Field Guide for Sharks of the Genus *Carcharhinus*

Matthias Voigt & Dietmar Weber

Field Guide for Sharks of the Genus *Carcharhinus*

Matthias Voigt & Dietmar Weber

Verlag Dr. Friedrich Pfeil
München 2011 · ISBN 978-3-89937-132-1

Impressum

Bibliographic information published by the Deutsche Nationalbibliothek

The Deutsche Nationalbibliothek lists this publication
in the Deutsche Nationalbibliografie;
detailed bibliographic data are available
in the Internet at http://dnb.d-nb.de.

This project was supported by

Deutsche
Elasmobranchier-
Gesellschaft e.V.

Deutsche Elasmobranchier-Gesellschaft e. V.
Initiative for the conservation of sharks, skates, rays and chimaeras
The German Elasmobranch Society (Deutsche Elasmobranchier-Gesellschaft, D.E.G.) was founded in 1995 by a group of scientists, journalists and interested citizens as a non-profit, voluntary organisation based in Hamburg. The Society is dedicated to promote the study, management and conservation of sharks, skates, rays and chimaeras (cartilaginous fishes) in German, European and international waters.

Front cover
Underside of a silky shark *(Carcharhinus falciformis)* on the Bahamas.
Photographer: Perrine Doug
Rights manager: Photolibrary Group Ltd.

Pre-Press: Verlag Dr. Friedrich Pfeil, München
Press: Advantage Printpool, Gilching

Printed in the European Union

ISBN 978-3-89937-132-1

Printed on permanent and durable acid-free paper

Verlag Dr. Friedrich Pfeil
Wolfratshauser Straße 27
81379 München, Germany
phone: +49 89 742827-0
fax: +49 89 7242772
e-mail: info@pfeil-verlag.de
www.pfeil-verlag.de

Table of Contents

Foreword

When I first saw the illustrations of sharks by Dietmar Weber many years ago, during a meeting of the German Elasmobranch Society, I was completely baffled how someone can depict a shark so realistically and with such clarity in detail. The pictures, which are scientific absolutely correct portrayed, really screamed at one's face, use me for a shark identification book!

Since then, his pictures have illustrated conservation reports, calendars, posters, conference banners and now, for the first time, they have found their way into a field guide. But these are not only pictures of sharks; they are complete tables with featured details of specific body parts, sometimes juvenile specimens, rows of teeth and other highlighted characteristics that help the reader to identify the species.

Marine biologist Matthias Voigt has collected over several years all the nitty-gritty data on species of the genus of *Carcharhinus*. He has visited many museums, took pictures, contacted scientist around the world to send him information and photos to sum it up into one big picture of this genus. This book describes 33 species of a group of sharks that are very hard to identify due to their similar appearance. Together with a wealth of information on biology, life history and distribution, the identification key leads the reader to the species information.

Almost all sharks of the genus *Carcharhinus* are coastal to pelagic sharks that occur in the tropical fisheries world-wide as targeted catch or by-catch. Their fins are primary goals for the fin market in Asia. Correct identification of these similar looking species is essential to establish species-specific management plans and for reported landing data. May this book by this help to actively support the conservation of these magnificent creatures of the sea.

Heike Zidowitz
Chair of the German Elasmobranch Society

Preface

The appearance of a new field guide identifying all recent species of sharks in the renowned genus *Carcharhinus* arouses various expectations among hobbyists and divers as much as it does for ichthyologists and paleontologists and probably also for commercial fishermen and conservationists. In me, it awakens memories of long-standing interests, and it makes me happy that the authors asked me to publish this guidebook in my publishing house.

In 1973, I was a young student of biology and paleontology searching the literature to compare the teeth of fossil sharks. It was then that "The genus *Carcharhinus* (Carcharhinidae)" by Bass, D'Aubray and Kistnasamy appeared in the series, "Sharks of the East Coast of Southern Africa" published in the "Investigational Reports" in Durban, South Africa. Their illustrations of the seventeen species occurring in the region were the best pictures of modern *Carcharhinus* teeth) available at that time.

In 1979/1980 I was in Wellington, New Zealand as a doctoral student. My supervisor there, Professor J.A.F Garrick, worked on his monograph of the genus *Carcharhinus*, which was published in 1982, as "Sharks of the genus *Carcharhinus*". Traditionally – following Bigelow & Schroeder 1948 and Bass, D'Aubrey & Kistnasamy 1973 – the emphasis of the work was to compare as much morphometric data as possible. But Garrick saw the great importance of a more detailed investigation of the tooth morphology: *"If one takes an overview of shark systematics in general, the best single feature that could be cited for determining similarities and differences between taxa at all levels, but perhaps predominantly at the generic level, is the shape of the teeth."* (Garrick 1982, p. 20). I was fascinated by the detailed drawings of the teeth. Garrick recognized the great variability of the teeth within a species, and therefore he illustrated the tooth rows of both a male and a female specimen of the species, *C. brachyurus* and *C. dussumieri*, showing a pronounced sexual dimorphism.

Unfortunately, drawings and/or photographs of *Carcharhinus* teeth in more recent publications, e.g. in the FAO Species Catalogue "Sharks of the World" by Compagno 1984, in "Sharks of the order Carcharhiniformes" Compagno 1988, Last & Stevens 1994 and 2009, "Field Guide to the Sharks of the World" Compagno, Dando & Fowler 2005, and Naylor & Marcus 1994, were in many cases too much "idealized" and less suitable for identification. The Illustration of the whole animals varied from simple black-and-white contour drawings to more realistic and sometimes highly artistic colored drawings, e.g. in Last & Stevens 1994 and 2009.

Nowadays, all species of *Carcharhinus* could be differentiated by using DNA analysis in the laboratory. But how could this work in the field? How does one identify the shark seen while diving or visiting an aquarium or caught by rod or net or found in the fish market? For such determination, one needs meaningful, colored illustrations and a key to the most important characters, above all those of the teeth.

Some expect to find all information in the Internet but anyone who is experienced knows that what he will find there is not satisfying. If one looks for taxonomic publications about sharks in the genus *Carcharhinus*, one must read the scientific literature. This kind of literature, however, is not accessible to everybody, and sometimes not easy to understand. Over the last three decades, researchers have learned much about the physiology of sharks, their taxonomy, however, has remained where it was more than 30 years ago.

When the two authors of this book, Matthias Voigt and Dietmar Weber asked me to publish their book, I needed little persuasion to agree to the project. At last, here were two, enthusiastic, young marine biologists, who would continue the tradition of earlier scientific descriptions of sharks and would supplement it with very high quality illustrations that would result in a field guide useful for everyone. They combine the necessary experience with the insight that makes the sober, morphometric data accessible even for the layman and allows the identification of all presently known, recent species of *Carcharhinus*. Dietmar Weber's illustrations of sharks are as realistic as one could wish. The digitally remastered photographs of the teeth follow the tradition of Bass et al. 1973, and Garrick 1982, but show this time substantially more details. Especially for teeth of the holotypes, the illustrations shown here could not be any better.

I connect this field guide with the hope that further volumes may follow. It is a substantial step towards a comprehensive representation of all species of sharks. As a paleontologist, I would like to see even more of the variations that show up in the heterodonty of young and adult individuals and in the differences of males and females.

Hopefully, this field guide will contribute to improved recognition and better understanding of these animals, hunted and endangered by humans, so that they may finally be protected before they become extinct.

Fritz Pfeil
Munich 2011

Acknowledgements

We would like to thank the following persons, societies and institutions for their kindly support:

Heike Zidowitz (current chairman) and Frank Velte (former chairman) of the German Elasmobranches' Society (D.E.G.) for supporting this book project financially and logistically. Without this help it would not have been possible for us to visit several museums and private collections in Europe, and to measure and to take pictures of many specimens.

Thanks to Andreas Schenck and Thomas Buetzow for taking many of the detailed photographs of species dentition. Moreover, gratitude to Bill Heim (London, UK) for preparing photographs and scans of species dentition and to Joe Herber (Strassen, Luxembourg) for the possibility to measure and to take photographs of specimens of his huge collection. Very large thanks to Dr. Gordon Hubbell (JAWS International, Miami, USA) for providing us with photographs of *Carcharhinus porosus*. Special thanks also to Dr. Michael Potthoff (DHI Group (Hørsholm, Denmark) for the help arranging the photographs of species dentition.

We express further our gratitude to: Dr. Martin van Oijen, collection manager and curator of the Nationaal Natuurhistorisch Museum (RMNH, Leiden, the Netherlands) for his kindly support. He has made specimens from the collection for examination, and to take many of the photographs of the detailed dentition (of holotype material and other specimens) available

Ernst Mikschi, Director of the 1st Zoological collection (Mammals) and Dr. Helmut Wellendorf, manager of the fish collection of the Naturhistorisches Museum (NMV, Vienna, Austria). With their help and permission we could take photographs of dentition of the *Carcharhinus leiodon* holotype specimen

Dr. Patrice Pruvost, Romain Causse and Guy Duhamel, Institute d'Historie de Naturelles (MNHN, Paris, France) for the kindly support and for enabling measurement and to take photographs of dentition of holotype and several other specimen of the famous collection

Mark McGrouther, Australian Museum (AMS, Sydney, Australia) and to Dr. Ronald Fricke, Staatliches Museum für Naturkunde (SMNS, Stuttgart, Germany) for the friendly partnership. Because of their effort we could take photographs of dentition of some Australian holotype specimen which has been loaned from the AMS and sent to the SMNS

Dr. James MacLaine, Dr. Colin McCarthy and Dr. Darrell J. Siebert, Natural History Museum (formerly the British Museum of Natural History, BMNH, London, UK) for courtesy of taking photographs of the *Carcharhinus hemiodon* (holotype) material dentition

Dr. Jacques Herman, Service Géologique de Belgique (Brussels, Belgium) for the critical reading of the draft part "Dental formula and tooth morphology". We appreciate his pieces of advice and thank him for checking the usage of technical terms corresponding to the newest findings

Dr. Matthias Stehmann, Institut für Seefischerei, (Ichthys, Hamburg, Germany) for giving invaluable hints to use dentition technical terms correctly and for useful critics. Dr. Michael George, Institute of Applied Fish Biology (Hamburg, Germany), Richard Pemberton, and Alexandra Heussner (University of Konstanz) for scrutinising the final version of the manuscript. Susan and Michael Sobtzick for many interesting discussions about sharks' dentition. These discussions have inspired us to include plates with detailed photographs of species dentition in this book

Thanks also to the following persons: Robert Thacker, Smithsonian Institute (Torras Bocas, Panama), Dirk Erpenbeck, Queensland Museum (Brisbane, Australia), Ross Robertson, Smithsonian Tropical Research Institute (Balboa, Ancon, Panama), William White, Shark Research Centre (Murdoch University, South St. Murdoch, USA), Jeff Johnson, Queensland Museum (Brisbane, Australia), Shelton P. Applegate, Instituto de Geologia (Ciudad Universitaria, Mexico City, Mexico), John Stevens and Alistair Graham, CSIRO Marine Research (Hobart, Tasmania, Australia), Phillip C. Heemstra, South African Institute for Aquatic Biodiversity (Grahamstown, South Africa), Jack Randall, Bishop Museum (St. Bernice, Honolulu, USA) – for their help and hints in organising the availability of type material. Many thanks to Caroline Wahnbeck, IUCN Shark Specialist Group (Newbury, Berkshire, UK) for providing us with current information about shark catch statistics. Thanks also to Jenny Ovenden, Queensland Primary Industries and Fisheries (Queensland, Australia) for sharing her results about the occurrence and genetic relationships of *C. limbatus* and *C. tilstoni*. Very large thanks to Jeremy Stafford-Deitsch, The Shark Trust (UK) for dentition photographs and pictures of sharks in natural habitat.

The most sincere thanks to Dr. Kristina Barz (Federal Research Centre for Fisheries, Rostock, Germany). She was an invaluable help and we thank her for the encouragement to produce this field guide, for her suggestions, useful information and critical reading of the manuscript.

Further, we are very thankful for the help offered by all of these people and by others not mentioned. Without their help this book would not have been written. Many persons were extremely helpful in preparing this book. It was intended to be comprehensive and to list all of them, but unavoidably one or two names might be missing, please apologize for it.

Introduction

The interest in sharks and cartilaginous fishes has steadily increased over the last decades. Thus, research effort and knowledge about sharks concerning their sociobiology, reproductive biology, geographical distribution etc. have grown sustainably. This process can be related to the higher public interest in "man eating sharks" and their relatives. However, the fishing industry has the greatest interest in these animals. The massive threat to sharks from fishing industry impact has risen from year to year in the last five decades, visible on the dramatic increase in the number of caught individuals (IUCN).

Sharks are caught either selectively or as by-catch. The aims of selective catches are the fins of sharks, which are mostly amputated while the individuals are still alive. The fins are used e.g. for preparing shark fin soup, a doubtful and quite expensive enjoyment of the Asian cuisine. After fins amputation the mutilated, still living individuals or carcasses are discharged into the sea.

Other serious problems which menace sharks are the non-selective long line fishery and the trawling fishery where many individuals are caught and killed as by-catch. Furthermore, sharks are endangered by trophy hunting fishermen, and are strongly affected by being charged with poisons caused by the human contamination of the sea with dangerous chemical substances.

Many shark populations are acutely threatened to extinction as the number of caught sharks has exceeded a critical value (IUCN). The catch-all number of sharks (and rays) seems to have been stable since 1996, but a reduction of the catches in statistics may be masked by the remarkable recent improvement in the species breakdown of reported catches. Such catches were previously mostly combined under the generic item "Elasmobranchii" or even classified as "Marine fishes not identified" following the efforts of the FAO and Regional Fisheries Bodies to improve the catch statistics. Official sources give the value of caught sharks to be 600 000–800 000 tons per year, equivalent to more than 100 million caught and killed individuals (SHARKINFO, IUCN), but the correct value could be much higher. This opinion based on the fact that often the dried fins are landed only. Hence, the biomass and number of the caught sharks is unknown. From those fins-only catches it is almost impossible to identify species exactly, and it cannot be excluded that (local) protected species, e.g. the Great White shark, *Carcharodon carcharias* were caught. To certify the species identity of such fin-catches it is necessary to examine the samples by extensive DNA-analyses, but such DNA fingerprint kits are available for few species only, e.g. *Carcharhinus brevipinna*, *C. limbatus* or *Sphyrna lewini* (CHAN et al. 2003, BLANDON et al. 2004, CHAPMAN et al. 2009). Recommendations and public laws to regulate and prohibit shark finning practice in parts were enacted to protect populations more effectively (e.g. Council Regulation of the European Union, EU, No. 1185/2003, or U.S. Public Law 106-557/Dec. 21, 2000). The Council Regulation of the EU demands a special fishing permission in accordance with Council Regulation No. 1627/1994 for the concession of finning of dead sharks on board of vessels, but almost all sharks are dead when they come on board. Furthermore, [...] "in order to ensure that all remaining parts of sharks are kept on board, after the removal of fins, masters of vessels which hold a valid special fishing permit should keep records of the amount of shark fins and of the remaining parts of sharks after evisceration and beheading".

Such records should be kept in the logbook as provided for by Council Regulation, No. 2847/1993 establishing a control system applicable to the common fisheries policy, or in a special register as appropriate (Council Regulation, No. 1185/2003, §8).

Frequently, catches of sharks are not registered in official statistics and violations of the mentioned regulations by fishermen are often not punished. Moreover, those protection measures are limited to sovereign country territories but are not an overall protection worldwide procedure. Hence, it is necessary to give a well based knowledge to people who come in contact with sharks or who decide about protection procedures to lead to an efficient protection. Sharks, rays and chimaeras need a strong lobby which will fight for maintenance, protection and regeneration of the populations. Otherwise the top predator of the sea will become extinct soon, and the side effects on marine environment are not predictable.

Aims of this book

This book reflects all of the known facts which are needed to identify members of the genus *Carcharhinus* and shall be understood as part of the broad campaign for wide-ranging protection of sharks and their relatives worldwide.

A further reason for composing a comprehensive book about all current known species of the genus *Carcharhinus* was to give an identification guide of this taxon with countless facts as a valuable tool to fishery and shark researchers, to people encountering sharks in the field, people who are active in aquatic sports, e.g. to divers and to the interested public.

Evolution of sharks – fossil records

Chondrichthyes are an evolutionary old taxon because the first appearance of earliest ancestors of the modern sharks reaches back to 400 to 500 million years ago. Carcharhinidae species have an abundant fossil record in the Cenozoic (mostly teeth but some whole-bodied specimens also) and occur in the Mesozoic from Upper Jurassic and Cretaceous deposits (COMPAGNO 1988).

The taxon Chondrichthyes is separated from bony fishes, taxon Osteichthyes, by having a skeleton of cartilage instead of calcified bones. Chondrichthyes can further be divided into Elasmobranches or Selachii (what means sharks and rays), and in Holocephali or Chimaeras (but rays and chimaeras are not further discussed here).

Taxonomic classification of sharks

The term 'shark' is in practice universally used as a catchall term in English language for all members of the taxon Chondrichthyes, which are not Batoids or Chimaeras. The French 'Requin', the Spanish 'Tiburon' and the German 'Hai' are comparable general terms. In the current zoological nomenclature system sharks are separated into eight orders (COMPAGNO 1984, 1991; COMPAGNO et al. 2005; ESCHMEYER 1998, 2007). That orders are: Hexanchiformes, Squaliformes, Squatiniformes, Pristiophoriformes, Heterodontiformes, Lamniformes, Orectolobiformes and Carchariniformes (Figure 1).

The next lower level in the nomenclature hierarchy is the separation in families. The family Carcharhinidae comprises 12 genera, more than 50 valid species, over 150 synonyms, dubious names and *nomina nuda*.

Short introduction to general biology and anatomy of sharks (of the genus *Carcharhinus*)

Sharks in general (and of the genus *Carcharhinus* in details) have mostly been classified in previous works, books and reports by external characters, e. g. of body, snout and fins length, their morphology, tooth counts and shape, and other characters. Many of these characters used in previous works that are still of value are used here too.

The total length (TL) is the most frequently cited measurement in shark systematics and the one usually used as the indepent variable for proportional dimensions of other structures. In some cases fishery biologists use precaudal length (abbreviated as PL or PCL) as measurement from the snout tip to the upper caudal origin or fork length (FL) from the snout tip to the notch in the postventral caudal margin as alternatives to TL (Figure 13).

Morphology

All members of the genus *Carcharhinus* have a smoothly arcuate or convex frontal head profile at eye level with its surface curving smoothly medial from the eye above and below. A few *Carcharhinus* species have distinct subangular brow bridge above the eye, and have more or less flattened or depressed, elongated snouts. The snout length (preoral length) varies from between half as long to about equal the mouth width.

The chondrocranium (also neuro- or endocranium) is an unsutured box of cartilage, forms the main component of skull and consists of an axial and medial braincase which contains the brain, lateral sense capsules for the olfactory sense (lateral line canals, and masses of Ampullae of Lorenzini), auditory organs and the eyes (COMPAGNO 1988, WHITEHEAD 2002, SCHUESSEL et al. 2008).

The mandibular arch and the hyoid arch are closely associated with the chondrocranium. The palatoquadrates (or upper jaw) are suspended from the chondrocranium or fused to it and the Meckel's cartilages (or lower jaw) are supported from the upper jaws. The hyomandibulae or dorsolateral elements (epihyals) of the hyoid arch have their proximal ends attached to the otic capsules of the chondrocranium and their distal end to the posterior ends of the jaw cartilages (COMPAGNO 1988).

Vertebrae were routinely excised from the monospondylous precaudal region of the vertebral column below D1. Their calcification pattern can be studied either by slicing or sawing the centra at their midlengths or by radiographing them with their longitudinal axes parallel to the path of radiation from the X-ray head.

Eye shape varies from long, low, slit like, through intermediate shapes to the vertically oval or circular eyes of *Carcharhinus* species. The eye shape is correlated with the presence of subocular ridges and with nictating lower eyelid type:

- slit like eyes have strong subocular ridges beneath them and rudimentary or external nictating lower eyelids,
- oval or circular eyes have no subocular ridges and internal nictating lower eyelids, and
- transitional shapes have transitional external or internal nictation eyelids and variably developed subocular ridges (COMPAGNO 1988).

Gill openings vary in size relative to the body and one to another (3rd or 4th are usually the widest, while the 1st or 5th are the narrowest). Requiem sharks have usually relatively narrow gill openings approximately one third the head height at the branchial region or less but exceptionally up to one half (in *Carcharhinus isodon*). The last one or two gill openings are mostly over pectoral fin base and behind its origin, but the fifth is usually over fins origin (Figure 4).

Nostrils have incurrent and excurrent apertures arranged that water is flowing into the incurrent apertures, enters the blind olfactory sacks and leaves them through the excurrent apertures while the shark swims forward. The anterior margin of the nostril has usually a posteriorly directed anterior nasal flap that often has a longitudinal keel on its inner surface (Figure 6 a, b).

This keel separates along with an anteriorly directed projection of the posterior nostril margin partially the incurrent and excurrent apertures. The anterior nasal flaps are usually small to moderately large, and in some *Carcharhinus* species the incurrent aperture is expanded as a narrow to broad prenarial groove (COMPAGNO 1988).

Mouth shape varies from angular or arcuate, narrow, long to broad or short. Labial furrows are shallow to deep grooves or dermal pockets on the exterior surface of the upper and lower jaws that partly encircle and parallel the posterior corners of the mouth (Figure 7). Upper and lower labial furrows join each other at mouth corners or slightly behind them, and circumscribe a pleat of skin. The labial cartilage is the supporting structure of the labial folds.

An interdorsal ridge is present in most of the *Carcharhinus* species on the midline of the back between dorsal fin bases but can be absent in some species also. Further, they have caudal keels on each side of the caudal peduncle from the insertions of D2 and anal fin bases to the caudal bases (Figure 4 and 10). The precaudal pits are depressions (or notches) in the middorsal and midventral surfaces of the caudal peduncle at caudal fin origins, setting the caudal base apart from the peduncle. Precaudal pits of *Carcharhinus* species are usually straight or crescent notches with their long axis more or less

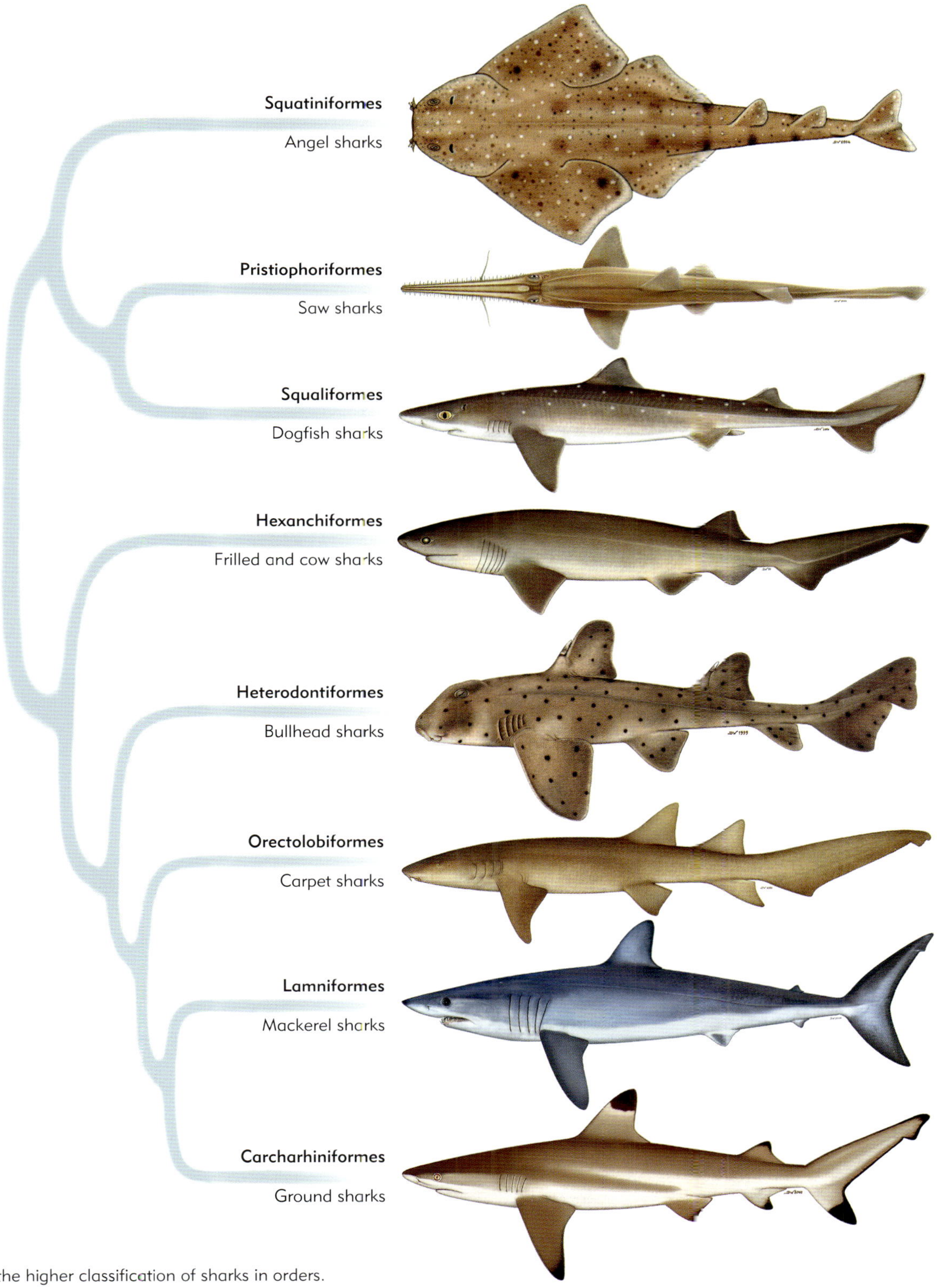

Figure 1. Cladogram showing the higher classification of sharks in orders.

transverse to the longitudinal axis of the individual. Upper precaudal pits are larger and more distinct than lower ones, and occasionally lower pits are absent.

Body and fins of sharks are covered by dermal denticles (or placoid scales), minute tooth-like structures which can be found in buccal cavity and pharynx also. They usually consist of a flattened plate-like crown attached by a narrow pedicle to an occasionally wide-spreading base. The denticle crown is above skin surface, and anchored to it by the skin-imbedded base. On crowns a species-specific number of longitudinal ridges and a corresponding number of posterior cusplets are present (Figure 12). The shape of denticles can vary widely on different parts of shark body to denticles, e. g. on leading edges of fins and on snout edges (COMPAGNO 1984, 1988).

All members of the genus *Carcharhinus* have paired pectoral and pelvic fins, an anal fin and caudal fin, and two dorsal fins without spines (Figure 4). For all precaudal fins the base is the part of the fins that merges with the body. The anterior end of the base is the fin origin, the posterior end fin's insertion (Figure 8–10). The exterior distal tip of the fins is called apex, the posterior extension of fins edge between origin and apex is the anterior margin, between apex and free rear tip the posterior margin, and between free rear tip and fin insertion its inner margin. Caudal fin upper lobe origin is located at the junction of its upper edge with the caudal peduncle and at the posterior edge of the upper precaudal pit (if present).

Senses

Sharks have several, considerably distinct senses to find either prey or a sexual partner, e. g. Ampullae of Lorenzini for measuring very small electrical impulses i. e. for orientating on earth magnet field. Further, they have an extreme visual faculty which is in comparison to the human visual faculty much stronger that enables sharks to see in muddy and turbid waters, the sensitive auditory system and the very sensitive lateral line that sharks use to detect changes in water pressure to find prey or to realise approach of other sharks (WHITEHEAD 2002).

Teeth

Shark teeth are of comparable material then dermal denticles and because of considerable difference between several families, genera and species they can be used for species determination. The teeth are embedded in gums and continuously replaced during whole lifetime of the individual. It is known that some species recycle the dentin of replaced teeth by swallowing them. Sexual heterodonty of tooth (gynandric heterodonty or dental sexual dimorphism) exists when the two sexes have differences in tooth morphology in similar positions in the mouth at the same life stages. This phenomenon has been recorded also for *Carcharhinus* species, e. g. *C. brachyurus*, *C. dussumieri* or *C. sealei* (HERMAN et al. 1991).

Buoyancy and breathing

Sharks lack a swim bladder and buoy up and tare by their huge, oil filled liver (containing Squalene in high concentration). A few species are able to swallow actively air and pump it into the stomach to tare therewith. Moreover, some species need not to swim permanently to conduct water through the gill filaments for breathing but those species, e. g. Sandtiger shark *(Eugomphodon taurus)* or Whitetip reef shark *(Triaenodon obesus)* are able to pump water through the gills actively while lying on the sea bed.

Reproduction

In contrast to Osteichthyes, sharks have an internal fertilisation. The copulation occurs by insertion of male claspers (a cylindrical extension visible as paired grooved extensions of posterior bases of the pelvic fins with their internal skeletons formed rearward extensions, Figure 5) in female's cloaca. Claspers of mature males are elongated and rigid, whereas immature males have short and soft claspers which could be easily overlooked. The female sexual organ, the cloaca, is visible as a short slit-like opening between the pelvic fins. The onset sexual maturity in males can be determined from condition and length of the claspers. Claspers of juveniles are short and flexible whereas claspers of adult males are rigid, elongated and calcified (BASS et al. 1973, CAPAPE et al. 2005). In contrast, the sexual maturity of females can be determined from the condition of ovaries and the morphology of the reproductive tract. Immature individuals have small flaccid and undeveloped ovaries and oviducts with no discernable eggs. Maturing females have enlarged ovaries and oviducts with visible eggs (approximately 5–10 mm in diameter), mature individuals have ovaries containing large, heavily yolked eggs (approximately 50–60 mm in diameter), and reproductive active individuals (can) have fertilized eggs and i. e. or embryos in the oviducts. Recent post-partum individuals have flaccid and stretched oviducts (NATANSON & CAILLIET 1986). Three reproduction strategies are known for sharks: oviparity, ovoviviparity and viviparity but in genus *Carcharhinus* viviparity is present only (DULVY & REYNOLDS 1997).

Distribution and behaviour

All members of the genus *Carcharhinus* are distributed in saltwater. Only one species, *Carcharhinus leucas* is known from verifiable records from (brackish) river estuaries and lakes (TAN & LIM 1998). Freshwater records of other members of the genus, e. g. *C. melanopterus* or *C. hemiodon*, may be from semi-brackish areas indicating they are tolerant of reduced salinities but are not truly euryhaline.

Several *Carcharhinus* species show distinct seasonal migrations, following prey or tend to avoid areas with rising or sinking oxygen saturation of the water, salinity or water temperature, e. g. *C. obscurus* and *C. limbatus* (COMPAGNO 1984, LIESKE & MYERS 1994, HEUPEL & SIMPFENDORFER 2002).

Distinct territorial behaviour patterns are mainly known from smaller reef associated living species, e.g. *C. amblyrhynchos* or *C. melanopterus*, but has also been reported from larger, more pelagic living species, e.g. *C. falciformes* (RANDALL & HELFMAN 1973, GARRICK 1982, COMPAGNO 1984, MCKIBBEN & NELSON 1986).

Sexual dimorphism

Sexual dimorphism of features e.g. dentition and colouration can be important for identification. Moreover, sexual and age segregation in geographical and depth distribution within many species are proven. Variances of appearance, e.g. body-shape or TL or of diet can occur as adaptations of the local conditions of habitat. Changes of internal features, e.g. number of vertebrae can occur as natural mutations or as adaptation on environmental setting when individuals grow up to less TL (BASS et al. 1973; GARRICK 1982; COMPAGNO 1984, 1988).

Preparation of this book

With exception of *Carcharhinus acarenatus* and *Carcharhinus* sp. A type material of almost all species of this shark genus was examined for the preparation of this book. Characteristics used here to enable identification and diagnosis of the species are the traditionally most applied features, e.g. morphometrics and body colouration, as well as internal issues, e.g. tooth formula and counts of precaudal and caudal vertebrae. The book connect and combine these facts with the newest findings of shark research and thus can be used as informative field guide with further notes to sociobiology, geographical distribution, reproductive biology, preferred food spectrum of the species, as well as link to the taxonomic literature for sharks of this genus.

Generally, in structure this book is orientated to the works of BASS et al. (1973), GARRICK (1982, 1985), COMPAGNO (1984, 1988), ESCHMEYER (1990, 1998/2007), and COMPAGNO et al. (2005). Similarities of plates, maps of geographical distribution and graphs were intended to enable easy comparisons between all mentioned works. A main intention in preparing this book was to exclude errors and inaccuracies of former works, to summarize the information of those works, and to combine all of them with the newest results of shark research in one book. Thereby, many statements made here refer to the mentioned works.

Based on the fact that the taxonomic classification of the family Carcharhinidae and the genus *Carcharhinus* have almost not changed in the last 20 years, the used nomenclature is according to COMPAGNO (1984), with some supplements. For a good and widespread taxonomic discussion about the genus see BASS et al. (1973), GARRICK (1982, 1985), COMPAGNO (1984, 1988) and others. In contrast to COMPAGNO et al. (2005) and most of the other works about the genus *Carcharhinus* all known facts about 32 species considered to be valid members of the genus and one species which is still in question (*Carcharhinus* sp. A) are discussed more detailed in this book.

Material and Methods

Key to the species

The presented identification key is a tool to determine individuals to species level. Identification keys are normally dichotomous. They follow in steps of alternative choices to single out a class, order, family, genus or species covered. Diagnostic features are presented as comprehensive lists of characters with the choice generally limited to clear distinctive patterns. Following the key the mentioned diagnostic patterns refer to a single species of the genus *Carcharhinus*. Subjective patterns in dependence of author's point of view, e.g. the description of fin shape and internal characters which are not easy to use, e.g. the beginning of mono- and diplospondylous vertebrae in the vertebral column are not listed in this key. However, for a definitely and doubtless of a species identification it is mostly necessary to count the vertebrae of the present individual. These counts can be easily made in the field with the help of a large, sharp, wide-bladed butcher knife, according to the proposed method of COMPAGNO (1984). However, this method should be limited to dead specimens only because no individual should be killed for this purpose!

[...] Precaudal counts are taken to the anterior edge of precaudal pit and can be easily done by placing the specimen on its side and starting at the tail filleting it by running the knife along the vertebral column continuing forward right into or through the chondrocranium. For counting the caudal vertebrae it is quite important not to damage the delicate terminal vertebrae (COMPAGNO 1999).

The alternative method of counting vertebrae of living individuals by X-ray photography especially for small and newborn specimens is applicable in laboratories only!

Systematic arrangement

The systematic arrangement used here is based on works of BASS et al. (1973, 1986), GARRICK (1982, 1985) and many other authors but in major parts it is those of COMPAGNO (1984, 1988) and COMPAGNO et al. 2005) with supplements. The latest agreements for the status of genera and species in the currently used zoological system are included but some relationships are still controversial at present and disagree with those of previous workers. In such cases these disagreements are discussed in species characterisations and the relevant reference is given.

Generic synonyms

In previous study reports, guides and species catalogues about sharks of the genus *Carcharhinus* (BLAINVILLE 1816), different generic synonyms were stated. Some of these generic synonyms used instead of this denomination are mentioned in Table 1. These synonyms are the previously most used and established names, listed with author name(s), year of publication and a short explanation for the term. Other levels of relation between genera and species are not discussed here. For a comprehensive overview of all known generic and species synonyms consult BASS et al. (1973, 1986), GARRICK (1982), COMPAGNO (1973, 1988), ESCHMEYER (1990, 1998/2007) and FROESE & PAULY (2010).

Examination of specimens

In some cases morphometric data were taken from other works e.g. BASS et al. (1973, 1986), GARRICK (1982, 1985), COMPAGNO (1984, 1988) or COMPAGNO et al. (2005) and combined with own measurements on species type material of zoological collections as well as of private collections. The number of examined specimen, i.e. used data sets from peer reviewed sources, to create mean values and graphs for every species are shown in Table 2.

Various body proportions were measured on a large number of specimens for almost all of the introduced 33 species of the genus *Carcharhinus*. Type material of *Carcharhinus* sp. A and *C. acarenatus* was unavailable. In case of *C. leiodon* a half upper and lower jaw were available only. Methods for taking measurement of specimen are illustrated in Figures 13–17. Hence, values for body proportions can be given from published literature only. The measured, derived and calculated values for body proportions are presented in Table 4.

Species characterisations

Every species is introduced in a separate characterisation chapter. The chapters are arranged in alphabetical order in regard of the current valid scientific name of the species. In each species characterisation chapter the name of the author and year for publication of species first description, and species vernacular English (En), French (Fr), Spanish (Sp) and German (Ge) species name(s) are mentioned.

Table 1. Overview about generic synonyms, author's names, year of publication, and explanations for the terms.

Generic synonym	Author	Explanation for the term
Aprion	MÜLLER & HENLE 1839	subgenus, type species *Carcharias (Aprion) isodon*, VALENCIENNES (1839)
Aprionodon	GILL 1861	used for *Carcharhinus brevipinna*
Bogimba	WHITLEY 1943a-c	subgenus of *Galeolamna*, used for *C. leucas*
Carcharinus	CLOQUET 1817	an erroneous spelling of *Carcharhinus*
Carcharias	CUVIER 1817	not *Carcharias*, designated by RAFINESQUE (1810)
Eulamia	GILL 1862	restricted to ridge-backed species, used for *Squalus longimanus*, POEY (1861)
Galeolamna	OWEN 1853	species dubia, but genus refers to *Carcharhinus*
Galeolamnoides	WHITLEY 1934	type species by original designation
Gilisqualus	WHITLEY 1934	type species by original designation
Gymnorhinus	HEMPRICH & EHRENBERG 1899	synonyms for *Carcharias (Prionodon) menisorrah* and for *Gymnorrhinus pharaonis*
Hypoprion	MÜLLER & HENLE 1839	subgenus, type species *Carcharias (Hypoprion) macloti*, MÜLLER & HENLE (1839), GILL (1862)
Hypoprionodon	GILL 1862	subgenus, type species *Carcharias (Prionodon) hemiodon*, VALENCIENNES (in MÜLLER & HENLE 1839)
Isogomphodon	GILL 1862	current genus, within family *Carcharhinidae*
Isoplagiodon	GILL 1862	type species *Carcharias (Prionodon) sorrah*, VALENCIENNES (in MÜLLER & HENLE 1839)
Lamiopsis	GILL 1862	current genus, within family *Carcharhinidae*
Lamnarius	WHITLEY 1943a-c	current genus of *Galeolamna*, used for *C. leucas*
Longmania	WHITLEY 1944a, b	mainly used in Australian literature for *Carcharias (Aprion) brevipinna*
Loxodon	MÜLLER & HENLE 1837-1839	current genus, within family *Carcharhinidae*
Mapolamia	WHITLEY 1934	mainly used in Australian literature for *Carcharias melanopterus*
Nasolamia	COMPAGNO & GARRICK 1983	genus of *Nasolamia velox* (formerly *C. velox*)
Platyposdon	GILL 1862	equals to *Carcharias (Prionodon) falciformes*
Prionodon	MÜLLER & HENLE 1841	junior homonym of a genus of mammals
Pterolamia	SPRINGER 1950	a homonym of a genus of beetles
Pterolamiops	SPRINGER 1951	used for *C. longimanus*
Uranga	WHITLEY 1943a-c	type species by original designation, equals to *Aprionodon*
Uranganops	WHITLEY 1943a-c	subgenus of *Galeolamna*

Older generic synonyms (see paragraph species synonyms) such as *Carcharias*, *Galeolamna* or *Prionodon* are given in brackets to show the previous generic affiliation.

Coloured figures of species

Coloured figures are included in the species characterisation chapters showing lateral view and underside, a detailed view of the left nostril, of the dentition of the upper and lower jaw (commencing at the teeth next to the symphysis towards the corners of mouth), and an illustration of dermal denticles for most of the species (Figure 18–50). Every figure is marked with a legend to indicate total length (TL), sex and locality where the shown specimen was caught. All figures were created accordingly to real percentage of body proportions i. e. morphometric data.

In most detailed illustrations of dentition a magnified pair of single tooth of upper and lower jaw is inset presented with a scaling bar and specification of tooth position in the jaw. In case of known sexual dentition dimorphism a second magnified image for the other gender is added. Moreover, the lateral view of a juvenile is shown for some of the species.

The current institutional deposition of the type material (Holo-, Syn-, Para- or Lectotype) is stated with short mark and collection catalogue number. In case material from private collections was used and examined such catalogue number(s) is missing.

Location and depth where the type material specimen was caught are stated (if available), and TL and sex of the specimen are denoted.

Species synonyms

Most of the known species synonyms for the introduced species (considered to be not valid accordingly to current taxonomical system regulations) and which are occasionally still used in other publications are listed in this section to avoid confusion by the usage of different scientific names for one species. For example, *Carcharhinus dussumieri* was previously often termed *C. menisorrah* but the term *menisorrah* is actually considered to be part of a synonym for *C. falciformes*.

However, the presented list of synonyms is considered to be incomplete and literature sources of synonyms are not necessarily listed in the bibliography. An extended and detailed discussion about synonyms status and usage can be found for instance in Compagno (1984, 1988), Garrick (1982, 1985) and Eschmeyer (1990, 1998/2007).

Diagnostic features

The mentioned diagnostic features for the identification of the species refer to obvious and clear distinctive patterns and are arranged hierarchically in the section. Condition for listing these features was the easy usage of the characters for a doubtless identification purposes under simple application of technical equipment in the field. All mentioned morphometric values are average values given as percentage of total length [% TL]. The number of used data sets (means of examined specimens combined with published morphometric data) for calculation of the average and Minimum-Maximum values are given in Table 2. Furthermore, these data sets were used to create graphs (Figure 51–59) for most of the measured morphometric features to enable easy comparisons of the species one to another (Figure 13–17).

The first considered diagnostic feature for identification is the body-shape, further the presence or absence of an interdorsal ridge and of a precaudal keel. In addition, the ratio of internarial width to preoral length is given. Further, shape and diameter of the eyes, shape of nostrils, and length and shape of the nasal flaps are listed. Since some *Carcharhinus* species have enlarged hyomandibular pores alongside corners of mouth, this obtains a miscellaneous aspect for identification. Enlarged hyomandibular pores are a discrete group of comparatively large and noticeable pores of the hyomandibular series opening near the upper labial furrow and extending dorso-posteriorly a short distance.

Frequently, these pores are arranged as a straight line, sometimes irregularly biserial. Difficulties in counting can occur when pores are clogged or when they had not broken through the dermal denticles, especially in smaller and young specimens.

Next features shape, height and position of fins origin are mentioned. To describe fin length, shape and position of fin origins the following abbreviation for the fins were used: First dorsal fin (D1), second dorsal fin (D2), pelvic fins (P), pectoral fins (PP), anal fin (A).

Table 2. Number of used specimens for calculation of morphometric mean values, and for creation of the graphs.

Scientific species name	Specimens [n]	Scientific species name	Specimens [n]
C. acarenatus	14	*C. leiodon*	1
C. acronotus	5	*C. leucas*	24
C. albimarginatus	20	*C. limbatus*	28
C. altimus	18	*C. longimanus*	23
C. amblyrhynchoides	12	*C. macloti*	10
C. amblyrhynchos	16	*C. melanopterus*	24
C. amboinensis	19	*C. obscurus*	21
C. borneensis	24	*C. perezii*	18
C. brachyurus	21	*C. plumbeus*	23
C. brevipinna	19	*C. porosus*	10
C. cautus	7	*C. sealei*	14
C. dussumieri	11	*C. signatus*	16
C. falciformes	18	*C. sorrah*	23
C. fitzroyensis	4	*C. tilstoni*	12
C. galapagensis	22	*C. wheeleri*	20
C. hemiodon	5	*C. sp. A*	3
C. isodon	9		

No. of rows left side ( ) – No. of central rows ( ) – No. of rows right side ( )	upper jaw
No. of rows left side ( ) – No. of central rows ( ) – No. of rows right side ( )	lower jaw

Figure 2. Schematically distribution of teeth of the left and right upper and lower half of jaw.

An auxiliary criterion for species identification is the number of precaudal and caudal vertebrae (also centra). Counts of precaudal centra give all mono- and diplospondylous centra to the upper caudal fin lobe origin, whereas the number of caudal centra gives all vertebrae of caudal fin (Table 4).

At last, shape and arrangement of the dermal denticles are described in details but examinations of this feature must be done in a laboratory using an appropriate microscope and photographic equipment.

Dental formula and tooth morphology

To the diagnostic features the attribute 'Dental formula' is added including a description of the tooth-shapes to enable a nearly doubtless identification of an individual. Members of the genus *Carcharhinus* are characterised by having the following dentition characters:

- roots neo-holaulacorhizid,
- with broad, shallow median grooves which are flatting of the inner root surface and / or rising of its bottom,
- the pulp cavity is greatly expanded internally, especially in upper teeth,
- root lobes are expanded and generally obtuse angled,
- the crowns of upper teeth are serrated on mesial and distal cutting edges, cusplets are absent,
- the crowns of lower teeth are slender, often smooth-edged,
- strong dignathic heterodonty is distinct.

Some intermediate forms between holaulacorhizid and neo-holaulacorhizid are known in the taxa Carcharhiniformes (HERMAN et al. 1991).

Tooth counts refer to the number of functional rows of teeth, following in general the nomenclature and counting methods of HERMAN et al. (1991) and NAYLOR & MARCUS (1994). This nomenclature is completed by some technical terms of the used nomenclature of APPLEGATE (1965) and COMPAGNO (1984).

The dental formula for each species comprehend the number of tooth rows around outer margin of the upper (above the line), and the lower jaw (below the line) (Figure 2). Each upper and lower jaw count is subdivided into the number of rows on the left side, followed by the number of small central or symphysial tooth rows (if present), then the number of tooth rows on the right side.

Each tooth of a row is the outer element of a file of teeth aligned inwards and derived from the same tooth primordium. Thus, each row is made up of one functional tooth together with their developing replacements (DIAZ 1979).

In some cases more than one dental formula is given in the species characterisations. The first is the dental formula of the holotype specimen the second the most common dental formulae of this species listed with known and frequently present alternative numbers of teeth (given in brackets).

To specify tooth-position in the jaw (beginning at the symphysis toward angle of jaw, see Figure 3) the following abbreviations were used:

PS	pseudo- or parasymphysial position
S	symphysial
A	anterior
AL	antero-lateral
L	lateral
LP	latero-posterior
P	posterior
CP	commissural-posterior
C	commissural
X	further teeth

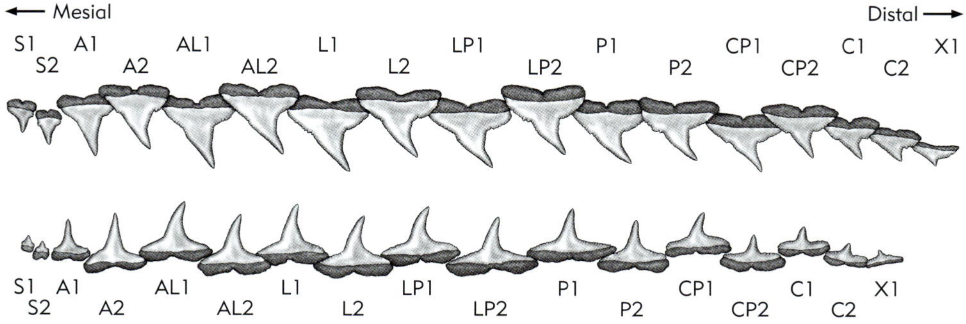

Figure 3. Used abbreviation for tooth position, on example of the dentition of *C. brachyurus*. Symphysis (in the figure) is situated to the left, rising terminology from left to the right, towards angle of mouth (right). Shown are the upper and lower halves of jaws.

The tooth-positions PS, S and X are exceptionally composed by a varying number of teeth. All other tooth-positions are formed by a pair of tooth rows. Tooth-shape descriptions follow the guidelines of HERMAN et al. (1991), GARRICK (1982, 1985), COMPAGNO (1984) and COMPAGNO et al. (2005). Thereby, the method to count and to compare tooth shape is doubtless an appropriate instrument to compile a present set of jaws with reference jaws (if available) of adequate and doubtless identified specimen. In case of sexual dimorphism in dentition, facts on variability of tooth-shape or serration are mentioned and explained in the section tooth morphology too. The tooth descriptions include the following aspects:

- Size and shape of teeth (erect, oblique, high, narrow ...), with specification of tooth position in the jaw for this appearance.
- Description of the distal and mesial median keels shape (straight, concave, convex ...).
- Whether crowns have serrated or smooth-edged median keels (because very often serration is very weak and not visible without microscope).
- Description of the root-shape.
- Number of symphysial and anterior rows.
- Which kind of dentition arrangement is present?

Colouration

The description of specimen's body-colouration is commonly a reliable and useful diagnostic feature but sometimes the exact determination of colour pattern is difficult especially for post mortem individuals. Only clear and pronounced colouration patterns are discussed for the species characterisations in this book but no pattern is all-exclusive! Variations in colouration are possible in all of the introduced species. Different colour patterns for juveniles and sub-adults were frequently reported and hence, the presented descriptions of body colouration must be seen as the generally most distinct patterns only. In addition, after death of the specimen colour patterns often move into grey or drab brown. After conservation in alcohol colour can occasionally disappear and then a precise description of colouration is almost impossible.

Size

The determination of specimen's total length (TL) is the most used independent variable in identification and is a standard measurement of specimens in fishery biology. However, among shark researchers no agreement for a standardized method of measuring specimen TL exists yet. Hence, values for TL of different sources in the literature may not be strictly comparable. To get an impression for this problem take a look at *C. tilstoni* species first description (WHITLEY 1950) where a fairly different measurement system was used. However, in the present book the method of COMPAGNO (1984) as standard method for measurements is preferred and advocated.

All size data are given in centimetres [cm]. Length can vary depending on how the upper caudal lobe is positioned. TL is measured as a straight line (not over the body curve) from snout tip to the apex of the upper caudal fin lobe. Usually, the lobe is pulled back while still maintaining a "normal" as per in living form tail position (COMPAGNO 1984, 1988). For the illustration of how measurements were taken see Figure 4–17.

The section 'Size' includes the maximum and average TL for each species which is in some cases correlated to a specific specimen age. In case of rare and poorly known species, or if immature individuals are available only, the hypothetical maximum size is listed also. The main intention of giving maximum and average size was to exclude old and unverifiable undersized or exaggerates size records of other sources.

In addition, in some species characterisations a maximum weight correlated to a specific length is mentioned. However, this feature is not reliable for identification purposes because it is very variable depending on season, state of pregnancy, degree of gut filling and other circumstances and thus, should be used as additional information only.

Misidentifications

Similarities among species of the genus *Carcharhinus* (and possible confusions with other genera or species) are mentioned to avoid misguiding in identification. Explanations are given to underline the differences between closely related or very similar species. If mistakes have been occurred in identification possible alternatives are discussed to still determine the correct species.

Geographical Distribution

Although sharks are primarily marine organisms a few of the introduced species are able to enter brackish estuaries, lagoons and bays or even freshwater areas, and occur far up in rivers and freshwater lakes with connections to the sea. Hence, the overview about the geographical distributions for all species is given by listing countries names, of their coasts, and in some cases rivers and river systems where the species occur. All data about geographical distribution were taken from published literature and should be seen as a review of the current state of knowledge.

In the maps of geographical distribution areas with proven occurrence of the considered species are marked dark blue along coastlines. Not proven or dubious areas are mentioned and sometimes discussed in the text only. Exceptions were made for the species *C. falciformes*, *C. longimanus*, *C. melanopterus* and *C. leucas*. In case of the first three mentioned species it is supposable that they occur more widely distributed then presently known. In previous works the distribution of such wide-ranging pelagic living species was limited to island groups and continental shelves only. Here, the geographical distribution for these species is given as covering huge oceanic areas all over, although proven reports and catches exist from islands only. In case of *C. leucas* the distribution of the species in Indian rivers was not given because of a fairly probable confusion with the related species *Glyphis gangeticus*.

A remarkable amount of freshwater records exist and may be assigned on species as *C. melanopterus* or *C. hemiodon* but the identity of the species was mostly not established correctly. Furthermore, many species of the genus *Carcharhinus* have been reported from brackish regions but mostly the wrongly identified species are incapable to exist in freshwater far from the sea. Thus, these freshwater or brackish area records are considered to be not valid for species distribution. It must further be noted in many cases gaps in data presentation of geographical distribution may not indicate absence of a species but lack of knowledge.

Habitat preference

This section covers information about physical conditions of the considered species. The known depth range of the species is given in meters [m], as well as their preferred position in the water column at different daytimes. Furthermore, the preferred type of substrate and the preferences of distance in relation to the coastline are noted, e.g. inshore, offshore, pelagic, oceanodromous. In some cases physical parameter data on salinity, oxygen content, and specific water temperature of species mainly preferred habitat are presented – if known.

General biological issues

General biological issues, e.g. facts about population structure and dynamics, reproduction, ecology and sociobiology, behaviour and locality at different life stages are given. Some facts of special hunting or defence strategies are registered also, and for some species the theoretical calculated longevity is quoted.

Reproductive Biology

All currently known facts about the reproductive biology, birth strategy, the mean and the range of young per litter, as well as the size at birth are given. Facts of known sexual or age segregation are added, and some facts about the development and growth rate of the newborn and juveniles are listed.

Food spectrum

This section is supplied as a comprehensive overview of the chiefly preferred diet of the considered species. Changes in species prey spectrum depending on specific age or as adaptation on different habitats are listed.

Table 3. Overview about abbreviations used to classify the IUCN Red List status of species.

EX	Extinct	NT	Near Threatened
EW	Extinct in the Wild	LC	Least Concern
CR	Critically Endangered	LR/nt	Low Risk/near threatened
EN	Endangered	NE	Not Evaluated
VU	Vulnerable	DD	Data Deficient

Red list status

The IUCN Red List status is given in Table 3 to show the degree of exposure of every species accordingly to the current statements.

Tables and Graphs

The number of data sets for the calculation of morphometric mean values for each species is given in Table 2. Values of proportional measurements taken on the specimens are listed in Table 4. This table provides data in the form of mean values and ranges (Minimum and Maximum values). The values were used to create graphs and to compare and to visualise morphometric differences between the species directly (Figure 51–59). The Minimum-Maximum values are presented to show differences in ratio of body proportions of the considered species outgoing from embryos and newborn individuals (or the smallest caught specimen) to the biggest specimen caught ever. In the graphs the mean values for all species are given as dots with range (Min-Max values as vertical line).

For some of the introduced species e.g. *C. leiodon* (GARRICK 1985) only a very limited number of specimens or data was available. This fact is the reason for the appearance of single dots in the graphs without giving a range (Figure 51–59). Two species, *Carcharhinus acarenatus* (MORENO & HOYOS 1983a) and *Carcharhinus* sp. A (COMPAGNO et al. 2005) were not considered in creating graphs. For the mentioned two species values for body proportions from literature can be given only.

For *C. tilstoni* (WHITLEY 1950) a few body size features cannot be given because of the different kind of measurement. A conversion of the published data set was attempted but for several morphometric data it was not successful. Hence, these values are marked in Table 4 with an asterisk to sign their different character. Moreover, for some specific characteristics no data were available, e.g. the clasper length for *C. perezii* and hence, those features were also not considered in preparing graphs (Figure 51–59).

Detailed photographs of dentition

Detailed photographs of dentition are presented for nearly all of the introduced species to include an additional tool and to enable a doubtless identification of specimen of the genus *Carcharhinus*.

Adequate sample material was unavailable for *Carcharhinus acarenatus* and *Carcharhinus* sp. A, and for few species, *C. cautus* and *C. fitzroyensis*, an incomplete set of jaws of the holotype specimen was available only.

All photos of dentition have a constant composition: the upper part always show upper teeth and the lower parts lower teeth. The choice to show a left or right jaw half depends on the preservation quality of specimens tooth file. Teeth of different positions are shown magnified to provide a better impression of the small structures of serration, and a scaling bar shows the proportions. Every picture is marked with a legend to indicate TL, sex and locality where the specimen was caught (particularly the institutional deposition is given also, in case the specimen is listed with a unique number in a museum's collection data base), whereas specimen without catalogue number were taken from private collections.

Measurements

Illustrations of taken measurements of type material specimens, about used abbreviations for the measurements, and explanations for the terms.

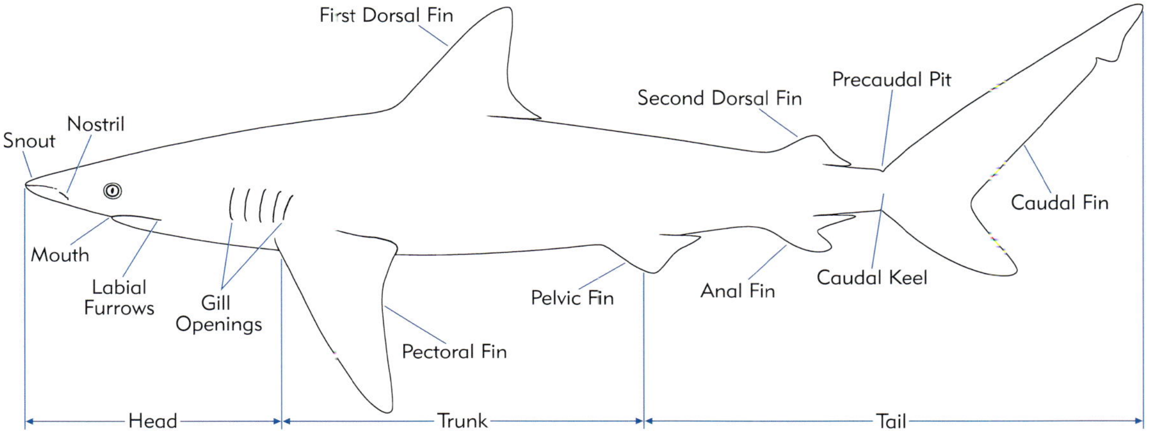

Figure 4. Lateral view of a shark.

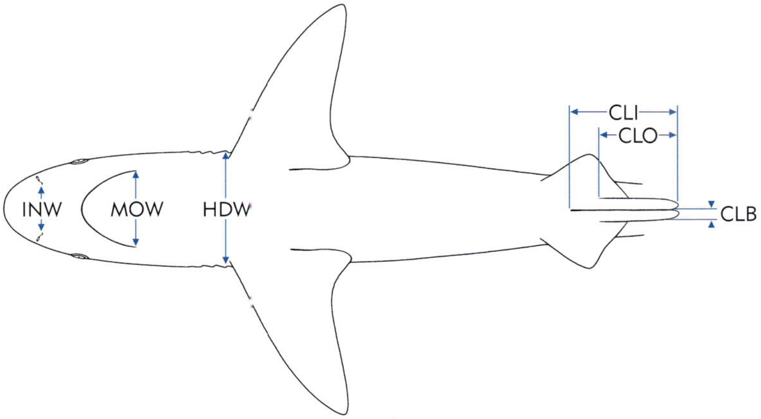

CLB Clasper Base Width
CLI Clasper Inner Length
CLO Clasper Outer Length
HDW Head Width
INW Internarial Space
MOW Mouth Width

Figure 5. Ventral view.

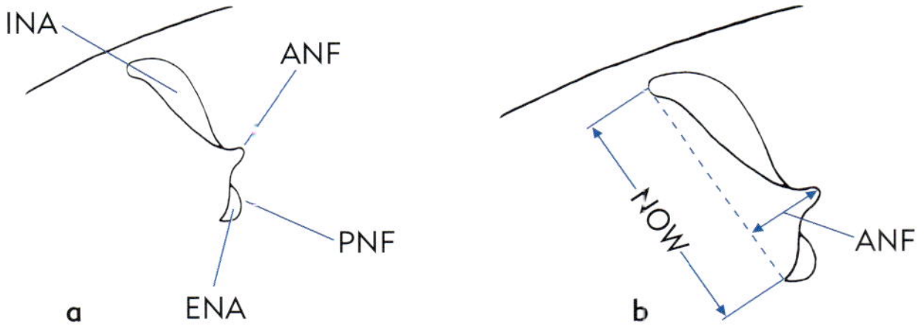

ANF Anterior Nasal Flap
ENA Excurrent Nasal Aperture
INA Incurrent Nasal Aperture
NOW Nostril Width
PNF Posterior Nasal Flap

Figure 6 a, b. Nostril.

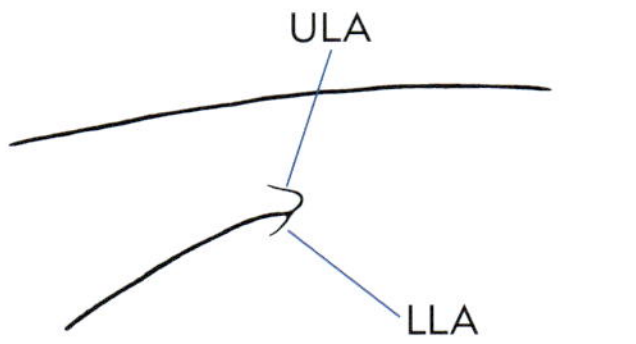

Figure 7. Mouth corner.

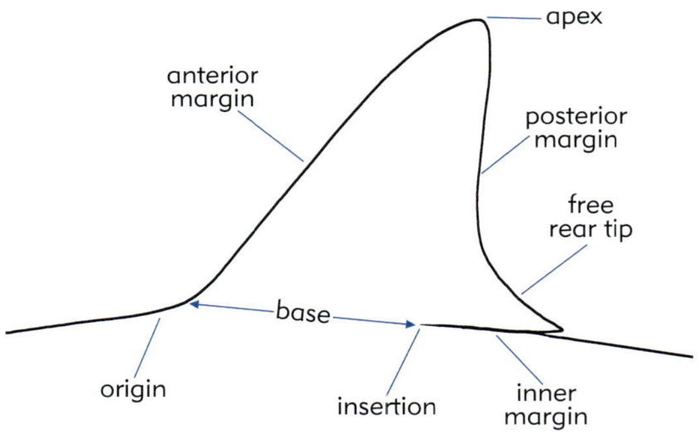

Figure 8. Dorsal fin.

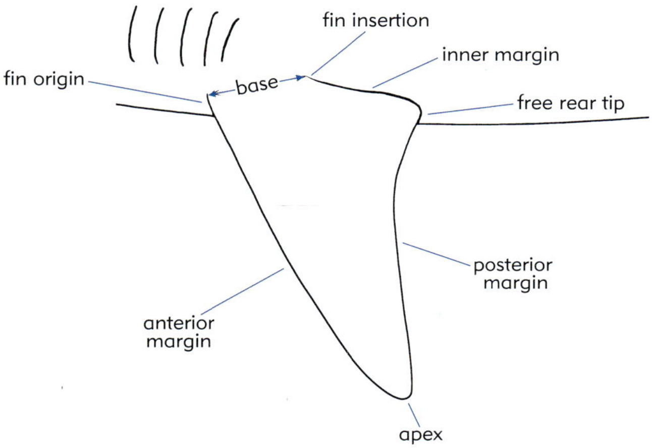

Figure 9. Pectoral fin.

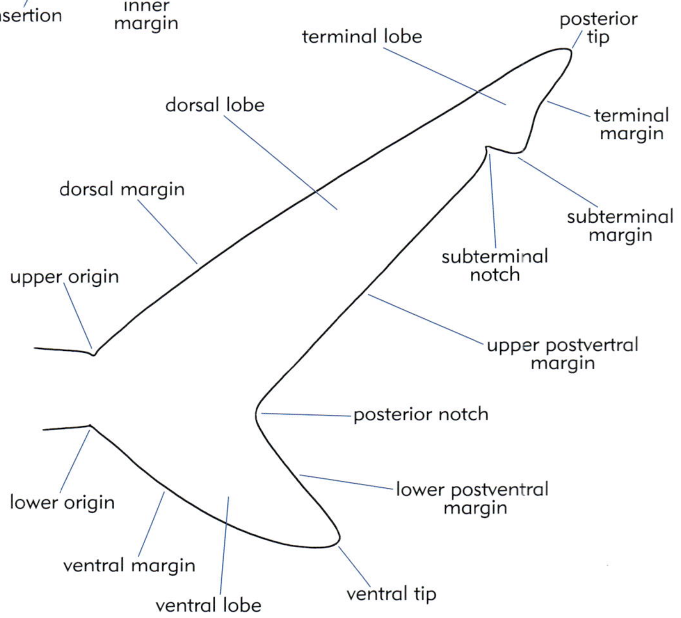

Figure 10. Caudal fin.

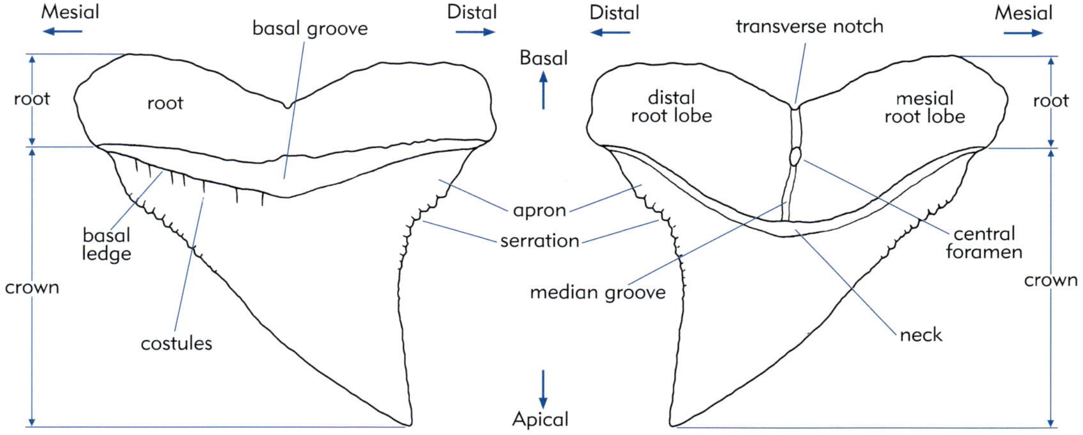

Figure 11. Tooth terminology (left upper anterolateral tooth); on the left side: labial view, and on the right side: lingual view. – (From COMPAGNO 1984, modified).

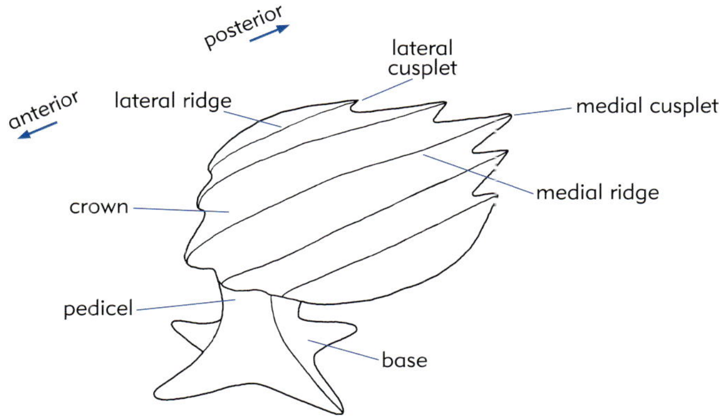

Figure 12. Oblique anterolateral view of dermal denticle (of lateral trunk). – (From COMPAGNO 1984, modified).

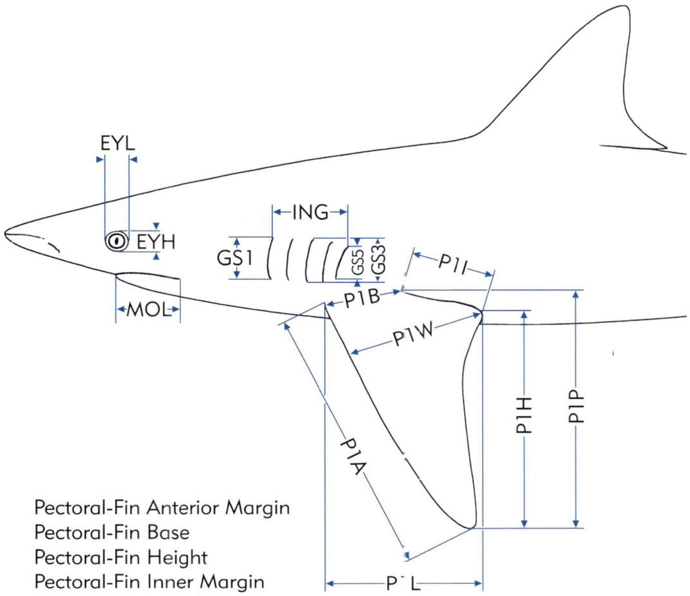

EYL	Eye Length	P1A	Pectoral-Fin Anterior Margin
EYH	Eye Height	P1B	Pectoral-Fin Base
ING	Intergill Length	P1H	Pectoral-Fin Height
GS1	First Gill Slit Height	P1I	Pectoral-Fin Inner Margin
GS3	Third Gill Slit Height	P1L	Pectoral-Fin Length
GS5	Fifth Gill Slit Height	P1P	Pectoral-Fin Posterior Margin
MOL	Mouth Length	P1W	Pectoral-Fin Width

Figure 14. Head and pectoral fin measurements.

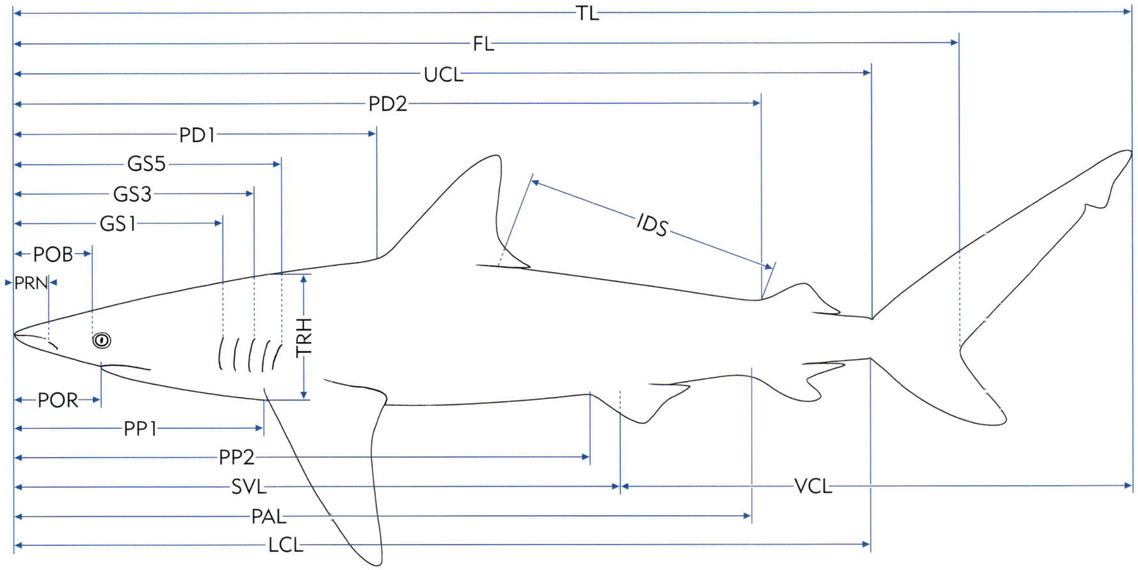

FL	Fork Length	PD1	Pre-First Dorsal-Fin Length	SVL	Snout-Vent Length
GS1	Snout To First Gill Slit Opening	PD2	Pre-Second Dorsal-Fin Length	TL	Total Length
GS3	Snout To Third Gill Slit Opening	POB	Preorbital Length	TRH	Trunk Height
GS5	Snout To Fifth Gill Slit Opening	POR	Preoral Length	UCL	Upper Precaudal-Fin Length
IDS	Interdorsal Width	PP1	Prepectoral-Fin Length	VCL	Vent Caudal-Fin Length
LCL	Lower Precaudal-Fin Length	PP2	Prepelvic-Fin Length		
PAL	Preanal-Fin Length	PRN	Prenarial Length		

Figure 13. Main longitudinal measurements.

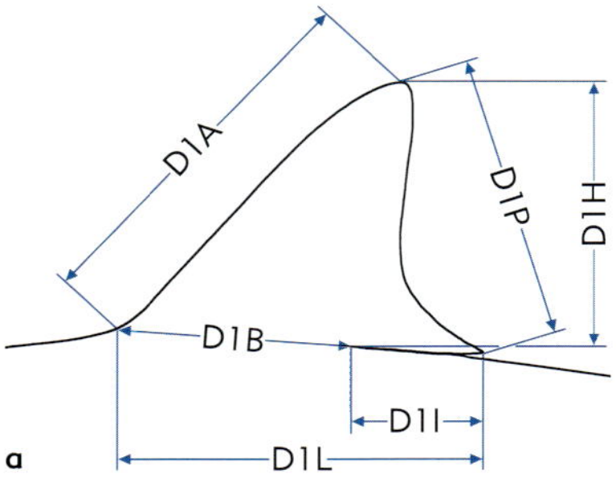

a

D1A	First Dorsal-Fin Anterior Margin
D1B	First Dorsal-Fin Base
D1H	First Dorsal-Fin Height
D1I	First Dorsal-Fin Inner Margin
D1L	First Dorsal-Fin Length
D1P	First Dorsal-Fin Posterior Margin

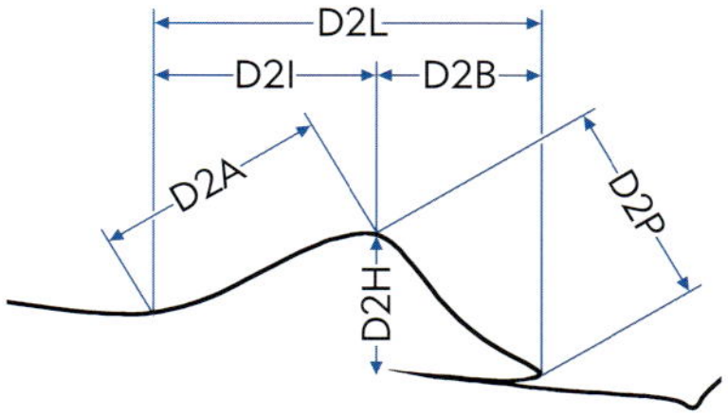

b

D2A	Second Dorsal-Fin Anterior Margin
D2B	Second Dorsal-Fin Base
D2H	Second Dorsal-Fin Height
D2I	Second Dorsal-Fin Inner Margin
D2L	Second Dorsal-Fin Length
D2P	Second Dorsal-Fin Posterior Margin

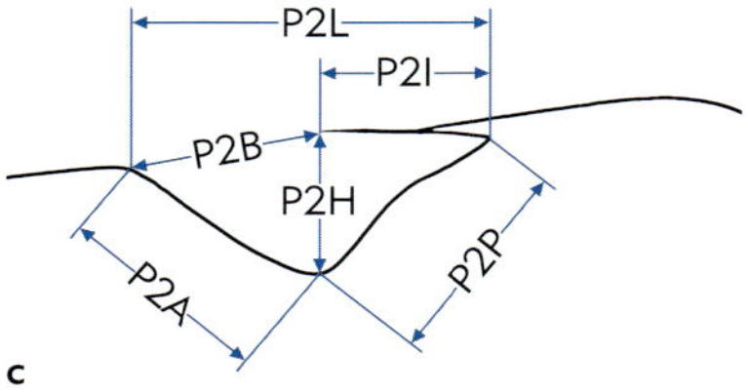

c

P2A	Pelvic-Fin Anterior Margin
P2B	Pelvic-Fin Base
P2H	Pelvic-Fin Height
P2I	Pelvic-Fin Inner Margin [Length]
P2L	Pelvic-Fin Length
P2P	Pelvic-Fin Posterior Margin [Length]

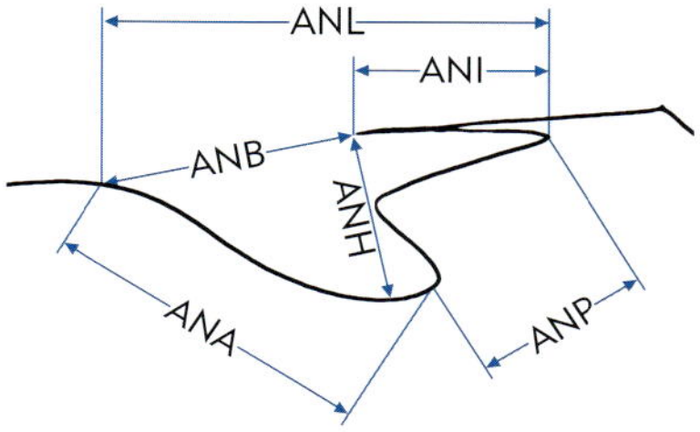

d

ANA	Anal-Fin Anterior Margin
ANB	Anal-Fin Base
ANH	Anal-Fin Height
ANI	Anal-Fin Inner Margin
ANL	Anal-Fin Length
ANP	Anal-Fin Posterior Margin

Figure 15. Measurements of dorsal (**a**, **b**), pelvic (**c**) and anal fins (**d**).

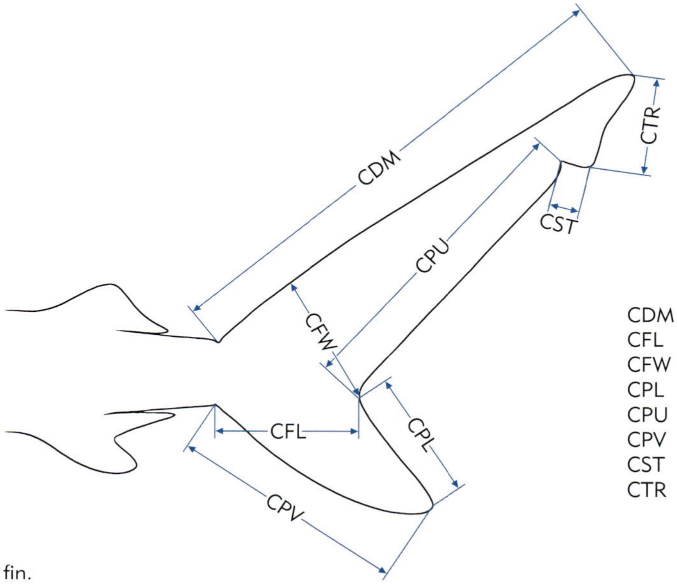

CDM	Dorsal-Caudal-Fin Margin
CFL	Caudal-Fin Fork Length
CFW	Caudal-Fin Fork Width
CPL	Lower Postventral Caudal-Fin Margin
CPU	Upper Postventral Caudal-Fin Margin
CPV	Preventral Caudal-Fin Margin
CST	Caudal-Fin Subterminal Margin
CTR	Caudal-Fin Terminal Margin

Figure 16. Measurements of caudal fin.

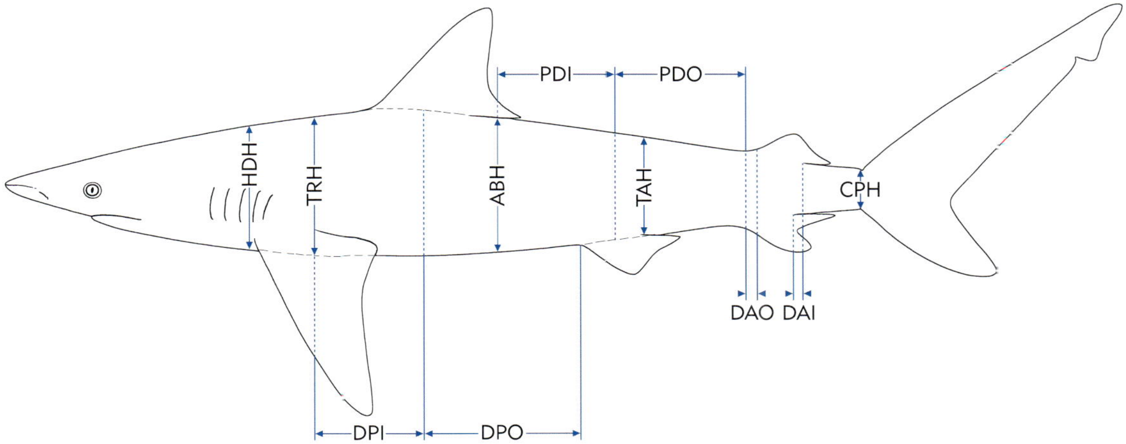

ABH	Abdomen Height	HDH	Head Height
CPH	Caudal-Fin Peduncle Height	PDI	Pelvic-Fin Midpoint First Dorsal-Fin Insertion
DAI	Second Dorsal-Fin Insertion Anal-Fin Insertion	PDO	Pelvic-Fin Midpoint Second Dorsal-Fin Origin
DAO	Second Dorsal-Fin Origin Anal-Fin Origin	TAH	Tail Height
DPI	First Dorsal-Fin Midpoint Pectoral-Fin Insertion	TRH	Trunk Height
DPO	First Dorsal-Fin Midpoint Pelvic-Fin Origin		

Figure 17. Other common measurements.

Technical Terms

For the preparation of the list of technical terms used in the field guide to describe anatomy and biology of sharks (and parts of them) explanations were taken (in parts) from COMPAGNO (1984, 1988), COMPAGNO et al. (1989, 2005), PURDY (1990), HERMAN et al. (1991) and STEVENS (2004), and modified.

Alternate overlapping dentition. Distal and mesial root lobe of tooth covering lobes of previous and succeeding tooth.

Ampullae of Lorenzini: Special sensing organs (electroreceptor's), forming a network of jelly-filled canals which help sharks to sense electric fields in the water as well as temperature gradients, e.g. to find hidden prey in sandy sea-bottom. Each ampulla consists of a jelly-filled canal opening to the surface by a pore in the skin and ending blindly in a cluster of small pockets. The ampullae are mostly clustered into groups visible on head as dark spots in the skin.

Anal fin: A single fin on the ventral surface of the body between the pelvic fins and caudal fin.

Anaulacorhizid: Vascularisation through scattered foramina of equal size on both outer and inner faces (e.g. within Hexanchidae).

Angular: The angularity of the shallow notch, which occurs about midway between the apex and base of the cutting edge, may be sub-angular.

Annular rings or **annuli:** In a vertebral centrum in cross section, rings of calcified cartilage separated by uncalcified cartilage that occupy the intermedialia only, or concentric rings that cross both the intermedialia and basalia.

Anterior: The front, forward, in the longitudinal direction of the snout tip to head end.

Anterior margin: In precaudal fins, the margin from the fin origin to its apex.

Anterior nasal flap: A flap on the front edges of the nostrils which serve to partially divide the nostril into incurrent and excurrent apertures.

Anterior teeth: Large, narrow teeth positioned near the symphysis.

Apex: In (precaudal) fins, the apical tip which can be acutely pointed to broadly rounded.

Apical: Towards cusp of oral teeth. Used to indicate the direction towards the tip of a fin also.

Apron: Expansion of the central part of the outer crown base.

Arcuate or arched: Forming an arc.

Basal: Bottom face concerned.

Basal groove: A deep groove proximal to the basal ledge on the labial surface of the crown and apical to root margin.

Basal ledge: A shelf-like projection basally on the labial surface of the crown.

Basals or basalia: In a vertebral centrum, the diagonal spaces below the attachment surfaces of the basidorsal cartilages above the basiventral cartilages, and between the two halves of the double cone. Basalia may be filled with uncalcified cartilage, may have diagonal calcifications penetrating the uncalcified cartilage, or may have calcified annuli or solid calcified cartilage that is continuous with calcification of the intermedialia.

Base: In precaudal fins, the proximal part of the fin between origin and insertion, extending distally, and supported by the cartilaginous fin skeleton. In caudal fin, that thickened longitudinal part of the fin enclosing the vertebral column and between the epaxial and hypaxial lobes or webs of the fin.

The anchoring structures to fix dermal denticles in the skin.

Basidorsals: Bilaterally paired cartilages on the dorsolateral side of the notochord, which are replaced by ossified neural arches or dorsal (arco)-centra, forming a continuous neural arch with the interdorsal cartilages to protect the spinal cord.

Basihyals: A small, thick, relatively immovable piece of cartilage that is found on the floor of the mouth of sharks ("sharks tongue").

Basiventrals: A pair of rounded or wedge-shaped cartilages on the ventro-lateral surfaces of a vertebral centrum that form the bases for attachment of ribs in monospondylous precaudal vertebrae. In diplospondylous precaudal and caudal vertebrae the basiventrals form haemal arches along with the interventral cartilages for protecting the caudal artery and vein.

Batoids: Rays and Skates.

Body: An entire shark sometimes restricted to the trunk and precaudal tail.

Branchial arch(es): The paired visceral arches behind the hyoid arch and just in front of the scapulocoracoid that support the gills. In elasmobranches the five to seven branchial arches primitively consist of a pair of dorso-medial wedge-shaped cartilages, the pharyngobranchials, closely situated against the roof of the pharynx. That are a pair of dorso-lateral and more cylindrical epibranchials connected dorsomedially to the pharyngobranchials, a pair of ventro-lateral cylindrical ceratobranchials that are connected ventro-laterally to the epibranchials, a pair of ventro-medial hypo-branchials that are connected ventro-laterally to the ceratobranchials, and unpaired ventro-medial basibranchials that are connected ventro-laterally to the hypobranchials. The hypobranchials and basibranchials along with the expanded ventral ends of the cerato-branchials form the basibranchial skeleton of the floor of the branchial pharynx. The branchial skeleton is variably modified in elasmobranches, with basibranchials and sometimes hypobranchials often lost, the last two pharyngobranchials and the last epibranchial often fused together, and the last basibranchial often expanded into a long, broad cupula with which the anterior hypo-branchials and posterior ceratobranchials articulate.

Calcified cartilage: Shark skeletons are formed of hyaline cartilage or gristle, but this is often reinforced with layers of calcified cartilage (cartilage impregnated with a mineral Hydroxyapatite, HAP) similar to that of bone but organized differently, in a hard, tile-like pavement of tiny tesserae, or more compactly as in the calcified structures of vertebral centra.

Carcharhinoid: Ground shark, member of the order Carcharhiniformes, including Carcharhinidae (requiem sharks), Hemigaleidae (weasel sharks), Leptochariidae (barbeled houndsharks), Proscylliidae (fin backed catsharks), Scyliorhinidae (catsharks), Triakidae (hound sharks), and Sphyrnidae (hammerhead sharks).

Carina: A low, blunt mesodistal ridge on tooth crowns replacing the cutting edge in sharks that eat hard-shelled invertebrate prey.

Caudal: Pertaining to the tail region.

Caudal fin: The tail fin.

Caudal keels: A fleshy ridge on each side of the caudal peduncle that may extend onto the base of the caudal fin, and may in other families and genera of shark extend forward as a body keel to the side of the trunk.

Caudal peduncle: The part of the precaudal tail extending from the insertions of the dorsal and anal fins to the anterior margin of caudal fin.

Centrum (pl.: **Centra**): A spool-shaped, partially or usually fully calcified structure that forms as a segmental constriction in the notochordal sheath of Neoselachians which is an articulated string. It forms the principal structural units of the vertebral column. Centra are primarily formed by the calcified double cones in the notochordal sheath, which may be their only calcification. In addition, secondary calcification may occur in the centrum between the outer surfaces of the calcified double cones, including calcified intermedialia, radii, annuli, and diagonal calcifications.

Ceratotrichia: From the Greek *keratos* (horn), and *trichos* (hair), slender soft or stiff filaments of an elastic protein, superficially resembling keratin or horn. Ceratotrichia run parallel and radial to the fin base and support the fin webs.

Ceratohyals: The ventral branches of the hyoid arch are situated within the elasmobranches (sharks) medially between the mandibular branches, ventro-medially connected due to basihyals. Pertaining to the cartilage they are situated below the epihyals in the hyoid arch.

Chimaera or chimaeroid: A member of the order Chimaeriformes, subclass Holocephali (see Holocephali).

Chondrichthyan: Referring to the class Chondrichthyes.

Chondrichthyes: The class Chondrichthyes, from Greek *chondros* (cartilage), and *ichthos* (fish), a major taxonomic group of aquatic, gill-breathing, jawed, finned vertebrates with primarily cartilaginous skeletons, 1–7 external gill openings, oral teeth in transverse rows on their jaws, and mostly small, tooth-like scales or dermal denticles. Chondrichthyes include the living Elasmobranchs, Holocephali and their numerous fossil relatives, and can be termed also shark-like fishes or simply sharks.

Chondrocranium: See neurocranium.

Claspers: The paired cylindrical copulatory organs, present on the pelvic fins of male sharks, for internal fertilization of eggs.

Classification: The ordering of organisms into groups on base of their relationships which may be originates from similarity or common ancestry.

Cloaca: The common chamber at the rear of the body cavity of Elasmobranches through which body wastes and reproductive products including sperm, eggs, and young pass, to be expelled to the outside through a common opening or vent.

Commissural: Tooth position near the end of jaw.

Costules: Short, vertical ridges sometimes present on inner and/or outer crown base.

Crown: The enamelated upper part of teeth with cutting edges.

Cusp: Usually a large, pointed projection of dermal denticles, sharp or blunt, associated with the medial and lateral ridges.

Cusplet: Small projection in association with a cusp.

Cutting edge: The compressed sharp, smooth or serrated longitudinal ridge on the mesodistal margins of oral teeth.

Dentine: The primary material of teeth, a hard tissue with numerous vascular and nonvascular canals.

Dermal denticle or placoid scale: Regular placoid scale, found in cartilaginous fishes, and it's various morphological modifications in certain areas. Basically a small tooth-like scale covered with enameloid. Consist on a core and base of dentine. Usually, small and often close set one to another covering the entire body.

Diagonal calcifications: In a vertebral centrum in cross-section, plate-like (diagonal calcified lamellae) or knob-like (diagonal calcified lobes) structures of calcified cartilage that partially fills the uncalcified basalia. They have a radial orientation from the centre of the centrum.

Dignathic: Heterodonty by having different tooth morphology in upper and lower jaws.

Diplospondylous vertebrae: Vertebrae of the tail with two centra and two basidorsal and basiventral elements per segment. Mostly, with a haemal arch which is formed by the basiventral and interventral elements. These include diplospondylous precaudal vertebrae between monospondylous vertebrae and caudal fin base, and diplospondylous caudal vertebrae in the caudal fin.

Distal: In any direction, at the far end of a structure.
Tooth edge or part toward angle of jaws.

Dorsal: The upper surface of the body or head, upwards, in the vertical direction of the back.

Dorsal fin: The unpaired fins (D1 and D2) along the upper surface of the back, between the head and caudal fin.

Dorsal lobe: In the caudal fin, the entire fin (including its base), epaxial and hypaxial webs but not including the ventral lobe.

Dorsal margin: In the caudal fin, the margin from the upper origin to its posterior tip.

Elasmobranch: Referring to the subclass Elasmobranchii.

Elasmobranchii: Subclass Elasmobranchii, from Greek *elasmos* (plate), and *branchos* (gill), in allusion to their plate-like gill septa. These are shark-like fishes, other than Holocephali (Chimaera), including the living non-Batoids (Rays and Skates), and hosts many fossil species. They differ from Holocephali in having 5–7 gill slits on each side of body, not protected by a hard gill cover (operculum), having separate oral teeth not formed as tooth plates, a fixed first dorsal fin with or without a fin spine, and a short spined or spineless second dorsal fin.

Embryo: An earlier development stage of the young of a live-bearing shark, ranging from nearly microscopic to moderate-sized but not like a miniature of the adults.

Enameloid: The shiny hard external coating of the tooth crowns.

Epaxial lobe or web: In the caudal fin, that part of the caudal fin between the base and dorsal margin supported by ceratotrichia.

Epaxial web: The entire fin web above the vertebral column and caudal fin base.

Epihyals: A segment next above the ceratohyal in the hyoid arch.

Euselachian: Referring to the Euselachii.

Euselachii: The cohort Euselachii, from Greek *Eu* (true, good or original), and *selachos* (shark or cartilaginous fish). The spined or 'phalacanthous' sharks, including the modern sharks (Neoselachii), and fossil shark groups, including Hybodonts, Ctenacanths, and the Xenacanths, all primitively with anal fins and having two dorsal fins with fin spines.

Excurrent apertures: The posterior and ventrally facing openings of the nostrils which direct water out of the nasal cavities and which are often partially covered by the anterior nasal flaps. These are usually medial on the nostrils and posterior-medial to the incurrent apertures but may be posterior to the incurrent apertures only.

Eye notch: A sharp, anterior or posterior indentation in the eyelid which is dividing the upper and lower eyelids.

File: Tooth row from symphysis toward end of jaw.

First dorsal fin: The anterior dorsal fin of two, ranging in position from over the pectoral fin bases to far posterior on the precaudal tail.

Foetus: A later development stage of the unborn young of a live-bearing shark which essentially resembles a small adult. Term foetuses are ready to be born, and generally have teeth and dermal denticles erupting. They have already a distinct colour pattern which is often more strikingly than in adults.

Free rear tips: All fins (pectoral, pelvic, dorsal, and anal) have a movable rear corner or flap which is separated from the trunk or tail by a notch and an inner margin. In some sharks the rear tips of some fins are very elongated.

Functional series: A series of teeth that are in functional position on the jaw.

Gill openings or **slits:** Long, narrow, paired rows of five transverse openings on the sides or underside of the head for the discharge of water through the gills.

Haemal arch: The arch ventral to the notochord or vertebral centra on tail vertebrae that is formed by basiventrals and interventrals and which houses the caudal artery and caudal vein in a haemal canal.

Head: That part of a shark from its snout tip to the last gill slits.

Hemiaulacorhizid: Vascularisation through a median groove and one or two lateral foramina on inner face (e.g. within the Squatinidae or Orectolobidae).

Heterodonty: Different tooth morphology within a tooth file. Two types of heterodonty are known: dignathic and monognathic.

Histotype: Type of internal tooth vascularisation.

Holaulacorhizid: Vascularisation through many small foramina concentrated in a median groove running from outer to inner face (e.g. within the Rajidae).

Holocephali: Covers the taxon Chimaera.

Holotype: Either the only specimen used and mentioned in an original description of a species, with or without a designation of such, or one of two or more specimens used and mentioned in an original description of a species and designated as such. This individual becomes the 'name-bearer' of the species, and is used to validate the species or scientific name by anchoring it to a single specimen.

Homodonty: Equal tooth morphology within a tooth file.

Hyoid arch: The visceral arch which supports the tongue and the rear of the upper jaws. The hyoid arch is between the mandibular arch and the first branchial arch, and has the spiracular pocket between it and the mandibular arch.

Hyomandibular pores: Series of pores, sometimes enlarged, extending posteriorly as a line directly behind corners of mouth.

Hypaxial web: The entire fin web below the vertebral column (vertebral axis) and the caudal base.

Hypercalcified structures: Parts of the skeleton that have developed extremely dense calcified cartilage, primarily during growth and maturation which sometimes swell to knobs that distort and engulf existing cartilaginous structures. The rostrum of the hardnose shark *(Carcharhinus macloti)* is a particularly impresssive hypercalcified structure.

Hypohyal: A ventral element of the hyoid arch which links ceratohyals and basihyals. The gill arches have two basic parts. Each part is associated with a main gill arch segment dorsally and ventrally. Thus, each of the branchial arches, the arches which actually function as respiratory arches in fishes, has an epibranchial and a ceratobranchial.

Imbricate overlapping dentition I: Distal root lobe of tooth covering mesial lobe of next tooth.

Imbricate overlapping dentition II: Distal root lobe of tooth covered by mesial lobe of next tooth.

Incurrent apertures: The anterior and ventral facing openings of the nostrils which direct water into the nasal cavities. Usually, they are laterally on the nostrils and anterior-lateral to the excurrent apertures but may be anterior to the excurrent apertures only.

Independent dentition: Teeth along a meso-distal series in which the roots do not overlap next tooth separated by a space.

Inner face: Directed toward the jaw cartilage.

Inner margin: In precaudal fins including the pectoral, pelvic, dorsal and anal fins, the margin from the fin insertion to the rear tip.

Insertion: The posterior or rear end of the fin base in precaudal fins. The caudal fin lacks insertions.

Interdorsal ridge: A ridge of skin on the mid-dorsal surface of back of sharks in a line between D1 and D2 (becomes an important feature for identification of sharks of the genus *Carcharhinus*, family Carcharhinidae).

Interventral cartilages: A pair of rounded or wedge-shaped cartilages fitting between the basiventral cartilages of each vertebra that in diplospondylous precaudal and caudal vertebrae form the haemal arches with the basiventral cartilages.

Jaws: See mandibular arch.

Keel: A fleshy or bony ridge/edge.

Labial: The outer face of teeth that is directed outside the mouth and towards the lips.

Labial cartilages: Paired internal cartilages that support the labial folds at the lateral angles of the mouth.

Labial folds: Lobes of skin at the lateral angles of the mouth. Usually, with labial cartilages inside them and separated from the sides of the jaws by pockets of skin (called labial furrows too).

Labial furrows or **labial grooves:** Grooves around the mouth angles on the outer surface of the jaws of many cartilaginous fishes isolating the labial folds. Primitively there is a distinct upper labial furrow above the mouth corner and a lower labial furrow below it.

Lateral or laterad: Outwards, refers to the sides, in the transverse direction towards the periphery of the body.
Tooth position half way along the jaw.

Lateral line: A sense organ to detect movement and vibration in the surrounding water. Lateral lines are usually visible as faint lines running lengthwise down each side, from head to base of tail. Sharks use the lateral line to detect magnetic fields as well (along with Ampullae of Lorenzini and other sense organs).

Lateral line canals: Canals in which neuromasts are not directly exposed to the environment but communicate with it via canal capsules. Additional neuromasts may appear individually at various locations on the body surface.

Lateral sense capsules: Capsules in the neurocranium for the nasal organs, eyes, and inner ears.

Lectotype: One of two or more specimens that were syntypes in an original description, designated as a lectotype by a subsequent writer. Then it becomes equivalent to a holotype, and anchors the name of the species to a specimen unless invalidated by a ruling of the International Commission on Zoological Nomenclature (ICZN) or a previous design-nation of a lectotype.

Lingual: The inner face of tooth that is directed inside the mouth and towards the tongue.

Live-bearing: A mode of reproduction in which female sharks give birth to their young, which are miniatures of the adults.

Longitudinal: Symphysial or commissural direction of a tooth file or structure.

Longitudinal ridges: In dermal denticles, the parallel ridges that extend anterior-posteriorly on the surface. These may be in the form of a single medial ridge (sometimes paired), paired lateral ridges, and may terminate in lateral cusplets.

Longitudinally: Apico-basally directed structuring on a tooth.

Lower eyelid: The ventral half of the eyelid, separated by a deep pocket (conjunctival fornix) from the eyeball.

Lower origin: In the caudal fin, the anteroventral beginning of the hypaxial or lower web of the caudal fin, at the posterior end of the anal-caudal or pelvic-caudal space.

Lower postventral margin: In the caudal fin, the lower part of the postventral margin of the hypaxial web from the ventral tip to the posterior notch.

Mandibular arch: The paired primary jaw cartilages of sharks, including the Palatoquadrates dorsally and Meckel's cartilages ventrally.

Meckel's cartilages: The paired lower jaw cartilages, articulating mesially with each other at the midline (or symphysis) of the lower jaw, and laterally with the distal ends of the Palatoquadrates. Meckel's cartilages are fused together at the symphysis in some sharks or are articulated to a symphysial cartilage in other fishes.

Medial: Inwards, in the transverse direction towards the middle of the body.

Median groove: Groove running from the inner root base to the inner crown-root junction dividing a holaulacorhizid type of root into two root lobes. It includes the main foramina of the vascularisation system.

Median keel: Transverse ridge dividing tooth crowns into inner and outer face.

Mesial: Tooth edge or part towards symphysis of left and right jaw halves.

Middorsal dermal ridge: Mostly, a low dermal ridge along the midline between the first and the second dorsal fin.

Monognathic: Heterodonty within one jaw only (appears as gradient or disjunction).

Monospondylous precaudal vertebrae: Vertebrae with one centrum and one pair of basidorsals, basiventrals, and ribs per body segment (myotome). Generally, extending from the occiput to the end of the body cavity and to over the pelvic girdle. Much variation in the position of the monospondylous-diplospondylous transition occur which can range well in front or behind the pelvic girdle.

Monospondylous-diplospondylous transition: The position on the vertebral column where monospondylous centra end and diplospondylous centra begin. In lateral view the transition often appears as an abrupt decrease in length of the diplospondylous centrum compared to the last monospondylous centrum but can be obscure in some species with very numerous, very short centra. Often, a centrum of intermediate length appears between a long monospondylous centrum and a short diplospondylous centrum. In a few shark species a transition zone of alternating long and short centra is present that marks the transition. Basidorsals and basiventrals have foramina for the spinal nerves on every other vertebra rather than on each vertebra as in monospondylous vertebrae. The transition from long to short centra is generally coordinated with the transition of vertebrae with free ribs and without haemal arches to those no ribs and with haemal arches. However, in some sharks the two transitions can be anterior or posterior to each other.

Nasal aperture: An aperture on the neurocranium, in the anterior-ventral surface of each nasal capsule through which the nostril directs water into and out of the nasal organ.

Nasal capsules: A pair of spherical, oval or trumpet-shaped, thin-walled structures on the neurocranium behind the rostrum (if present), in front of the orbits, cranial roof and basal plate. They serve as containers for the nasal organs (organs of smell), and have passages into the cranial cavity to connect the nasal organs with the brain.

Nasal curtain: Anterior nasal flaps that are expanded medially and posteriorly and have fused with each other. Nasal curtains are found in some Carcharhinoid sharks and in many Batoids.

Nasal flap: Skin flaps extending from the nostrils serving direct water into and out of them including the anterior, posterior, and mesonarial flaps.

Neck: A narrow band of finely porous dull tissue (possibly orthodentine) encircling the proximal end of the crown of a tooth, and apparently covered with dental membrane.

Neo-holaulacorhizid: Modification of the holaulacorhizid type of root combining a shallow median groove and an extremely expanded pulp cavity.

Neoselachian: Referring to Neoselachii.

Neoselachii: From Greek *neos* (new), and *selachos* (shark). This group of modern sharks consists of the living Elasmobranches and their immediate fossil relatives.

Neotype: A specimen, not part of the original type series for a species, which is designated by a subsequent author. Particularly, if the holotype or other types have been destroyed, if they were never designated in the original description or when they are presently useless.

Neural arch: In shark vertebrae, a dorsal arch formed by basidorsal and interdorsal cartilages above the centrum and forming a neural canal containing the spinal cord.

Neurocranium: In sharks, a box-shaped complex cartilaginous structure (chondrocranium) at the anterior end of the vertebral column, containing the brain, protecting and supporting the nasal organs, eyes, ears, and other sense organs, and which is supporting the visceral arches (or splanchnocranium).

Neuromasts: Receptors in the lateral line, each of which is composed of a group of hair cells. The hairs are surrounded by a protruding jelly-like cupula, typically 0.1–0.2 mm long. Hair cells and cupolas of the neuromasts are usually at the bottom of a visible pit or groove.

Nictitating lower eyelid: A movable lower eyelid that has special posterior eyelid muscles that lift it and, in some species, completely close the eye opening (or palpebral aperture). Often, the term is used incorrectly as nictitating membrane, a different, non-homologous structure in terrestrial vertebrates.

Nictitating upper eyelid: The upper eyelid has anterior eyelid muscles that pull it down and close the eye opening analogous to the nictitating lower eyelids of the Carcharhinidae.

Nomenclature: In biology, the application of distinctive names to groups of organisms.

Notch, shallow: A notch located on the cutting edge of tooth about the halfway between apex and base of the cutting edge.

Notch, deep: The distal cutting edge of the crown of a tooth which is more or less vertical and ends at the shoulder.

Nostrils: The external openings of the cavities of the nasal organs (organs of smell).

Notochord: In embryonic sharks a fluid-filled tube below the spinal cord that has a connective-tissue notochordal sheath surrounding it. The notochord forms the primitive developmental base of the chondrichthyan vertebral column. Chimaeroids retain the notochord and its sheath without constriction (although some have ring-like centra in the sheath), but in Neoselachii it is constricted by the development of double-cone calcifications of the centra within the sheath into biconical chambers between each centrum.

Olfactory sense: The sense of smell, a form of chemoreception is mediated by specialized sensory cells of the nasal cavity.

Origin: The anterior end of the fin base in all fins. The caudal fin has upper and lower origins but no insertion.

Orthodont: Histotype of vascularisation in which a tooth is supplied primarily by an internal pulp cavity radiating into numerous tiny canals penetrating the orthodentine layer.

Osteodont: Histotype of vascularisation, in which a tooth is supplied without any pulp cavity by scattered tiny cavities and canals penetrating the osteodentine layer of the root and the internal crown material.

Otic capsules: On the neurocranium a pair of complex thick-walled capsules containing the inner ears. They are located between the orbits and the occiput, partially separated medially by the cranial cavity.

Outer face: Directed outward the jaw cartilage.

Overlapping dentition: Mesodistal series of teeth in which the roots overlap (imbricate overlapping dentition), not separated by a gap.

Oviparity: Production of eggs inside the mother. The fertilised eggs are attached on substrate by adhesive filaments which are hardening after a while. The developing young in the egg capsules are nourished by their yolk-sac (adjective oviparous).

Ovoviviparity: Production of eggs, hatched inside the mother but without a placental connection of the embryos to the oviduct or uterus, and so they are not nourished by the mother. The embryos feed on the yolk in the yolk-sacs. The yolk is gradually depleted and the yolk-sacs reabsorbed until the young are ready to be born as miniature adults, free-swimming and feeding.

Paired fins: Refers to pectoral and pelvic fins.

Palatoquadrates: The paired upper jaw cartilages articulating mesially with each other at midline or symphysis of the upper jaw, and laterally with the distal ends of the Meckel's cartilages. The palatoquadrates of Neoselachians are divided into cylindrical antero-medial sectors or palatine processes, which are connected to each other at the symphysis. Sometimes the processes are variably modified conical to flattened articular structures or orbital processes on the middle of the palatoquadrates for attachment to the neurocranium at the orbital notches.
The posterodistal quadrate processes are elevated often. They form an articulation with the distal ends of Meckel's cartilages and are loosely or firmly attached to the distal ends of the hyomandibulae. In few living Neoselachii and many fossil Elasmobranches, the quadrate processes have postorbital articulations with the rear surfaces of the postorbital processes of the neurocranium.

Palpebral aperture: The eye opening, defined by the upper and lower eyelids.

Parasymphysial: First anterior tooth row, if a symphysicl tooth row is absent.

Paratype: Each specimen of a type series other than the holotype. Specimens other than the holotype become automatically paratypes unless the author designates them as referred specimens that are not part of the type series.

Pectoral fins: A symmetrical pair of fins on each side of the trunk just behind the head and in front of the abdomen.

Pedicle: In (lateral trunk) denticles a narrow stalk separating crown from the base.

Pelvic fin: A symmetrical pair of fins on the sides of the body between the abdomen and precaudal tail.

Placenta: See yolk-sac placenta.

Placental viviparity: Live-bearing in which the young develop a (yolk-sac) placenta which is apparently confined to the Carcharhinoid sharks.

Placoid scale: See dermal denticle.

Polyaulacorhizid: Vascularisation of a tooth root through many small foramina concentrated in several grooves running parallel from the outer to the inner face (e. g. in Myliobatidae).

Pores, pigmented: In few shark species the pores for the lateral line and Ampullae of Lorenzini are conspicuously black-pigmented, and look like little black specks.

Posterior: Rearwards, the hind or tail-end, in the longitudinal direction of the caudal-fin tip or tail filament.
Tooth position towards angles of jaw.

Posterior margin: In precaudal fins, the margin from the fin apex to either the free rear tip (in sharks with distinct inner margins) or the fin insertion (for those without inner margins).

Posterior nasal flaps: Low flaps or ridges arising on the posterior edges of the excurrent apertures of the nostrils.

Posterior notch: In the caudal fin, the notch in the postventral margin dividing it into upper and lower parts.

Posterior tip: The posterior corner or end of the terminal lobe of the caudal fin.

Postorbital processes: On the neurocranium, posterior-lateral projections of the supraorbital crests below which the postorbital walls originate.

Postventral margin: In the caudal fin, the margin from the ventral tip to the subterminal notch of the caudal fin.

Preanal ridges: A pair of low, short or long, narrow ridges on the midline of the caudal peduncle extending anteriorly from the anal fin base.

Precaudal fins: All fins in front of the caudal fin.

Precaudal pit: A depression at the upper and sometimes lower origin of the caudal fin where it joins the caudal peduncle.

Precaudal tail: That part of the tail from its base at the vent to the origins of the caudal fin.

Precaudal vertebrae: Vertebrae from the occiput to the dorsal origin of the caudal fin.

Predorsal ridge: A low, narrow ridge of skin on the midline of the back anterior to the first dorsal fin base.

Preorbital processes: On the neurocranium, anterior-lateral projections of the supraorbital crests below which the preorbital walls originate.

Preventral margin: In the caudal fin, the margin from the lower origin to the ventral tip of the caudal fin.

Proximal: In any direction, at the middle of a structure.

Pseudo-apron: Apron-like vertical ridge that appears sometimes on lateral and posterior teeth.

Pseudo-osteodont: The former pulp cavity of an originally orthodont histotype of tooth being filled secondarily with osteodentine.

Pseudosymphysial: One of the parasymphysial tooth rows placed in the position of the symphysial tooth row (symmetry).

Pulp cavity: Cavity inside the tooth from which the vascularisation is spread via canaculi (see central foramen).

Recurved: The apical portion of the crown is curved.

Replacement series: A series of teeth lingual to the functional series and not in a functional position on the jaw.

Rhomboidal: In the form of a rhombus or diamond.

Root: Non-enamelated tooth part which forms the junction with the jaw and provides vascularisation of the tooth.

Root lobe: Often, tooth roots are divided into separate lobes at their midlengths which are termed mesial and distal root lobes.

Rostrum: On the neurocranium, a rigid projection of the snout, including lateral line canals and masses of ampullae, and is located in front of the nasal capsules and anterior fontanelle.

Rostrum, tripodal: The rostrum of the neurocranium in Carcharhinoid sharks is primitively tripodal. With a pair of dorsolateral lateral rostral cartilages that arise from the posterior-laterodorsal surfaces of the nasal capsules or from the preorbital wall, and a medial rostral cartilage that arises from the anteromedial surface of the internasal septum. The medial and lateral rostral cartilages extend anteriorly and articulate or fuse at the rostral node.

Row: A single replicating line of teeth, from inner to outer face of jaw approximately transverse to the longitudinal jaw axis, which includes functional teeth and their replacements.

Secondarily anaulacorhizid: Median groove of a holaulacorhizid type of root, totally overgrown to form a closed tube, internally connected or merged with the pulp cavity.

Secondarily hemiaulacorhizid: Median groove of holaulacorhizid type of root, overgrown to various extents but terminally groove or pores still open.

Second dorsal fin: The posterior dorsal fin of two, ranging in position from over the pelvic-fin bases to far posterior on the precaudal tail.

Secondary lower eyelid: The eyelid below or lateral to the nictitating lower eyelid separated from it by a subocular groove or pocket. In many Carcharhinoid shark species with internal nictitating lower eyelids it is functionally replacing them as lower eyelids.

Series: A line of teeth along the jaws, parallel to the jaw axis and which includes teeth from all rows present.

Serration: Minute teeth formed by the cutting edge of the crown that enhance the slicing abilities of the teeth.

- **weak:** can been seen clearly under usage of a magnifying glass or microscope only.
- **very fine:** not easily to see with naked eye but can be felt by running finger nail along the cutting edge.
- **fine:** visible with the naked eye but the number of serrations is not easily countable with naked eye.
- **coarse:** number of serrations is easily countable with the naked eye.

Sexual maturity: The age or stage when an organism can reproduce, and is sometimes considered synonymous with adulthood.

Shark: Generally, used to name cylindrical or flattened cartilaginous fishes with 5–7 external gill openings on the sides of their heads, pectoral fins that are not attached to the head above the gill openings, and a large, stout tail with a large caudal fin, that is, all living Elasmobranches except Batoids.
The term shark is also used loosely for fossil Chondrichthyans that are not Neoselachians but have a shark-like form, even for 'spiny sharks' (Acanthodians) and for certain teleosts.

Skull: Head skeleton of sharks which includes the neuro- and splanchnocranium (or visceral arches). The visceral arches articulate with and are associated with to neurocranium but are not fused to it.

Snout: That part of sharks in front of its eyes and mouth including the nostrils.

Spiracle: A small to large opening between the eyes and first gill opening representing the modified gill opening in the pharynx between the jaws and hyoid (tongue) arch.

Splanchnocranium: That part of shark skulls which is including the visceral arches. These include the jaws or mandibular arch, the tongue or hyoid arch, and the five to seven gill or branchial arches (called also viscerocranium).

Squalene: A long-chain oily hydrocarbon present in the liver oil of cartilaginous fishes.

Striae: Vertical ridges running from crown base toward apex.

Subterminal margin: In the caudal fin the margin from the subterminal notch to the ventral beginning of the terminal margin.

Subterminal mouth or ventral mouth: Mouth located on the underside of the head behind the snout. It is termed as inferior mouth also, in reference to its ventral position, but not in its function.

Subterminal notch: On the caudal fin of most sharks, the notch in the lower distal end of the caudal fin between the postventral and subterminal margins, and defining the anterior end of the terminal lobe.

Sulcus: Groove developed by the primary vascularisation canals leading from root base to the main foramina in anaulacorhizid root type. It differs from the median groove in which several foramina are concentrated of the holaulacorhizid root type and the parallel grooves of the polyaulacorhizid root type, respectively, in that a sulcus lacks foramina.

Symphysial or symphysials: Teeth at the junction (symphysis) of both halves of a jaw.

Symphysis: The midline (centre) of the upper and lower jaws where the paired jaw cartilages are connected with each other to form an articulation.

Syntype: Two or more specimens used and mentioned in an original description of a species where was no designation of a holotype or a lectotype and paratype(s) by the describer of the species.

Systematics: Scientific study of the kinds and diversity of organisms including relationships between them.

Tail: That part of sharks from the cloacal opening or vent to the tip of the caudal fin or caudal filament, including the anal fin, furthermore the second dorsal fin (if present) and caudal fin.

Taxon (pl.: **taxa**): A taxonomic group at any level in a classification.

Taxonomy: Often used as synonym of systematic or classification but narrowed by some researchers to the theoretical study of the principles of classification.

Term foetus: See foetus.

Terminal lobe: In the caudal fin, the free rear wedge-shaped lobe at the tip of the caudal fin, extending from the subterminal notch to the posterior tip.

Terminal margin: In the caudal fin, the margin from the ventral end of the subterminal margin to the posterior tip.

Tongue arch: See hyoid arch.

Transversal: Mesio-distally directed. The term is used to describe outer or inner direction of a tooth row.

Transverse groove: In oral teeth, a deep groove transverse on the lingual root surface, transecting it into mesial and distal root lobes.

Transverse notch: A distinct notch in the proximal labial edge of the root of teeth at about its midlength.

Transverse ridges: Small narrow ridges on the labial and lingual surfaces of tooth crown, apico-basally oriented and sometimes exten-ding to the cusp edges.

Trunk: That part of sharks between its head and tail from the last gill openings to the vent, including the abdomen, back, pectoral and pelvic fins, and often the first dorsal fin.

Umbilical cord: A modified yolk stalk in placental-viviparous sharks, carrying nutrients from the placenta to the foetus.

Unpaired fins: Means D1, D2, anal, and caudal fin.

Upper eyelid: The dorsal half of the eyelid, separated by a deep pocket (conjunctival fornix) from the eyeball. The upper eyelid fuses with the eyeball and the pocket is absent in all Batoids.

Upper origin: In the caudal fin, the anterior-dorsal beginning of the epaxial or upper web of the caudal fin at the dorso-caudal posterior end.

Upper postventral margin: In the caudal fin, the upper part of the postventral margin of the hypaxial web from the posterior notch to the subterminal notch.

Uvula: Lobate extension of the inner crown base.

Vascularisation: Growth of blood vessels into a tissue with the result that the oxygen and nutrient supply is improved.

Vent: The opening of the cloaca on the ventral surface of the body between the inner margins and at the level of the pelvic fin insertions.

Ventral: Downward, refers to the lower surface or underside of the body or head, in vertical direction of the abdomen.

Ventral fin: See pelvic fin.

Ventral lobe: In the caudal fin, the expanded distal end of the preventral and lower postventral margins, defined by the posterior notch of the caudal fin.

Ventral margin: In the caudal fin, the entire ventral margin from lower origin to posterior tip. Either, distinct as a continuous margin or is variably subdivided into preventral, postventral, subterminal and terminal margins.

Ventral tip: In the caudal fin, the ventral apex of the caudal fin where the preventral and postventral margins merge.

Vertebra (pl.: **vertebrae**): A single unit of the vertebral column, including a vertebral centrum and associated cartilages that form neural arches and ribs or haemal arches.

Vertebral column: The entire set or string of vertebrae from the rear of the chondrocranium to the end of the caudal base.

Visceral arch: See splanchnocranium.

Viviparity: Live-bearing, in which a placenta is developed, and the young are nourished by the mother. Placental viviparity is confined to the Carcharhinoid sharks.

Yolk sac or yolk sack: Almost all sharks start embryonic development as a large spherical yolky egg inside an elongated shell, the egg shell. A small disk of dividing cells represents the pre-embryo or blastula atop the huge yolk mass. The blastula expands around the sides and ventral surface of the yolk mass, and differentiates into an increasingly shark-like embryo: The yolk sac or bag-like structure is containing the yolk, and a narrow tubular yolk stalk, between the abdomen of the embryo and the yolk sac.

Yolk stalk: The connecting passage between embryo or foetus and yolk sac which allows yolk to pass from the sac into the embryonic gut.

Yolk sac placenta: An organ in the uterus of sharks (Carcharhinoid sharks), formed from the embryonic yolk-sac of the embryo and maternal uterine lining through which maternal nutriment is passed to the embryo. It is analogous to the placenta of live-bearing mammals. There are several forms of yolk-sac placentas in Carcharhinoid sharks, including entire, discoidal, globular, and columnar placentas (see COMPAGNO 1988, 2005).

Yolk sac viviparity: Live-bearing in which the young are nourished primarily by the yolk in the yolk sacs which is gradually depleted and reabsorbed until the young are ready to be born.

Key to Genera and Species

The keys to genera and species cover primarily external characteristics. However, features as nuances of fin shape are not listed because such descriptions are limited to the identifier subjective point of view. To avoid to key out some genera or species more than once internal features for identification such as descriptions of teeth are included. Due to teeth wide inter- and intraspecific variation, overlap in shape and in dental formula further internal characters, e.g. vertebral counts are given for an unambiguous identification of a genus or species. In addition, in the key to species the maximum total length (TL) and the wider range of the spatial distribution of the species are given in parenthesis.

Key to genera

Ia First and second dorsal fin are nearly of equal size. II

Ib First dorsal fin considerably higher than second dorsal fin. III

IIa Snout long and parabolic, labial furrows very short. Upper precaudal pit longitudinal, crescentic. D1 origin over pectoral inner margin, pectoral origin under 4th gill slit or between 4th and 5th. Upper teeth broad, triangular, with serrated cusps, lowers with narrow, smooth cusps. .... *Lamiopsis*

IIb Snout short, rounded to wedge-shaped, labial furrows very short. Upper precaudal pit longitudinal but not crescentic. D1 origin over pectoral rear tips, pectoral origin between 3rd and 4th gill slit opening. Upper and lower teeth with narrow, unserrated cusps. .... *Negaprion*

IIIa Spiracles present. IV

IIIb Spiracles absent. V

IVa Spiracles large, slit-like, very long labial furrows, extending to front of eyes. Large eyes without posterior notches, prominent interdorsal ridge and lateral keels on caudal fin peduncle. Vertical black or dusky bars on back, obscure or absent on adults. Upper and lower teeth cockscomb-shaped. .... *Galeocerdo*

IVb Spiracles normally very small (occasionally absent) present as tiny, slit-like pores. Eyes small with posterior notches, expanded anterior-nasal and mesonarial flaps forming a tube for excurrent aperture. Dorsal fins highlighted with white tips. Teeth in both jaws with high proximal and distal cusplets. .. .. *Triaenodon*

Va Without papillose gillrakers on gill arches, no lateral keels on caudal fin peduncle. VI

Vb Papillose gillrakers on gill arches present. Body brilliant dark blue dorsally. Eyes fairly large without posterior notches, weak distinct lateral keels on caudal peduncle. Precaudal pit transverse and crescentic. D1 origin clearly behind pectoral rear tips, teeth slender and triangular, serrated, without cusplets. ....*Prionace*

VIa Nostrils fairly small, internarial space moderately large. VII

VIb Nostrils very large, transverse and close-set. Snout long and very narrow, eyes fairly large, without posterior notches. Internarial width small, interdorsal ridge absent, without dermal keels. Upper precaudal pit transverse and crescentic, D1 origin over pectoral inner margin, teeth without cusplets. ... .*Nasolamia*

VIIa Teeth of lower jaw not protruding when mouth is closed. IIX

VIIb Lower teeth prominently protruding when mouth is closed, occasionally with or without cusplets. Snout broadly parabolic or rounded, eyes extremely small without posterior notches. Upper precaudal pit longitudinal, not crescentic. D1 origin far anterior, over last third of pectoral base. ....*Glyphis*

IIXa Eyes small to mid-sized, without posterior notches. IX

IIXb Eyes large, with posterior notches. Snout narrowly parabolic, very long. Labial furrows very short or rudimentary, without dermal keels on caudal peduncle, upper precaudal pit transverse and crescentic. D1 origin behind pectoral rear tips, teeth in both jaws without cusplets. .... *Loxodon*

IXa Anal fin higher than D2. X

IXb Anal fin (much) smaller than D2. XI

Xa Labial furrows varying from short (confined to mouth corners) to fairly long (falling behind eyes) but can be reduced in a few species. Snout narrow, broadly parabolic to obtusely wedge-shaped. Interdorsal ridge absent or rudimentary, anal fin with very long preanal ridges, teeth serrated, without cusplets. .... *Rhizoprionodon*

Xb Labial furrows very short or rudimentary, snout parabolic or bell-shaped, very long. Interdorsal ridge absent or rudimentary, anal fin with short preanal ridges. D1 origin over or behind pectoral rear tips, pectoral fins very broad, teeth without serration and cusplets. .... *Scoliodon*

XIa Snout long, acutely triangular. Eyes extremely small without posterior notches. Labial furrows very short but prominent. D1 origin halfway over pectoral inner margin. Anal fin with very short preanal ridges. Dentition varies from 49 to 61 tooth rows in upper and 49 to 56 rows in lower jaw, without cusplets. .... *Isogomphodon*

XIb Snout shape varies from narrowly parabolic or subangular to bluntly rounded or truncate, short or long. Labial furrows short, confined to mouth corners, end of uppers falling far behind eyes. Interdorsal ridge variably, it can be absent but occasionally present and prominent, sometimes vestigial. Upper precaudal pit transverse and crescentic, no dermal keels on caudal peduncle. Teeth shape is highly variable, serration is usually present, with or without cusplets, most species of the genus do not exceeding 33 tooth rows. .... *Carcharhinus*

Key to species of the genus *Carcharhinus*

1a Interdorsal ridge present. 2

1b Interdorsal ridge absent. 14

2a First dorsal fin with white tips or white trailing margins. 3

2b First dorsal fin not white, nor with white markings. 5

3a First dorsal and most of the other fins white mottled or with white margins (adults), distinct black dorsal saddles on caudal peduncle (juveniles), interdorsal ridge very low or occasionally absent, 125–128 precaudal centra (up to 400 cm TL; world-wide). .... *longimanus*

3b First dorsal fin not white mottled. 4

4a First dorsal fin and all other fins have white tips and posterior trailing margins, second dorsal fin dusky, 12–14 broad, triangular teeth in each upper half of jaw 116–122 precaudal centra (up to 300 cm TL; Indo-Pacific and eastern Pacific). .... *albimarginatus*

4b First dorsal fin without white margins, only dusky or plain. 5

5a Second dorsal fin with prominent black margins. 6

5b Second dorsal fin without black margins, only dusky or plain. 7

6a Second dorsal fin (occasionally), ventral caudal lobe and pectorals black tipped. First dorsal fin with a very narrow black trailing margin, anal fin plain, coarse serration on basal parts of tooth, 66–73 precaudal centra (up to 160 cm TL; Indo-Pacific). .... *sorrah*

6b Second dorsal fin with a prominent black or dusky tip, other fins with light margins, upper teeth strongly oblique, teeth on basal parts coarsely serrated, 56–74 precaudal centra (up to 100 cm TL; Indo-Pacific). .... *dussumieri*

7a First dorsal fin origin behind pectoral free rear tips, tips on fins other than first dorsal fin dusky but not black, upper teeth strongly serrated, 99–105 precaudal centra (up to 330 cm TL; worldwide distributed, but not in the Mediterranean). .. .. *falciformis*

7b First dorsal fin origin in front or above of pectoral free rear tips. 8

8a Origin of the first dorsal fin anterior to pectoral fin insertion, first dorsal fin very high, upper teeth noticeably long, triangular in shape, interdorsal ridge sometimes fairly low, 82–97 precaudal centra (up to 300 cm TL; worldwide). ... . *plumbeus*

8b Origin of the first dorsal fin posterior to pectoral insertion. 9

9a Not more than 13 upper and 12 lower teeth in each half of jaw, teeth narrow, triangular and serrated, black margins on anal and caudal fin, long expanded black tips on pectorals ventral side, 104–108 precaudal centra (up to 295 cm TL; worldwide). .... *perezii*

9b Dental formula different (usually, more than 13 upper and 12 lower teeth in each half of jaw). 10

10a Usually, 14 upper and 13 to 14 lower teeth in each half of jaw, teeth high and narrow, basally both tooth keels serrated, apically smooth, hard hyper-calcified rostrum, 68–71 precaudal centra (up to 110 cm TL; Indo-West Pacific). ....*macloti*

10b More than 14 teeth in each half of upper jaw. 11

11a Usually, 15 upper and 14 to 15 lower teeth in each half of jaw, upper teeth narrow and noticeably long, nasal flaps remarkable elongated, interdorsal ridge high, 108–110 precaudal centra (up to 300 cm TL; worldwide). .....*altimus*

11b Usually, 15 or more upper teeth in each half of jaw. 12

12a Upper teeth oblique, narrow, distal and mesial keels smooth (newborn), very coarse serration on both teeth keels (adults), snout very pointed, green eye colour, 98–104 precaudal centra (up to 280 cm TL; western and eastern Atlantic). .... *signatus*

12b Upper teeth oblique. 13

13a Lower teeth triangular, upper teeth narrow, basally strongly serrated and oblique, height of first dorsal fin 9.5–11.2 % TL, height of second dorsal fin 2.5–2.8 % TL, 104–109 precaudal centra (up to 370 cm TL; insular, worldwide). .... *galapagensis*

13b Lower teeth not triangular, slightly expanded distally in larger specimens, upper teeth finely serrated, height of first dorsal fin 6.0–7.9 % TL, and height of second dorsal fin 1.8–2.3 % TL, 88–93 precaudal centra (up to 400 cm TL; worldwide). .... *obscurus*

14a First dorsal fin white tipped and with trailing margins, black margins on second dorsal and anal fin, entire trailing margin of caudal fin, 111–115 precaudal centra (up to 170 cm TL; western Indian Ocean). .... *wheeleri*

14b First dorsal fin without white markings. 15

15a First dorsal fin with black tips or margins. 16

15b First dorsal fin without black tips or margins, only dusky or plain. 24

16a First dorsal fin, ventral lobe of caudal fin and anal fin with a prominent black blotch, brilliantly highlighted proximally with white, black tips on pectorals, 116–117 precaudal centra (up to 180 cm TL; Indo-Pacific, Mediterranean). .... *melanopterus*

16b First dorsal fin only with small black tips or margins. 17

17a Black tips on all fins, first dorsal fin origin over the middle of the pectoral inner margin, height of first dorsal fin 10.5 % TL, 115 precaudal centra, upper teeth margins smooth (up to 75+ cm TL; Red Sea). ....*leiodon*

17b Black tips not on all fins. 18

18a Conspicuous black tips on pectorals, second dorsal fin and ventral caudal lobe; other fins (except pelvics and anal fin) with small black tips, first dorsal fin with very narrow dark margin, long first dorsal rear tip (up to 200 cm TL; Indo-West Pacific). ....*hemiodon*

18b Conspicuous black tips on other fins or on few fin tips present. 19

19a Black tips on first and second dorsal fin, also on ventral lobe of caudal fin, outer tips of pectorals and pelvics are black tipped, dorsal lobe of caudal fin dusky, anal fin pale, 78–96 precaudal centra (up to 170 cm TL; Indo-West Pacific). ....*amblyrhynchoides*

19b Black tips on other fins or fewer black tips present. 20

20a Black tips on first and second dorsal, and on ventral lobe of caudal fin, apices of pectorals and anal fin are black, pelvic fins pale, height of first dorsal fin 6.0–8.8 % TL, 77-86 precaudal centra (up to 280 cm TL; Indo-Pacific, Atlantic). ... . *brevipinna*

20b Black tips on other fins or fewer black tips present, height of first dorsal fin different. 21

21a Second dorsal fin and ventral caudal lobe with black blotches, first dorsal fin and dorsal caudal lobe with black margin, pelvics and pelvic region black, height of first dorsal fin 8.2–12.4 % TL, 94–100 precaudal centra (up to 250 cm TL; worldwide). ....*limbatus*

21b Black tips on other fins or fewer black tips. 22

22a Black tips on second dorsal fin, and ventral caudal lobe, first dorsal fin and dorsal caudal lobe with black margins, pelvics plain, height of first dorsal fin 9.9–10.9 % TL, 86–91 precaudal centra (up to 200 cm TL, Indo-Pacific, off Australian shelves). .... *tilstoni*

22b Black tips on other fins or fewer black tips present. 23

23a Narrow black margins on first and second dorsal fin, black margins expanded to broad tips on both lobes of caudal fin and pectorals, 86–90 precaudal centra (up to 150 cm TL; Indo-South Pacific). ....*cautus*

23b Without narrow black margins on first or second dorsal fin, first dorsal fin plain. 24

24a Broad black margin on entire posterior margin of the caudal fin, other fins only with inconspicuous dark tips or dusky margins, origin of first dorsal fin over or just in front of pectoral free rear tips, 111–118 precaudal centra (up to 255 cm TL; Indo-Pacific). .... *amblyrhynchos*

24b Caudal fin without prominent black margins. 25

25a Snout tip with a prominent black or dusky blotch, second dorsal fin and dorsal caudal lobe dusky spotted to completely black, a trailing margin on the ventral lobe can be present, 81–88 precaudal centra (up to 200 cm TL; western Atlantic). .... *acronotus*

25b Snout tip without a black blotch. 26

26a Second dorsal fin with a conspicuous black blotch, other fins with pale posterior edges, but no black markings, teeth strongly serrated basally, but they are not serrated in themselves, 75–84 precaudal centra (up to 100 cm TL; Indo-Pacific). .... *sealei*

26b Without conspicuous coloration on all fins. 27

27a Hyomandibular pores forming a discrete series of five to eight enlarged pores directly behind corners of mouth, first dorsal fin and dorsal lobe of caudal fin dusky, other fins with light edges, 61–63 precaudal centra (up to 100 cm TL; Indo-West Pacific). ....*borneensis*

27b Hyomandibular pores are not enlarged and are not forming a discrete series of pores. 28

28a Pectoral fins very broad, upper teeth strongly serrated on both keels, finely to even smooth on apical part, coarser basally, 58 precaudal centra (up to 150 cm TL; Australia, western Pacific). ....*fitzroyensis*

28b Pectoral fins are not broad, no cusplets basally on teeth. ... .29

29a Pectoral fins fairly small, strongly serrated upper teeth, origin of first dorsal fin slightly posterior to inner pectoral corner, 41–67 precaudal centra (up to 150 cm TL; Indo-Pacific, western Atlantic). .... *porosus*

29b Pectoral fins not noticeably small. 30

30a Usually, 12 or more teeth in each half of upper jaw, height of first dorsal fin 11.1–12.0 % TL, origin of first dorsal fin over or just behind pectoral insertion, 90-94 precaudal centra (up to 280 cm TL; Indo-West Pacific, eastern Atlantic). .... *amboinensis*

30b Usually, more than 12 teeth in each half of upper jaw, lower height of first dorsal fin, higher number of precaudal centra. 31

31a Normally, 13 or more teeth in each half of upper jaw, height of first dorsal fin 7.0–10.8 % TL, origin over or slightly posterior pectoral fin insertion, nearer to inner pectoral corner, 110–119 precaudal centra (up to 350 cm TL; worldwide). . ... *leucas*

31b Normally, more than 13 teeth in each half of upper jaw. ... .32

32a Usually, 15 teeth in each half of upper jaw, fine or no serration on upper and lower teeth, pectoral fins are small, height of first dorsal fin 7.5–10.6 % TL, 78–80 precaudal centra (up to 200 cm TL; western and eastern Atlantic). ....*isodon*

32b Teeth serrated, distributed in other regions, shape of the upper labial furrows different. 33

33a Usually, 15 and more teeth in each half of upper jaw, upper teeth coarsely serrated, lower teeth finely serrated, height of first dorsal fin 6.8–9.7 % TL, 98–110 precaudal centra (up to 294 cm TL; worldwide). .... *brachyurus*

33b Usually, 15 and more teeth in each half of upper jaw, upper teeth coarsely serrated, lower teeth finely serrated, height of first dorsal fin 8.8–9.6 % TL, upper labial furrows typically Y-shaped (up to 285 cm TL; eastern Atlantic and Mediterranean). .... *acarenatus*

Species Characterisations

Carcharhinus acarenatus Moreno & Hoyos (1983a)

En: Strait shark; Fr: Requin du Détroit; Sp: Jaquetón del Estrecho; Ge: Seestraßenhai

Carcharhinus acarenatus Moreno & Hoyos (1983a). Holotype: VBCM 816. Type material is a male of 174.2 cm TL, caught near Casablanca-Rabat.

Synonyms: *Carcharhinus brachyurus* Günter (1870), designated by Garrick (1982).

Authors note. Hitherto, this species is listed in species catalogues and field guides about the genus *Carcharhinus* under the group synonym *C. brachyurus* (designated and discussed by Garrick 1982). In contrast to it, investigations by Moreno & Hoyos (1983a), Moreno (1995) and Barrull & Mate (2002) have revealed several distinct differences between *C. brachyurus* and *C. acarenatus*.

After studying species descriptions we are convinced *Carcharhinus acarenatus* should be separated from *C. brachyurus* and referred to a new single species as valid member of the genus *Carcharhinus*.

Several measurements usually taken from examined specimens to compare species one to another can not be given because the type material of the introduced was unavailable and thus, this species was not considered in creating graphs (Figure 51–59).

Diagnostic features. A large, slender species, without interdorsal ridge. Upper precaudal pit well developed, lower pit weak. Internarial width 1.2–1.4 times the preoral length. Rounded and flattened snout of moderate length, considered ventrally seems wide and parabolic shaped, eyes circular and small (diameter 1.1 % TL). Eyes anterior margin above or slightly anterior to the front of mouth. Nostrils oblique, with small, ovate, incurrent apertures. Anterior nasal flaps

Figure 18 *Carcharhinus acarenatus*: 174.2 cm TL male from the western Mediterranean; **a**, left side; **b**, ventral view; **c**, enlarged left nostril; **d**, left upper and lower teeth; **e**, detailed view of mouth corner.

with long, pointed lobes. Upper labial furrows short and typically Y-shaped. Gill slits short, hyomandibular pores alongside corners of mouth inconspicuous. First dorsal fin large and high (9.2 % TL), falcate, with a pointed or narrowly rounded apex. Its margin curving ventrally from fin apex, a short rear tip present. D1 origin over or anterior to pectoral free rear tips. D2 small and low (height 2.6 % TL), with a short rear tip and short inner margin. Fin origin over or slightly behind anal fin origin. Pectoral fins long (length of anterior margin 20.0 % TL). Number of vertebrae unknown (detailed investigations are not accomplished yet), what complicates comparisons with other species. Dermal denticles closed set, imbricate and sub-circular. With three low longitudinal ridges and a corresponding number of short posterior cusplets in smaller specimens, five posterior cusplets in larger ones.

Dental Formula:

$$\frac{15 - 2 - 15}{15 - 1 - 15} \quad \text{to} \quad \frac{15 - 2 - 15}{(14)\,15 - 1 - (14)\,15}$$

Upper teeth are narrow, crowns triangular, oblique. Distal median edges are concave on upper half of crowns, convex basally, deeply notched. Mesial edges are straight to convex, deeply notched halfway of the crowns. Both median edges finely serrated, basally more coarsely serrated, sometimes even with irregular serration. Roots arched, slightly notched. One to three rows of symphysial teeth are present. Lower teeth narrow, crowns erect to slightly oblique. Both median edges are concave and notched basally. Mesial edges convex apically, bent, fine and uniformly serrated, distal edges smooth-edged. Roots are arched, slightly notched. One to three rows of symphysial teeth are present. Files in upper and lower jaw are arranged as alternate overlapping dentition.

Colour. Dark or light grey above, with violet reflections. With an inconspicuous dark posterior trailing margin on D1, and on upside of pectoral fins. Ventrally white, dusky to black tips on lower side of pectorals, and a trailing white margin on the lower caudal lobe may be present.

Size. Maximum up to 285 cm TL, and an average of 200–230 cm TL (MORENO & HOYOS 1983a, MORENO 1995, BARRULL & MATE 2002). Largest caught specimen ever was a female of 235 cm TL, caught in August 1979 in the estuary of the Mouluya River between Morocco and Algeria (DEMADDALENA & BÄNSCH 2005).

Misidentifications. The sympatric species that might be mistaken with *C. acarenatus* is *C. brachyurus* but this species differs from the strait shark in having different shaped labial furrows (not Y-shaped), lacking an interdorsal ridge, and in having a different dental formula. Furthermore, confusion is possible with *C. falciformis*, *C. obscurus* and *C. galapagensis* but *C. acarenatus* differs from the mentioned species in tooth-shape, dental formulae, absence of an interdorsal ridge, and an in having a different D1 origin.

Geographical Distribution. Eastern Atlantic and the Mediterranean but detailed data are deficient. The species seems to be more abundant in the western Mediterranean. – (MORENO & HOYOS 1983a, DEMADDALENA & BÄNSCH 2005).

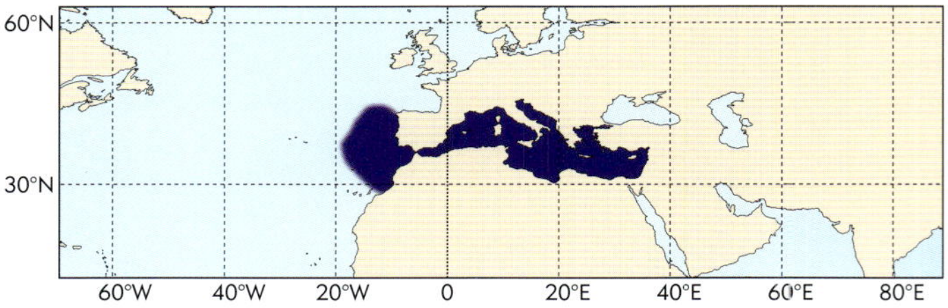

Habitat. *Carcharhinus acarenatus* is pelagic living and commonly found over continental and insular shelves in water depths of 10–600 m (MORENO & HOYOS 1983a, DEMADDALENA & BÄNSCH 2005).

Biology. It is a quick swimming species, more active in the night, and known to travel over long distances. The theoretical longevity for this species is unknown and little biological data for the species is available.

Reproduction. Males reach maturity at 190 cm TL. Females reach maturity at 230 cm TL. Reproductive mode is placental viviparous, number of young is up to 13 young per litter. At birth young have approximately 80 cm TL. Duration of gestation is unknown (MORENO & HOYOS 1983a, DEMADDALENA & BÄNSCH 2005).

Food. Feed on bony fishes, squid and octopus, rays and carrion.

Red List status: NE (IUCN 2010).

Carcharhinus acronotus (POEY 1860)

En: Blacknose shark; Fr: Requin nez noir; Sp: Tiburon Amarillo, Lombo-preto, Sarda; Ge: Schwarznasenhai

Squalus acronotus POEY (1860). Holotype: MNHN A-9661. Type material is a female of 113.5 cm TL, from the Antillean Islands.

Synonyms: *Prionodon curcuri* deCASTELNAU (1860), *Carcharias (Prionodon) remotus* DUMERIL (1865).

Diagnostic features. A small species, without interdorsal ridge. Upper precaudal pit well developed, lower pit weak, internarial width 1.3–1.8 times the preoral length. Rounded snout of moderate length, big eyes (diameter 1.9 % TL). Eyes anterior margin slightly anterior to front of mouth. Nostrils oblique, with broad, ovate incurrent apertures. Anterior nasal flaps with long, pointed lobes. Upper labial furrows short and inconspicuous. Gill slits short (2.3–2.9 % TL). Hyomandibular pores alongside corners of mouth inconspicuous. Apex of first dorsal fin sharply rounded (height 9.1 % TL). D1 origin over the free rear tip of the pectorals. D2 small (height 2.7 % TL), the origin over or slightly behind anal fin origin. Pectoral fins small (length of anterior margin 15.0 % TL). Vertebrae: 84 precaudal and 88 caudal centra on average. Dermal denticles close set, imbricate, with sub-circular main cusps. In adults and sub-adults five (sometimes seven) ridges and three to five posterior cusplets present. Newborn and half-grown specimens with three longitudinal ridges and three posterior cusplets.

Dental Formula:

$$\frac{12-2-12}{11-1-11} \quad \text{to} \quad \frac{12(13)-(1)2-12(13)}{11(12)-1-11(12)}$$

Upper teeth are narrow, crowns strongly oblique, except the symphysial and anterior rows. Distal median edges straight apically, convex basally, slightly notched. Mesial edges are straight to convex. Both median edges are fine and uniformly serrated, slightly coarser

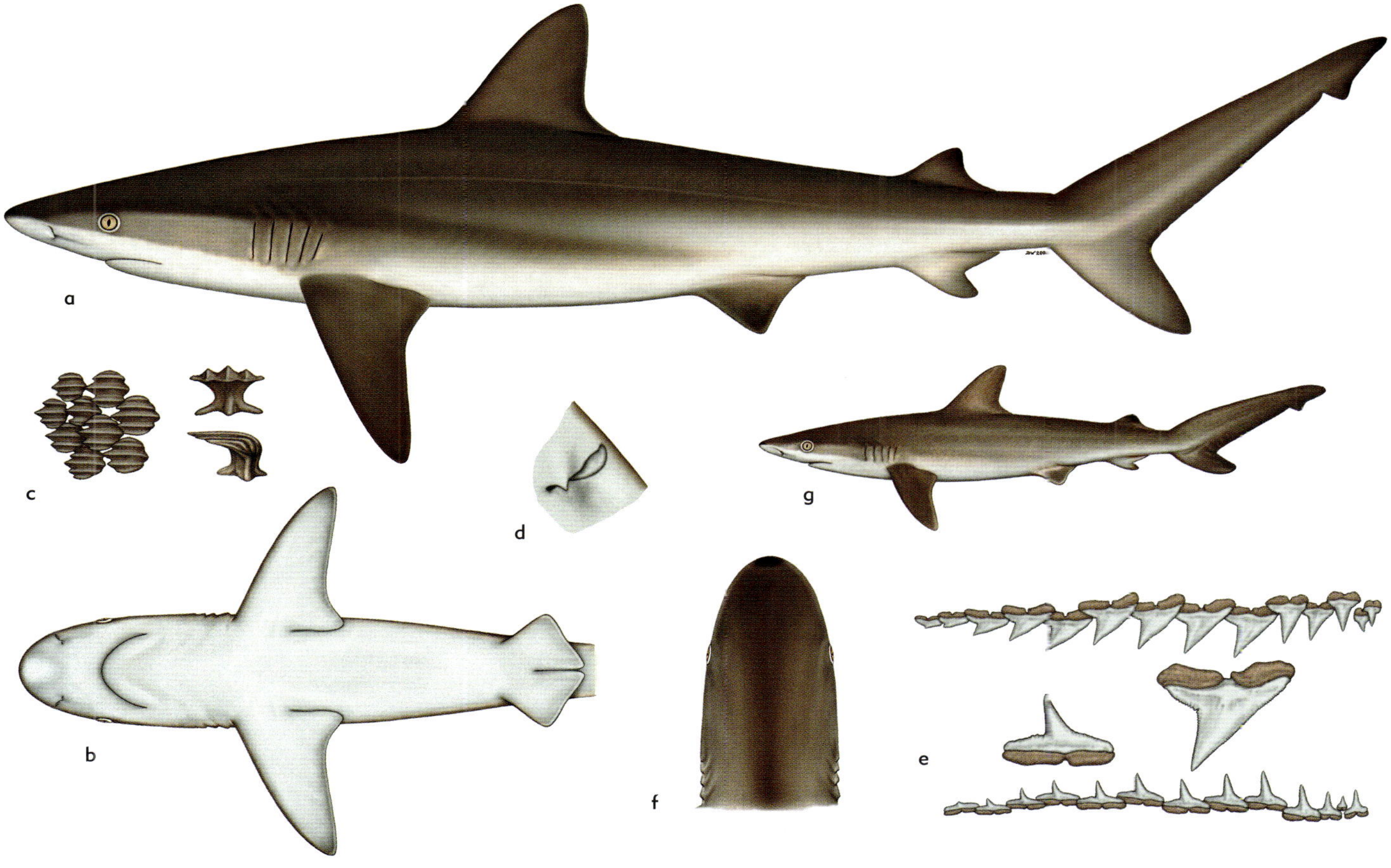

Figure 19. *Carcharhinus acrontus*: 148 cm TL female from the Bahamas; **a**, left side; **b**, ventral view; **c**, dermal denticles; **d**, enlarged left nostril; **e**, left upper and lower teeth (symphysis to the right), and inset are the enlarged fifth upper and lower teeth; **f**, upper side of head; **g**, juvenile female of 58 cm TL from Florida.

basally. Roots are straight (S and A rows) to oblique (all other rows), deeply notched, and divided into two root lobes. Usually, two rows of symphysial teeth are distinct but occasionally only one row can be present. Lower teeth are narrow, crowns are almost erect (S and A rows). Crowns of other rows are slightly oblique toward angle of jaw. Only one row of symphysial teeth is present. Distal median edges notched, apical part of crowns straight, mesial edges concave, both edges very fine and uniformly serrated basally, apically smooth-edged. Roots straight, slightly notched. Files in upper and lower jaw arranged as mixed alternate i.e. imbricate overlapping dentition.

Colour. Grey-brown above and with a prominent dusky to black blotch on snout tip, being more conspicuous in juveniles than in adults, ventrally white. D2 and upper caudal lobe are dusky spotted to completely black. Sometimes a trailing white margin on the lower caudal lobe may be present.

Size. Maximum up to 200 cm TL, an average of 100–140 cm TL, and a body mass of about 10 kg (COMPAGNO 1984). Maximum reported body mass of 18.9 kg (IGFA 2001).

Misidentifications. The only sympatric species that might be confused with *C. acronotus* is the small smooth-backed *C. porosus* but this species differs from the blacknose shark in the position of D2 origin, in dental formula, and in the average number of precaudal vertebrae (59 versus 84 in *C. acronotus*). The species may also occasionally be mistaken for *C. dussumieri* and *C. sealei* by body size features.

Geographical Distribution. Western Atlantic: North Carolina to Florida, Bahamas, Gulf of Mexico, Virgin Islands, Puerto Rico, Antilles, to southern Brazil. – (BRANSTETTER 1981, GARRICK 1982, COMPAGNO 1984, SCHWARTZ 1984, COMPAGNO et al. 2005).

Habitat. *Carcharhinus acronotus* is commonly in tropical and warm temperate water living, and typically found over continental shelves. The species has an affinity to grain-coarse sandy, i.e. coral/shell debris or rocky bottom, and occur in the western Atlantic Ocean in waters up to of 200 m depth.

Biology. Body size and sexual segregation is known within this species. Juveniles are typically living in shallower water while adults are located at greater depths. Blacknose sharks appear frequently in the coastal waters of North Carolina in summer and early autumn. Males dominate catches of local fishery in July, females from August to October and pregnant females occur in higher numbers from January to April off the south-western coast of Florida. Catches of this species vary over the years and are probably affected by seasonal

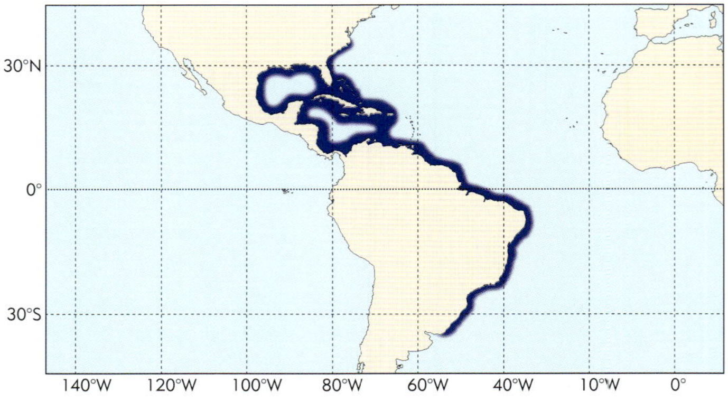

water temperature and salinity. It could indicate a local migration pattern but was not proven yet (CARLSON et al. 1999).

Carcharhinus acronotus is a quick swimming species, and is known to form large schools. Sometimes the species is associated with schools of mullets and anchovies. Blacknose sharks perform an agonistic or hunching display in captivity when confronted with divers or other disturbing, new and unknown things (CARLSON et al. 1999).

The theoretical longevity for the species was estimated to be 10–16 years for females, and 4.5–9.0 years for males, depending on specific geographical region the individuals are from (SCHWARTZ 1984, CARLSON et al. 1999, HAZIN et al. 2002).

Reproduction. Males reach maturity at 95–105 cm TL. Females reach maturity at 100–105 cm TL. Both sexes are mature at about 2 years of age (SPRINGER 1951). Reproductive mode is placental viviparous (DULVY & REYNOLDS 1997). At birth, young have 40–50 cm TL, and normal are 3–6 young per litter. In contrast to many other species of the genus, *C. acronotus* is a rapidly growing species.
Male blacknose sharks seem to produce sperm annually, in early June. Reproduction is biennially. Duration of the gestation is eight to eleven months. Females give birth during late May and early June, in protected inshore nursery grounds. Off Florida, gestation is from January to April (DRIGGERS 2002, DRIGGERS et al. 2004, SULIKOWSI et al. 2007).

Food. Feed on smaller bony fishes, e.g. pinfish, porcupine fish, croakers, porgies, anchovies and spiny box fish. Squid and octopus can also be part of the diet. Often blacknose sharks are eaten by other, larger shark species.

Red List status. NT (IUCN 2010).

Carcharhinus albimarginatus (RÜPPELL 1837)

En: Silvertip shark; Fr: Requin pointe Blanche, Aileron Blanc du récif; Sp: Tiburon de puntas Blancas; Ge: Silberspitzenhai

Carcharias albimarginatus RÜPPELL (1837). [...] Several specimens were taken, but no type material designated by RÜPPELL available (GARRICK 1982). Lectotype: SMF 3582. Type material is a stuffed immature male of 102.5 cm TL, caught at Ras Mehomet, Red Sea (KLAUSEWITZ 1960).

Synonyms: *Eulamia (Platypodon) platyrhynchus* GILBERT (1892).

Synonyms in recent use: *Carcharhinus platyrhynchus* GILBERT (1892).

Diagnostic features: Large, slim species, with a long and broadly rounded snout. Interdorsal ridge present, upper precaudal pit well developed, lower pit weak, internarial width 1.0–1.5 times the preoral length. Nostrils short, oblique and slit-shaped, elliptical incurrent apertures. Anterior nasal flaps poorly developed, rounded or pointed. Eye diameter large (2.5 % of TL), eyes anterior margin anterior to the front of mouth. Upper labial furrows short and inconspicuous. Hyomandibular pores alongside corners of mouth enlarged. Gill slits short (2.2–3.0 % TL, and less than a third of first dorsal base). First dorsal fin large and sickle-shaped (its height 8.8 % TL), apex weakly rounded (in adults) or pointed (in juveniles). D1 origin usually over or slightly anterior to pectoral rear tips, short inner margin. Second dorsal fin large (height 2.0 % TL), with rounded apex, its inner margin long. D2 origin over or shortly behind anal fin origin. Pectoral fins large and semi-falcate (length of anterior margin 18.6 % TL), with narrowly rounded or pointed apices. Vertebrae 120 precaudal centra and 103 caudal centra on average. Dermal denticles closed set, imbricate, mostly sub-circular to rhomboid or ellipsoid, with

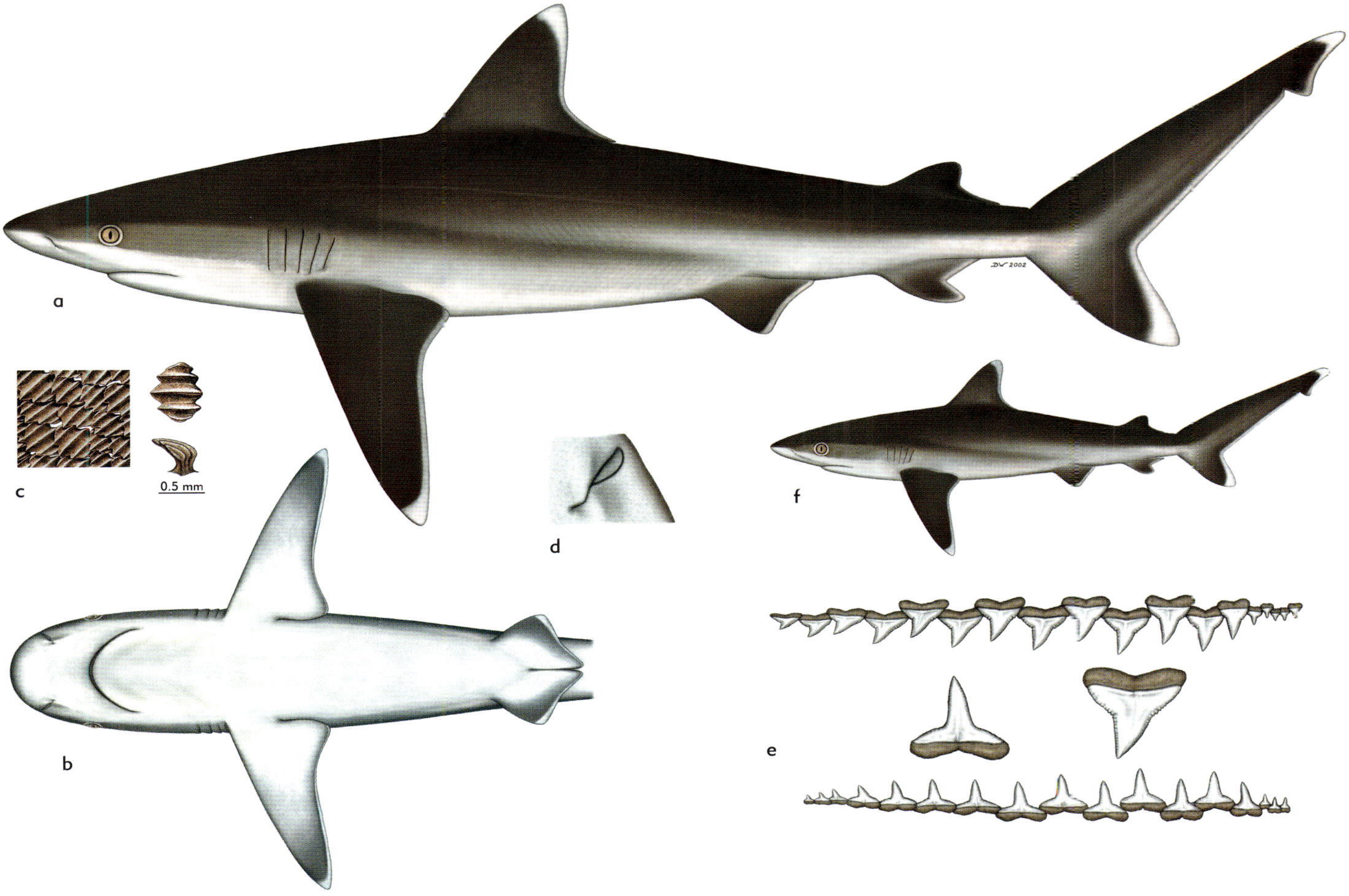

Figure 20. *Carcharhinus albimarginatus*: 190 cm TL female from the Red Sea; **a**, left side; **b**, ventral view; **c**, dermal denticles; **d**, enlarged left nostril; **e**, left upper and lower teeth (symphysis to the right), and inset are the enlarged fifth upper and lower teeth; **f**, juvenile female of 85 cm TL from Cocos Island.

three to five longitudinal ridges (and with short posterior cusplets in embryos and juveniles), seven to nine ridges and a corresponding number of posterior cusplets in adults.

Dental Formula:

$$\frac{13 - 1 - 13}{12 - 1 - 12} \quad \text{to} \quad \frac{12(14) - 1(2) - 12(14)}{12(14) - 1(2) - 12(14)}$$

Upper teeth are broad, crowns triangular, oblique, except symphysial and anterior rows. Both median edges are notched halfway longitudinally. Distal median edges straight apically, convex basally, deeply notched. Mesial edges straight basally to convex apically, slightly notched. Both median edges are serrated, coarser serration basally, roots arched. One or two rows of symphysial teeth are present. Lower teeth are narrow, crowns erect, median edges fine and uniformly serrated. Both median edges notched halfway longitudinally, apically straight, basal part of crowns convex. Serration in larger specimens (up to about 1.5 m TL) is present from base to apex of crowns, in smaller specimens on the apical third to the half of crowns only. Roots are arched, slightly notched only. One or two rows of symphysial teeth are present. Files in upper and lower jaw are arranged as alternate overlapping dentition, except LP to C rows in lower jaw. In contrast to the information of other sources, sexual dimorphism is not apparent in dentition of adult males and females (DIAZ 1979).

Colour. Dark grey dorsally (Pacific specimen with a reddish-brown tinge or sometimes bronze), often with a tinge of brown, white ventrally. Sometimes with an indistinct whitish band on flanks, and all fins have strikingly white tips and white posterior margins. D2 is often dark, and with a thin white posterior margin. The ventral surface of the pectoral and pelvic fins are darkened, particularly in the posterior parts, accentuating the white edging.

Size. Maximum up to 300 cm TL, and an average of 200–250 cm TL (RANDALL et al. 1990). Maximum reported body mass of 162.2 kg (IGFA 2001).

Misidentifications. *Carcharhinus albimarginatus* could be confused with the oceanic whitetip shark, *C. longimanus*, because of the white tips on fins but the spots are not mottled and the white does not extend regularly along the trailing margins in *C. albimarginatus*. In addition, the shape of first dorsal fin is broader and bluntly rounded in *C. longimanus*, and the tooth-shape is completely different. Both species differ in the number of vertebrae (120 in *C. albimarginatus versus* 126 in *C. longimanus*). Furthermore, confusion can occur with *C. wheeleri* because this species has white markings on fin apices and a trailing margin on D1 but markings of the other fins are dusky to black and are therewith in contrast to *C. albimarginatus*. Silvertip sharks can also be mistaken with *C. brevipinna* but this species has a more slender body-form, the snout is longer and pointed, and on average 82 precaudal vertebrae are present only.

Geographical Distribution. Western Indian Ocean: Southern Africa (occur probably regularly along of the northern Natal coast). Not recorded from Zanzibar but from Mozambique, Kenya, Madagascar, Mauritius, Seychelles, and Chagos Archipelago. – Western and central Pacific: Indonesian Archipelago (Macassar Strait), Taiwan, Philippines, Palau, Guam, New Caledonian, Marshall Islands, Solomon and Phoenix Islands, Tahiti. – Eastern Pacific: Southern Baja California, Clipperton, Cocos and Galapagos Islands. – (BASS et al. 1973, 1986; GARRICK 1982; COMPAGNO 1984; HEMIDA & LABIDI 2002; COMPAGNO et al. 2005).

Habitat. Coastal-pelagic and in tropical waters living, reported from inshore and offshore. This is a common to abundant species which is typically found from surface up to 400 m depth. FORSTER et al. (1970) gave a maximum depth of up to 800 m but this value seems to be doubtful. The species has a strong affinity for offshore islands but is not really oceanic. Usually, silvertip sharks are found near the edges of offshore banks or insular shelves, coral reefs and lagoons (COMPAGNO 1984).

Biology. Small and young specimens (occasionally adults too) migrate into shallower water of lagoons (up to 10 m depth) but only the juveniles are limited to these protected habitats (FOURMANOIR (1961, 1976a, b; WHEELER 1962). The species has been observed swimming along the edge of a group of other feeding shark species, dashing into the group to swipe some food. That is one reason why silvertip sharks are often caught as bycatch in fisheries of swarm fishes. The species is very aggressive one to another and dangerous to humans. Other, larger shark species can be predators for juvenile silvertip sharks but data about is very limited.

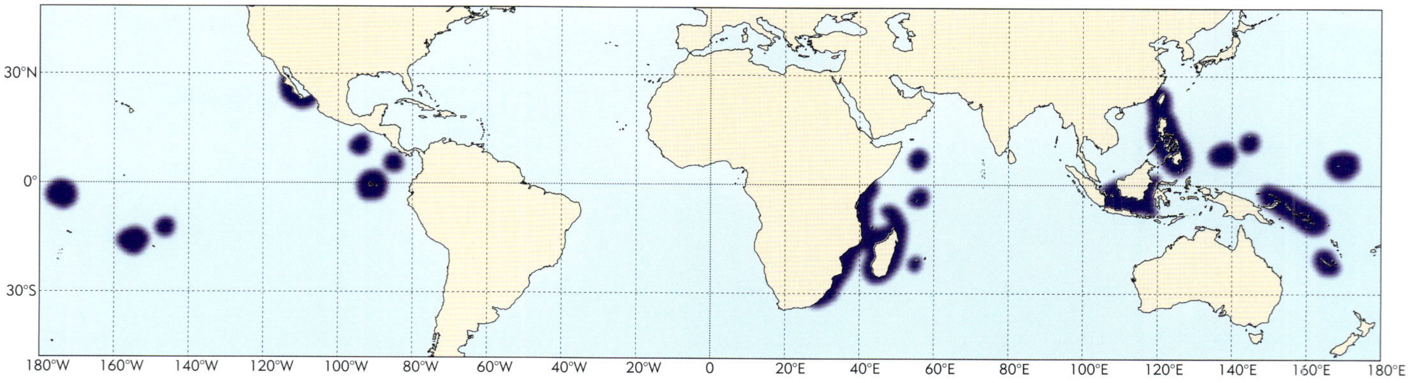

Reproduction. Males reach maturity at 165–170 cm TL, and are fully matured with 180 cm TL (WHEELER 1962). Females reach maturity at 160–200 cm TL, corresponding to an age of 3 years (GARRICK 1982). Reproductive mode is placental viviparous (DULVY & REYNOLDS 1997). Normal are 1 to 11 young per litter (alternatively 6–8) but an average of 5–6 young per litter was occasionally reported (GARRICK 1982). Young are born in summer after a gestation of about a year. Commonly, length at birth is 70–80 cm TL but occasionally shorter, with 55 cm TL (MYERS 1991).

Food. Adults feed mainly on bony fishes – pelagic and demersal (e.g. flying fishes, laternfish, gempylids, tunas, bonitos, wrasses, soles but also on eagle rays and cephalopods (FOURMANOIR 1961). FORSTER et al. (1970) have shown this species feeds also in midwater (for specimens caught at depths down to at least 400 m over deep water).

Juveniles that have fed on soles have sometimes coral fragments in the stomach probably because they have been chasing it along the bottom. This species feeds on a wide range of prey and occasionally even takes baits at the surface.

Red List status: NT (IUCN 2010).

Carcharhinus altimus (SPRINGER 1950)

En: Bignose shark, Knopp's shark; Fr: Requin babosse; Sp: Tiburón baboso, Tiburón narizón; Ge: Großnasenhai

Eulamia altima SPRINGER (1950). Holotype: USNM 133828. Type material is an immature female of 132 cm TL, from the Florida Keys, caught in 174 m depth.

Synonyms in recent use: *Carcharhinus radamae* FOUMANOIR (1961).

Diagnostic features. Large species, with a long, rounded or bluntly pointed snout (more pointed in eastern Pacific specimens). Broad interdorsal ridge present (midline of back between first and second dorsal fin with a low, but conspicuous broad, round-topped dermal ridge). Upper precaudal pit well developed, lower pit weak. Internarial width 1.2–1.6 times the preoral length. Eyes circular, moderately large (diameter 1.7 % TL), eyes anterior margin above the front of mouth. Nostrils oblique, with big, ovate incurrent apertures. Anterior nasal flaps high, with narrow, pointed lobes, excurrent apertures of nostrils small. Upper labial furrows short and inconspicuous. Hyomandibular pores alongside corners of mouth inconspicuous. Gill slits long (2.4–3.5 % TL). First dorsal fin high (10.2 % TL), falcate. D1 origin over pectoral fin insertion or behind it to almost 50 % of the length of the inner pectoral margin. D1 apex bluntly pointed. Second dorsal fin height 3.1 % TL, almost equal to anal fin height (1.1–1.4 times in length of its rear tip). D2 origin anterior of anal fin origin. Pectoral fins nearly straight and large (length of anterior margin 21.2 % TL). Vertebrae: 109 precaudal centra and 95 caudal centra on average. Dermal denticles wide-spaced (in specimens of > 1.5 m TL) but could be contiguous to imbricate. Pacific specimens usually with non-imbricate denticles. Younger specimens have small, ovoid denticles, wider than long, with three longitudinal ridges and a corresponding number of posterior cusplets in embryos and young, adults with some more.

Dental Formula:

$$\frac{15 - 1 - 15}{14 - 1 - 14} \quad \text{to} \quad \frac{14(16) - 1(2) - 14(16)}{14(15) - 1 - 14(15)}$$

Upper teeth are very long, crowns triangular, erect to slightly oblique, except symphysial row(s). Distal median edges concave, mesial edges straight to convex, both edges serrated. Serration is a little bit coarser on central part of distal median edges, roots arched. One row of symphysial teeth is present. In larger specimens the teeth of the ultimate row are noticeable smaller in proportion than the penultimate row and could easily be overlooked. Lower teeth are very narrow, crowns erect. Both edges straight apically, convex basally, finely serrated. Roots are arched, slightly notched. One row of symphysial teeth is present. Files in upper and lower jaw are arranged as mixed alternate overlapping dentition (except P to C rows, arranged as imbricate overlapping dentition type I).

Colour. Grey above, sometimes bronze to grey-brownish, a little bit dusky (more distinct in young than in adults). Ventrally white, with tips on fins, lower side of pectoral fins darker. With white markings on flanks but they are often inconspicuous. Sometimes a green sheen along the gills of freshly collected specimens can be present.

Size. Maximum up to 300 cm TL, an average of 240–260 cm TL, and a body mass of 170 kg (COMPAGNO 1984).

Misidentifications. The long upper teeth and the remarkable prominent nasal lobes are features in which *C. altimus* differs from all other members of the genus, e.g. *C. amblyrhynchoides*, *C. galapagensis*, *C. limbatus* *C. obscurus* and *C. plumbeus*. *Carcharhinus amblyrhynchoides* and *C. limbatus* have a stouter body-form, larger eyes, and a larger first dorsal fin than *C. altimus*. Moreover, *C. amblyrhynchoides* has a shorter snout than the bignose shark. *Carcharhinus altimus* differs from all other mentioned species in having on average 109 precaudal vertebrae *versus* 91 in *C. amblyrhynchoides*, 107 in *C. galapagensis*, 96 in *C. limbatus*, 91 in *C. obscurus* and 90 in *C. plumbeus*.

Geographical Distribution. Spanish Mediterranean coast. – Western Atlantic: Florida, Bahamas, Gulf of Mexico, Nicaragua, Costa Rica, Venezuela. – Eastern Atlantic: Cape Verde, Senegal, Gambia, Sierra Leone, Ivory Coast, Ghana. – Western Indian Ocean: Red Sea, Madagascar, South Africa, India. – Western Pacific: China, selectively on the coast of Australia. – Central Pacific: Hawaii. – Eastern Pacific: Gulf of California, southern Mexican coast, Colombia, Ecuador, Revillagigedo Islands. – (BASS et al. 1973, 1986; BARANES & BEN-TUVIA 1978 a, b, c; GARRICK 1982; MORENO & HOYOS 1983b; COMPAGNO 1984; HEMIDA & LABIDI 2002; COMPAGNO et al. 2005; LAST & STEVENS 2009).

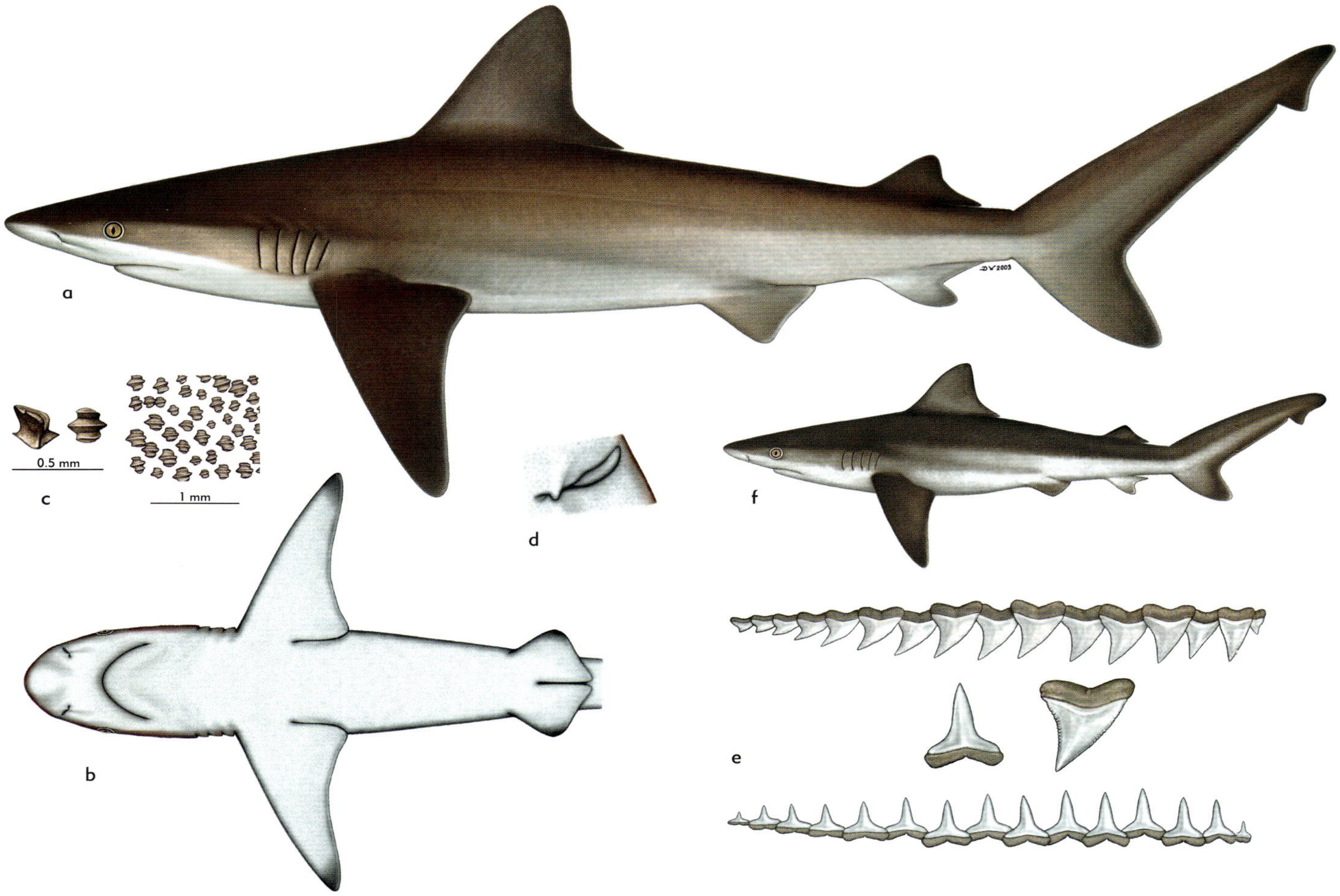

Figure 21. *Carcharhinus altimus*: 234 cm TL female from Cape Verde Islands; **a**, left side; **b**, ventral view; **c**, dermal denticles; **d**, enlarged left nostril; **e**, left upper and lower teeth (symphysis to the right), and inset are the enlarged fifth upper and lower teeth; **f**, juvenile female of 74.5 cm TL from South California.

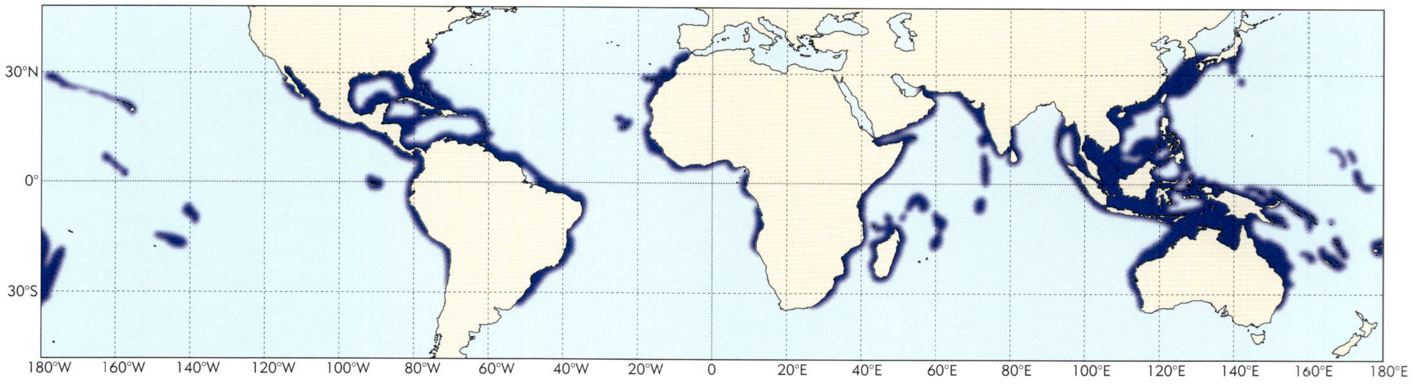

Habitat. Commonly, this species is distributed in warm, temperate waters, on continental shelves and uppermost slopes but is rare in shallow waters (COMPAGNO 1984, COMPAGNO & NIEM 1998).

Biology. *Carcharhinus altimus* is known to make diurnal vertical migrations. The species occurs in deep benthic waters near the bottom, in about 90–500 m by day. At night, individuals move into shallower waters or into the pelagic zone (ANDERSON & STEVENS 1996). The young occur often in shallower water up to 25 m depths (COMPAGNO 1984).

Reproduction. Males reach maturity at 215–270 cm TL. Females reach maturity at 230–280 cm TL. Reproductive mode is placental viviparous (DULVY & REYNOLDS 1997). Normal are 3–15 young per litter i. e. 7 and 8 young in two litters, reported from the coast of Florida (SPRINGER 1950). At birth, young have 70–90 cm TL.

Smallest pregnant female caught ever was a specimen of 205 cm TL, caught in the Hawaiian Islands (CROW et al. 1996). Different birth times are known depending on specific habitat: in the Mediterranean, bignose sharks giving birth in August and September, whereas in the western Indian Ocean (off Madagascar) between September and October.

Food. Mainly bony fishes, e. g. mackerels, soles, and batfish, but also small sharks (Squalidae, Scyliorhinidae), rays, cephalopods and cuttlefish.

Red List status: DD (IUCN 2010).

Carcharhinus amblyrhynchoides (WHITLEY 1934)

En: Graceful shark; Fr: Requin gracile; Sp: Tiburón grácil; Ge: Graziler Hai

Gillisqualus amblyrhynchoides WHITLEY (1934). Holotype: QM I. 2003. Type material is an immature female of 59.5 cm TL, from Cape Bowling Green, Queensland.

Synonyms in recent use: *Carcharhinus pleurotaenia* BLEEKER (1852) = *C. limbatus* VALENCIENNES (in MÜLLER & HENLE 1839).

Diagnostic features. Mid-sized to large species, snout short, wedge-shaped and moderately pointed. Brawny or robust body shape, without interdorsal ridge. Upper precaudal pit well developed, lower pit weak. Internarial width 0.8–1.3 times the preoral length. Eyes circular, fairly large (diameter 1.7 % TL), eyes anterior margin above or usually slightly anterior to the front of the mouth. Nostrils oblique, incurrent apertures large. Anterior margin of each aperture has a low, pointed lobe. Upper labial furrows short and inconspicuous. Hyomandibular pores alongside corners of mouth inconspicuous. Gill slits large (3.2–4.8 % TL). First dorsal fin triangular, with a short rear tip, apex sharply rounded or pointed (height 10.8 % TL). D1 origin over or slightly posterior to pectoral insertion. Second dorsal fin large (3.4 % TL), with a short rear tip. D2 origin over or slightly anterior to anal fin origin, length of rear tip 1.0–1.2 times its height. Pectoral fins long, falcate and sharp pointed distally. Origin of pectorals approximately below the fourth gill opening (length of anterior margin 18.3 % TL). Vertebrae: 91 precaudal centra and 95 caudal centra on average. Dermal denticles closed set and imbricate, almost rhomboid, with five ridges and posterior cusplets. Small specimens have almost or just contiguous, sub-circular denticles, with three longitudinal ridges and three weakly developed cusplets.

Dental Formula:

$$\frac{15 - 1 - 15}{14 - 1 - 14} \quad \text{to} \quad \frac{15 - 1(3) - 15}{14(15) - 1(3) - 14(15)}$$

Upper teeth are narrow, crowns erect (A to L rows) otherwise slightly oblique (LP to C rows). Distal median edges notched, concave, but mesial edges are not notched and slightly concave only, both median edges are fine serrated. The serration is slightly coarser basally, sometimes even weak or irregular. Roots are arched, deeply notched. One to three rows of symphysial teeth are resent. Lower teeth are narrow, crowns erect. Both median edges concave, basally convex, very fine and uniformly serrated. Roots are arched, deeply notched. One to three rows of symphysial teeth are present. Files in upper and lower jaw are arranged as mixed alternate/imbricate overlapping dentition.

Colour. Grey or brownish-grey dorsally. Apices of D1 and D2, and the lower lobe of caudal fin are black. On upper caudal fin lobe dusky margins can occasionally be present, and a prominent dusky or black mark on pectorals is distinct. Ventrally white or pale, with a white band along the mid-level of side, from pelvics forward to below the first dorsal fin. Often pelvic fins with a small dusky mark, anal fin normally pale coloured. Adults can be more pale coloured.

Size. Maximum of at least 167 cm TL, adult females with an average of 130–150 cm TL, and males of 135–140 cm TL (COMPAGNO 1984).

Misidentifications. *Carcharhinus amblyrhynchos*, *C. limbatus* and *C. brevipinna* are very similar to the graceful shark, and hence confusion is conceivable. All three mentioned species are moderate to large sized in appearance and share the following features: absence of a mid-dorsal ridge, presence of narrow, black tips on all fins, pointed snout and D1 origin is over the pectoral axis, upper teeth. Grey reef sharks have a dark caudal fin margin but in relation to *C. amblyrhynchoides* a shorter snout and having a higher second dorsal fin and both species have a different body colour. The blacktip shark, *C. limbatus*, seems to be distributed worldwide, whereas *C. amblyrhynchoides* is restricted to the Indo-western Pacific. Specimens of *C. limbatus* have a more slender body form than *C. amblyrhynchoides*. The spinner shark, *C. brevipinna*, is also a worldwide distributed species but differs in tooth-shape (smooth-edged lower teeth), in dental formula and D1 origin which is over the inner pectoral corner and the species has generally smaller fins than *C. amblyrhynchoides*. Furthermore, sub-adult and adult *C. brevipinna* have black-tipped

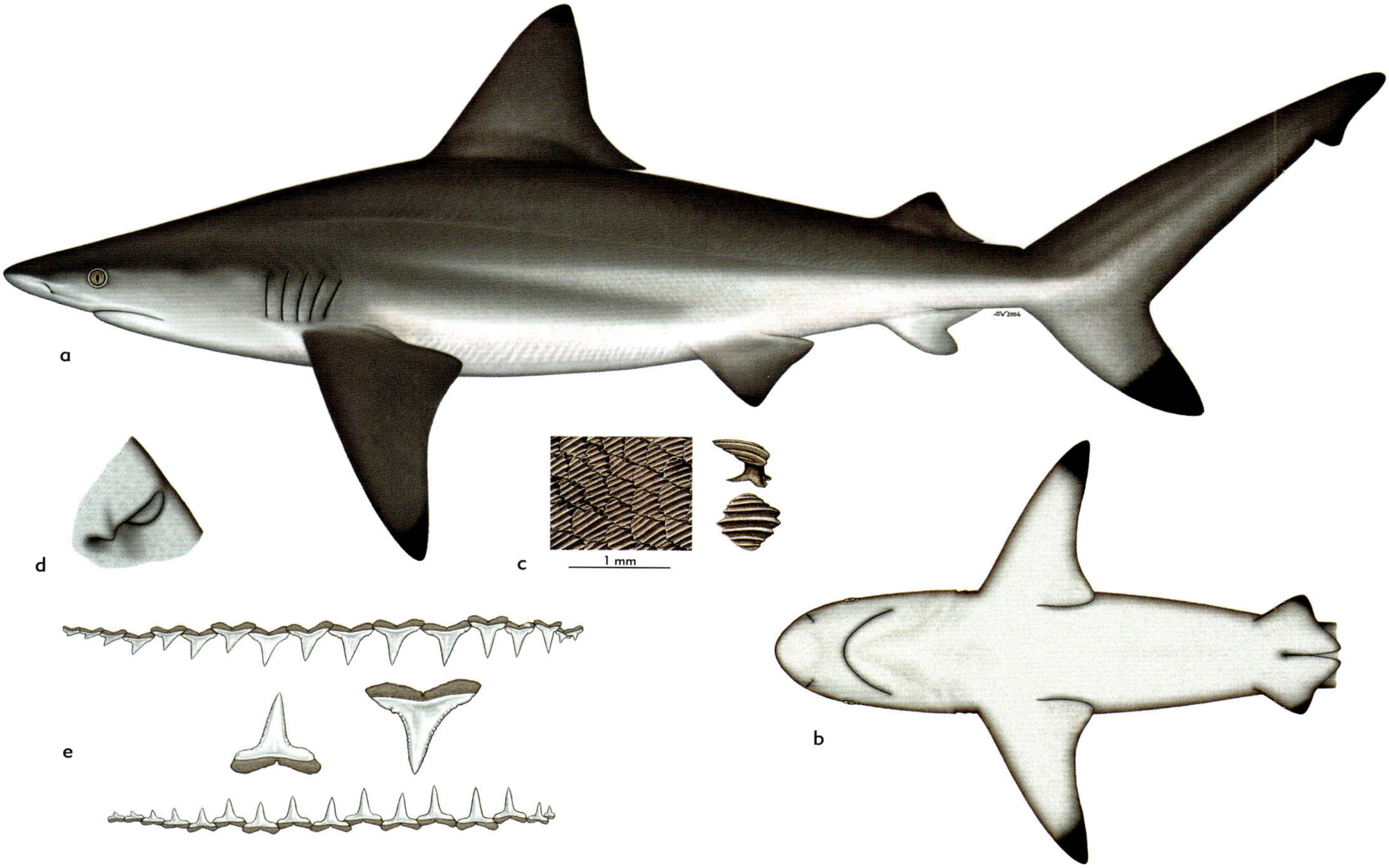

Figure 22. *Carcharhinus amblyrhynchoides*: 136 cm TL female from the Gulf of Aden; **a**, left side; **b**, ventral view; **c**, dermal denticles; **d**, enlarged left nostril; **e**, left upper and lower teeth (symphysis to the right), and inset are the enlarged fifth upper and lower teeth.

fins, newborn and juveniles lack this feature. The mentioned species are distinguishable from *C. amblyrhynchoides* by the average number of their precaudal vertebrae (91 in *C. amblyrhynchoides versus* 96 in *C. limbatus* and 82 in *C. brevipinna*).

Geographical Distribution. Indo-West Pacific: Gulf of Aden, southwestern coast of India, Sri Lanka, Gulf of Thailand, along the eastern coast of China, Vietnam, Philippines, Java, Borneo, Australia (Queensland and northern to north-western coast). – (Whitley 1934–1967, Garrick 1982, Compagno 1984, Lyle 1984, Compagno et al. 2005).

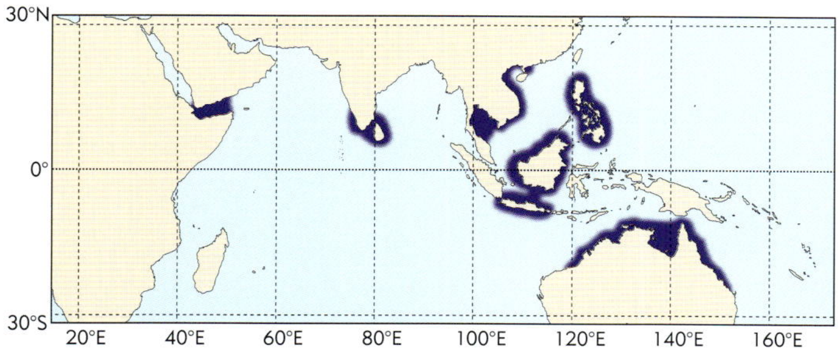

Habitat. Probably, it is a common species in tropical waters on Australias coast, inshore or coastal pelagic living, typically found over continental and insular shelves in depths of 50 m but little is known about it (Stevens & Wiley 1986).

Biology. No relevant data available.

Reproduction. Size at reaching maturity is for both sexes 110–115 cm TL. Reproductive mode is placental viviparous (Dulvy & Reynolds 1997). At birth young have 51–55 cm TL, and normal are three young per litter (Compagno 1984). Information about duration of gestation is unavailable.

Food. Mainly carangids and mugilids, also cephalopods and crabs. Probably, the species has the same prey spectrum as *C. limbatus* and *C. brevipinna*.

Red List status: NT (IUCN 2010).

Carcharhinus amblyrhynchos (BLEEKER 1856)

En: Gray reef shark, Longnose blacktail shark, Blacktail reef shark; Fr: Requin dagsit; Sp: Tiburón de arrecifes, Tiburón coralero rabinegro; Ge: Grauer Riffhai, Langschnäuziger Schwarzschwanzhai

Carcharias (Prionodon) amblyrhynchos BLEEKER (1856). Holotype: FMNH 7377. Type material is a female of 154 cm TL, caught in the Java Sea near Salambo Islands.

Synonyms: *Carcharias nesiotes* SNYDER (1904), *Galeolamna fowleri* WHITLEY (1944 a, b), *Galeolamna tufiensis* WHITLEY (1949), *Galeolamna coongoola* WHITLEY (1964 a, b), JOHNSON & NELSON (1973), *Carcharhinus wheeleri* GARRICK (1982).

Synonyms in recent use: *Carcharhinus menisorrah* VALENCIENNES (in MÜLLER & HENLE 1839) = *Carcharhinus falciformes* BIBRON (in MÜLLER & HENLE 1839) as restricted by GARRICK (1982).

Diagnostic features. Large species without interdorsal ridge (a weakly developed dermal ridge can occasionally be present). Upper precaudal pit well developed, lower pit weak. Snout long and rounded, internarial width 1.1–1.5 times the preoral length. Eyes circular and fairly large (diameter 2.5 % TL), eyes anterior margin slightly anterior to the front of mouth. Nostrils large, slit-shaped to ovate. Anterior nasal flaps small, with low triangular lobes, incurrent apertures long and narrow. Upper labial furrows short and inconspicuous. Hyomandibular pores alongside corners of mouth inconspicuous. Gill slits moderate sized (2.5–3.3 % TL). D1 semi-falcate, large (10.3 % TL), with narrowly rounded or acute apex. D1 origin over or just in front to pectoral free rear tips. Inner margin of first dorsal fin long, but less than half of dorsal base. Second dorsal fin large and long (height 3.1 % TL), with a fairly long inner margin of 1.2–1.6 times its height. D2 Origin over anal fin origin.

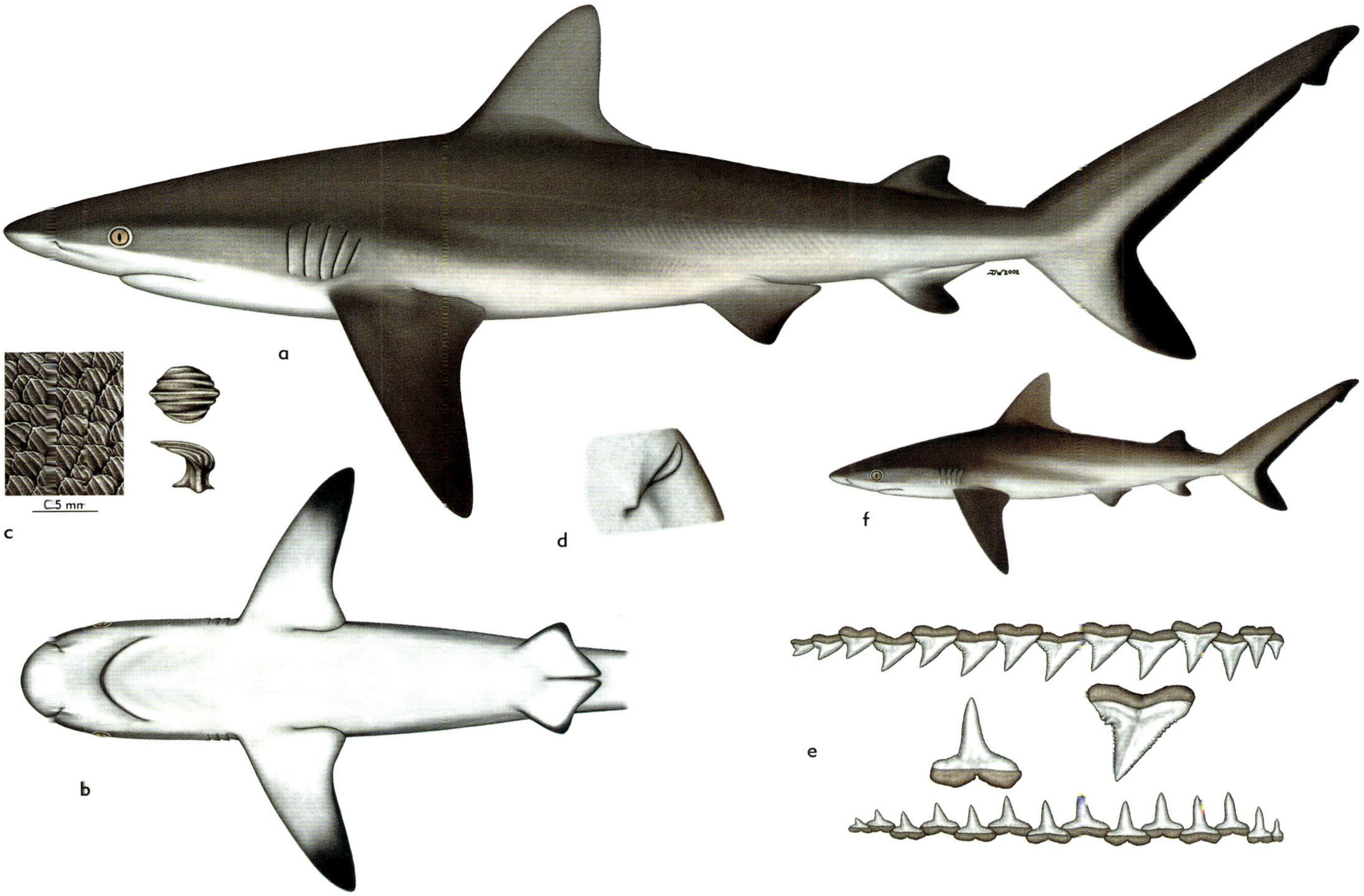

Figure 23. *Carcharhinus amblyrhynchos*: 178 cm TL female from French Polynesia; **a**, left side; **b**, ventral view; **c**, dermal denticles; **d**, enlarged left nosetril; **e**, left upper and lower teeth (symphysis to the right), and inset are the enlarged fifth upper and lower teeth; **f**, juvenile male of 66 cm TL from Great Barrier Reef, Australia.

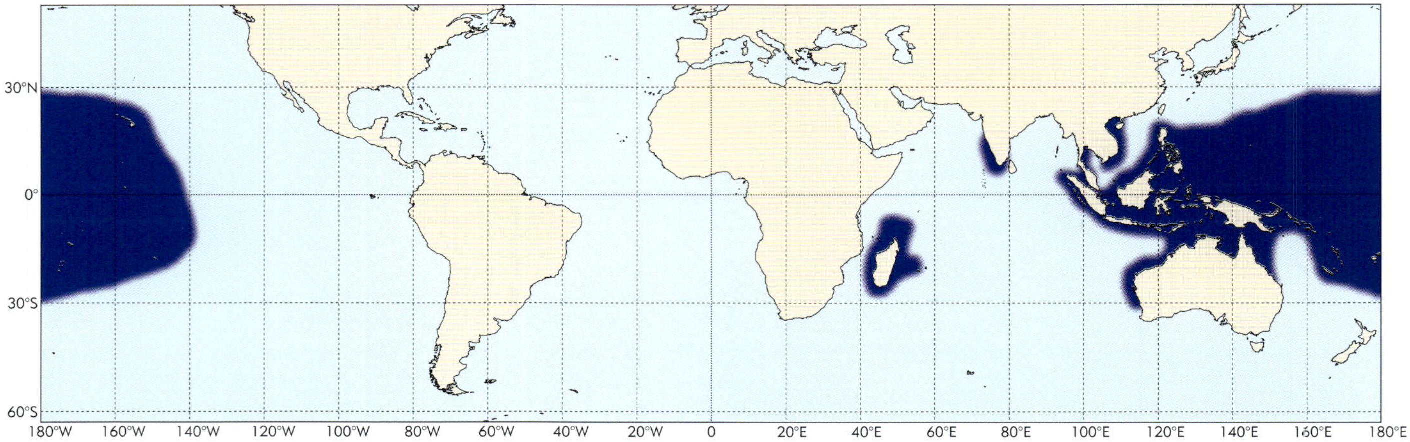

Pectorals large (length of anterior margin 19.3 % TL), narrow and falcate, with rounded or pointed apices. Vertebrae: 115 precaudal centra and 100 caudal centra on average. Dermal denticles closed set, imbricate and sub-circular in small, almost rhomboid in larger specimens. Each denticle with three to five strong longitudinal ridges and a corresponding number of sharp-pointed (but shorter) posterior cusplets in smaller specimens. Larger specimens have seven longitudinal ridges with a corresponding number of cusplets.

Dental Formula:

$$\frac{14 - 1 - 14}{13 - 1 - 13} \quad \text{to} \quad \frac{13(14) - 1(2) - 13(14)}{13(14) - 1 - 13(14)}$$

Upper teeth are narrow, crowns erect to oblique. Distal median edges concave, notched. Mesial edges weakly concave or almost straight, both median edges are coarsely serrated. Serration is obviously coarser basally on distal edges, roots are arched. One, seldom two rows of symphysial teeth are present. Lower teeth narrow, crowns erect, both edges concave, slightly notched basally. Median edges of smaller specimens smooth-edged (L to P rows), serration of larger specimens (LP to P rows) is very weak. Roots are arched, deeply notched. One row of symphysial teeth is present. Files in upper and lower jaw are arranged as alternate overlapping dentition.

Colour. Grey above and white to greyish ventrally. Normally, D1 is plain, but can be occasionally irregularly coloured and white-edged. On the entire posterior margin of caudal fin (terminal, pre- and postventral margins) a conspicuous, narrow (from terminal to upper postventral margin) to a broad (on lower postventral margin to ventral tip) black margin can be distinct. All other fins (pectorals, second dorsal and anal fin, pelvic fins) are apically blackish or dusky but not with deep black tips. Pectorals are above greyish-dusky, central and basal parts of the body are more pale (COMPAGNO & NIEM 1998).

Size. Maximum up to 233 cm TL (possibly even up to 255 cm) for males, i.e. 172 cm TL for females (RANDALL et al. 1990, FISCHER et al. 1990). With an average of 130–185 cm TL and a body mass of 34 kg (IGFA 1991).

Misidentifications. Two species can be confused with *C. amblyrhynchos*; the shorter snouted *C. wheeleri* and *C. sealei*. In contrast to *C. amblyrhynchos*, *C. wheeleri* has a white apical tip and a trailing margin of the first dorsal fin. In contrast to it *C. sealei* has a black apical tip on the second dorsal fin. The snout of *C. sealei* specimens is usually narrower than in *C. amblyrhynchos*, and on upper teeth smooth-edged lateral cusplets are present. The first dorsal fin is more falcate, and its posterior margins are deeply concave. All three species differ in the average number of their caudal vertebrae (100 in *C. amblyrhynchos versus* 105 in *C. wheeleri* and 76 in *C. sealei*).

Geographical Distribution. Indian Ocean: Madagascar, Mauritius, Seychelles and India. – Western and central Pacific: Singapore, Thailand, Vietnam, South China, Philippines, Indonesia, New Guinea, Australia (Queensland and Western Australia), east to Hawaii and the Tuamotu Atoll. The species is distributed on most of the Pacific atolls and island groups, e. g. New Caledonia and Lord Howe Islands, Palau, Caroline Islands, Marshall and Line Islands, Guam, Solomon and Phoenix Islands, Gilberts and Pitcairn Islands, Johnston Atoll, Wake Islands and Tahiti. – (WHEELER 1960; BASS et al. 1973, 1986; FOURMANOIR 1976 a, b; DECROSTA 1981; GARRICK 1982; COMPAGNO 1984; ECONOMAKIS & LOBEL 1998; COMPAGNO et al. 2005).

Habitat. Commonly, coastal living, distributed in shallow tropical and subtropical waters, inshore and offshore. Often found near coral atolls and lagoons adjacent to reef habitats, from surface to 280 m depth but has also been observed in waters down to 1000 m. Most sharks were caught at depths between 20–60 m, and the depth distribution of both sexes is similar, although females are more common at shallower depths (WETHERBEE et al. 1997).

Biology. [...] Over a four year period, free-ranging sharks were studied by ultrasonic telemetry and direct underwater observation. Movement patterns suggested that: (A) sharks tagged near the ocean reefs were generally nomadic, moving relatively long distances along the reefs and (B) sharks tagged on lagoon reefs or pinnacles had a more home-ranging pattern with different day and night areas, often returning day-after-day to the tagging site. Observed daytime grouping patterns were of the following three intergrading types: (1) polarized schools were seen close to bottom over level areas, (2) loose aggregations were usually near ocean-reef drop offs and (3) lone individuals were usually over shallower reefs and lagoon pinnacles (McKIBBEN & NELSON 1986). The species shows an agonistic display and is responsible for attacks on humans (JOHNSON & NELSON 1973). It is a more nocturnally active species and can lie during the day on the bottom for longer time. Together with the blacktip reef shark, *C. melanopterus*, and the whitetip reef shark, *Triaenodon obesus*, it is the most common species in tropical coral reefs.

Reproduction. Males reach maturity at 120–150 cm TL. Females reach maturity at 100–190 cm TL, and are normally fully matured at a length of 125–175 cm TL (WETHERBEE et al. 1997). Other sources give size for reaching maturity for both sexes with 130–145 cm TL, corresponding to an age of 7–8 years (SHARKINFO). Reproductive mode is placental viviparous (DULVY & REYNOLDS 1997). Gestation duration of about 12 months was reported (MYERS 1999). Litter size varies from 1–6 with a mean of 4 pups. At birth, young have 45–60 cm TL (WETHERBEE et al. 1997).

Food. This species (≈80 %) is known to consume mainly teleosts but feed on cephalopods, crustaceans and shrimps also. For adult specimens, the frequency of consuming teleosts is declining, whereas the occurrence of cephalopods in diet increases (HOBSON 1961, WETHERBEE et al. 1997). Adults will prey on young individuals of their own species also.

Red List status: NT (IUCN 2010).

Carcharhinus amboinensis (MÜLLER & HENLE 1839)

En: Java- or Pigeye shark; Fr: Requin balestrine; Sp: Tiburon baleta; Ge: Javahai

Carcharias (Prionodon) amboinensis MÜLLER & HENLE (1839). Holotype: RMNH 2582. Type material is a stuffed female of about 74 cm TL, from Amboina.

Synonyms: *Carcharias henlei* BLEEKER (1853), *Prionodon henlei* BLEEKER (1853), *Carcharias brachyrhynchos* BLEEKER (1859), *Prionodon brachyrhynchos* BLEEKER (1859), *Triaenodon obtusus* DAY (1873–78).

Diagnostic features. Large to stout species, without interdorsal ridge. Upper caudal pit well developed, lower pit weak, internarial width 0.8–1.1 times the preoral length. Snout very short, broad and bluntly rounded. Eyes small and circular (diameter 1.4 % TL), eyes anterior margin slightly anterior to the front of mouth. Nostrils strongly oblique-rounded, incurrent apertures large and ellipsoid to slightly triangular. Anterior nasal flaps mid-sized. Upper labial furrows short and inconspicuous. Hyomandibular pores alongside corners of mouth inconspicuous. Gill slits mid-sized (2.5–3.2 % TL). First dorsal fin large (11.6 % TL), broadly triangular to falcate, its apex pointed or sharply rounded. The posterior margin curving ventrally or posterior-ventrally from fin apex. D1 origin over or halfway behind pectoral insertion. Inner margin of the first dorsal fin short, less than a third of dorsal base. Second dorsal fin height 3.1 % TL, its inner margin short. D2 origin anterior to anal fin origin. Pectoral fins large and broad (length of anterior margin 20.7 % TL), triangular to semi-falcate, with narrow, pointed apices. Vertebrae: 92 precaudal centra and 96 caudal centra on average. Dermal denticles relatively large, lose-spaced in smaller specimens, but fairly closed-set in larger specimens, each with three to five longitudinal ridges and five posterior cusplets.

Dental Formula:

$$\frac{12 - 1 - 12}{1 - 1 - 1} \quad \text{to} \quad \frac{11(13) - 1 - 11(13)}{10(12) - 1 - 10(12)}$$

Upper teeth are broad, crowns triangular, erect to slightly oblique. Distal median edges slightly concave, slightly notched distally. Mesial edges straight to weakly convex on the basal half of crown. Both edges are serrated, merge smoothly with slightly coarser serration basally, roots are arched (A to L rows) to almost straight (LP to C rows). One row of symphysial teeth is present. Lower teeth are broad, narrower than upper crowns erect to semi-oblique. Both edges are concave, the mesial edges weakly convex apically. Serration of lower teeth is finer than those of upper, the roots are arched. One row of symphysial teeth is present. Files in upper and lower jaw are arranged as mixed alternate i.e. imbricate overlapping dentition.

Colour. Grey above, white ventrally, tips on D2 and lower caudal lobe are dusky, more distinct in juveniles but they are not strikingly marked. An inconspicuous white band on the flanks can be present (COMPAGNO et al. 1989).

Size. Maximum of 280 cm TL, and an average of 190–250 cm TL (RANDALL et al. 1990).

Misidentifications. Often *C. amboinensis* is confused with the sympatric species *C. leucas* and *Glyphis gangeticus* but the species are (well) distinguishable on from another. The pigeye shark has broader cusped lower teeth and a bigger first dorsal fin. All mentioned species are differentiable in dental formula, in the average number of precaudal centra (92 in *C. amboinensis*, whereas 116 in *C. leucas*), and other morphometric dimensions. In *Glyphis* sp. anterior teeth have elongated, strongly curved and hooked cusps which usually protrude from the mouth even when closed, but in *Carcharhinus* sp. the lower cusps are shorter, straighter, and do not protrude. In addition, the eyes of *Glyphis* sp. are smaller than in *Carcharhinus* species (with negative allometry of eye diameter with ongoing growth, large individuals of small-eyed *Carcharhinus* species such as *C. leucas* overlap small *Glyphis* sp. in eye diameter) (COMPAGNO 1988).

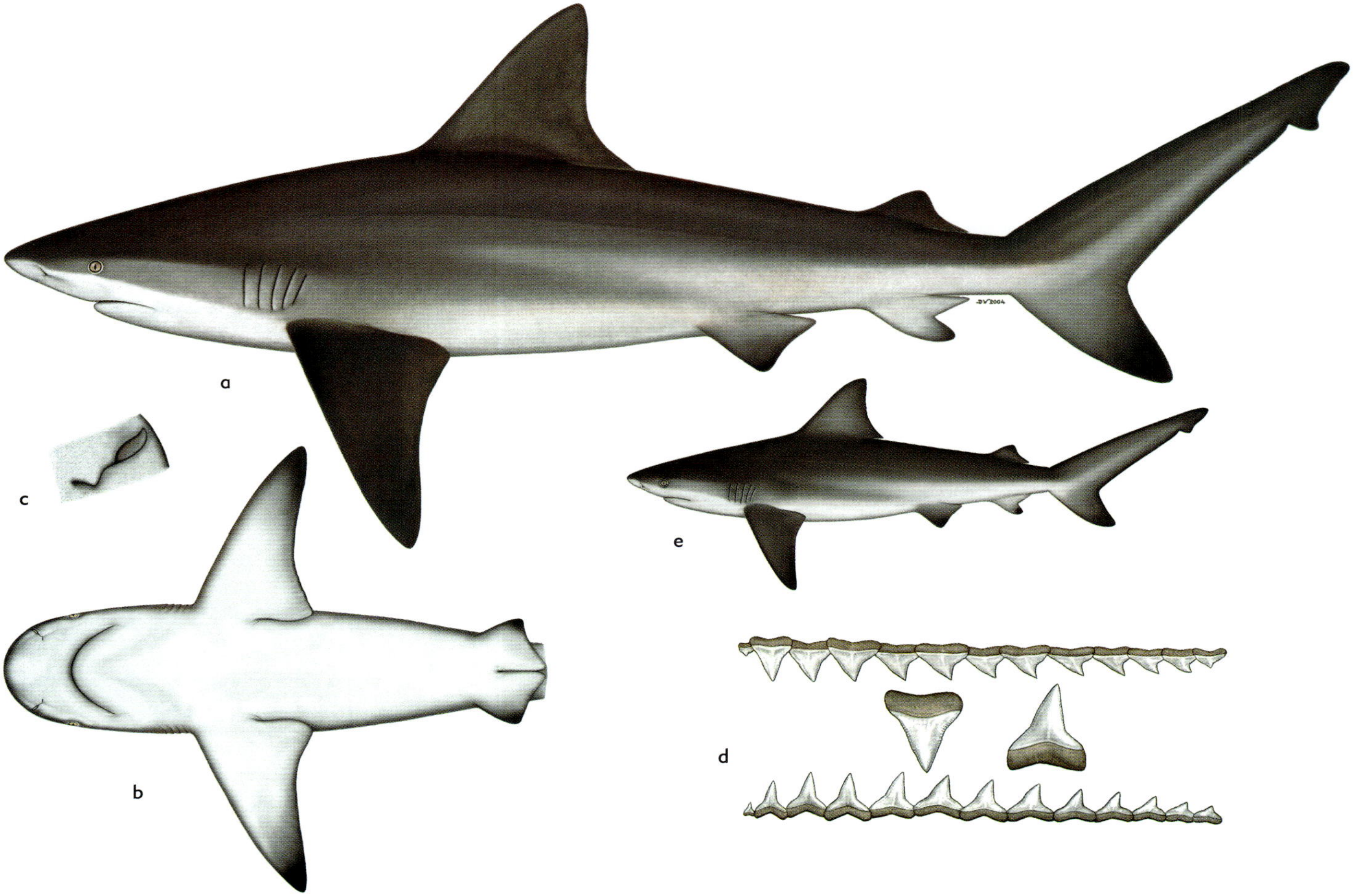

Figure 24. *Carcharhinus amboinensis*: 260 cm TL female from Queensland, Australia; **a**, left side; **b**, ventral view; **c**, enlarged left nostril; **d**, left upper and lower teeth (symphysis to the left), and inset are the enlarged second upper and fifth lower teeth; **e**, juvenile female of 73 cm TL from Natal, South Africa.

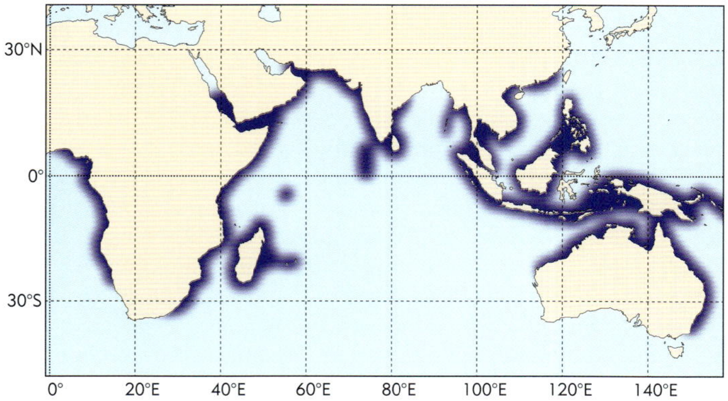

Geographical Distribution. Eastern Atlantic: Cape Verde, Senegal to Nigeria, and Gabon to Namibia. – Indian Ocean: Southern Red Sea, Gulf of Aden, Bahrain, Djibouti, Oman, Somalia, Mozambique, Seychelles, Mauritius, Madagascar, east coast of South African, Maldives, Pakistan, India, Sri Lanka, Andaman Sea, Gulf of Thailand, South China Sea, southern China, Philippines, Indonesia (Java, Amboina, Aru, Lomblen Islands), Borneo, Celebes Sea, Papua New Guinea and north-western to north-eastern Australia, and on several Micronesian Atolls. – (Bass et al. 1973, 1986; Garrick 1982; Compagno 2005; Compagno et al. 2005; Last & Stevens 2009).

Habitat. This species has an affinity to continental and insular shelves, and occur mostly in shallow waters, close inshore (but not truly ascending rivers), as well as off the open coast, along surf line and beaches, in depths of 150 m up to the surface (Compagno 1984, Compagno & Niem 1998, Last & Stevens 1994).

Biology. It is a rare species in the south-eastern Indian Ocean but more abundant on the eastern coast of Madagascar and other parts of its spatial distribution area. The species is able to tolerate for a short while brackish waters off Madagascar estuaries but is normally not abundant distributed in freshwater (BASS et al. 1986).

Reproduction. Males reach maturity at 190-195 cm TL. Females reach maturity at 200–225 cm TL. Reproductive mode is placental viviparous (DULVY & REYNOLDS 1997). At birth, young have 70 cm TL (COMPAGNO 1984). Little is known about the reproduction biology of this species.

Food. Mainly, prey is taken near the sea bed living and consist on bony fishes, e.g. croakers, soles, other sharks (*Rhizoprionodon* sp.), skates, sea snails and whale carcasses, crustaceans (shrimp), molluscs and cuttlefish.

Red List status: DD (IUCN 2010).

Carcharhinus borneensis (BLEEKER 1859)

En: Borneo shark; Fr: Requin-tigre houareau; Sp: Tiburon de Borneo; Ge: Borneohai

Carcharias (Prionodon) borneensis BLEEKER (1859). Holotype: RMNH 7386. Type material is an immature male of 24 cm TL, from Singkawang, Borneo.

Diagnostic features. Small, slender species, without interdorsal ridge. Upper precaudal pit well developed, lower pit weak. Snout long and pointed, internarial width 1.1-1.5 times the preoral length. Eyes circular, large (diameter 2.1 % TL), eyes anterior margin slightly

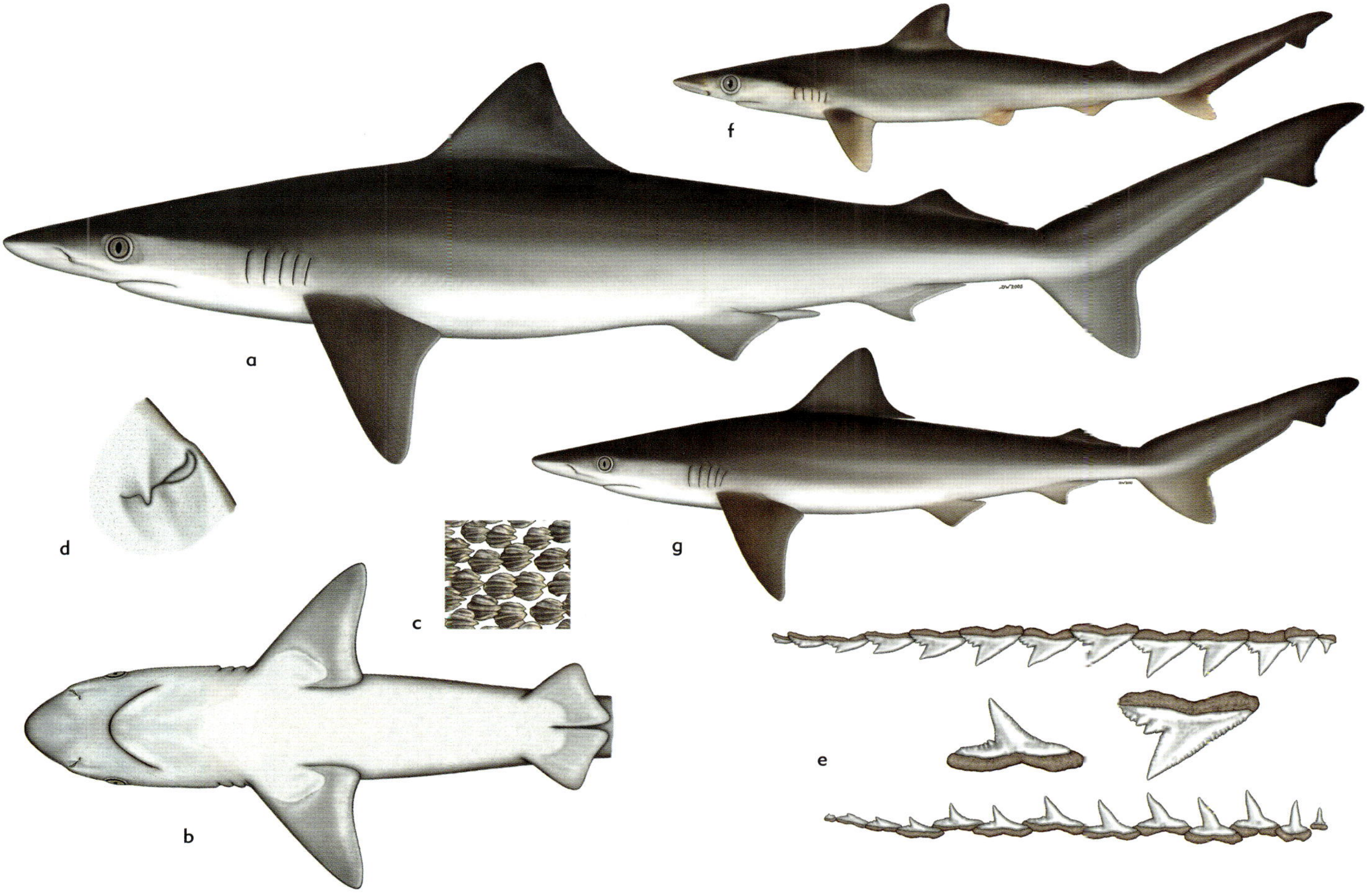

Figure 25. *Carcharhinus borneensis*: 57.4 cm TL male from Sarawak, Malaysia; **a**, left side; **b**, ventral view; **c**, dermal denticles; **d**, enlarged left nostril; **e**, left upper and lower teeth (symphysis to the right), and inset are the enlarged fifth upper and lower teeth; **f**, left side of a male of 32.2 cm TL from Sabah, Malaysia; **g**, left side of a female of 46.9 cm TL from Chu Shan Island, China.

anterior to the front of mouth. Nostrils small, incurrent aperture slit-shaped, oblique to falcate. Anterior nasal flaps high and narrow, nipple-shaped. Upper labial furrows short and inconspicuous. Hyomandibular pores alongside corners of mouth discrete and conspicuously enlarged. Gill slits short (2.2–2.7 % TL). D1 moderately large (height 8.0 % TL), triangular, with a bluntly pointed apex. The posterior margin is curving ventrally from apex, with a short rear tip. D1 origin slightly anterior to inner pectoral rear tips, inner margin relatively long. Small, low second dorsal fin (height 2.0 % TL), with short rear tip. D2 origin above or slightly posterior to anal mid-base. Small pectoral fins (length of anterior margin 13.7 % TL), falcate, narrowly rounded or pointed apices. Vertebrae: 61 precaudal centra and 58 caudal centra on average. Dermal denticles spaced apart, slightly imbricate and sub-circular, a little bit longer than wide, with three longitudinal ridges, and a corresponding number of posterior cusplets.

Dental Formula:

$$\frac{12 - 1 - 12}{11 - 1 - 11} \quad \text{to} \quad \frac{11(12) - 1 - 11(12)}{11(12) - 0(1) - 11(12)}$$

Upper teeth are quite narrow, crowns high and oblique. Distal median edges deeply notched, almost straight apically, convex basally. Mesial edges almost straight to slightly convex. Both edges are finely serrated on apical parts of crowns, coarser on basal part. Basally on distal median edges serration is very coarse and deeply notched. Roots are almost straight or slightly oblique (A rows) to arched (for the other tooth rows). One row of symphysial teeth is present. Lower teeth are narrow, crowns oblique, except the symphysial row. Distal median edges are concave, deeply notched, coarsely serrated, mesial edges concave. Both edges of smaller specimens are smooth to finely serrated, only slightly notched. Roots are arched, slightly notched. One or even none row of symphysial teeth is present. Files in upper and lower jaw are arranged as mixed alternate i.e. imbricate overlapping dentition.

Colour. Brown above to slate-grey, ventrally white. Usually, fins are without conspicuous markings but sometimes a dark brown or dusky tip on D1 apex can be present. The margin of the upper caudal lobe is dusky or with white markings, the pectorals (and sometimes the pelvic fins too) have light white edges (WHITE et al. 2010).

Size. Maximum of 100 cm TL for both gender. The average seems to be 60–70 cm TL (GARRICK 1982, COMPAGNO 1984, WHITE et al. 2010).

Misidentifications. *Carcharhinus borneensis* is easily to confuse with its congeners of the genus, *C. macloti*, *C. porosus* and *Carcharhinus* sp. A, but differs from them in having D2 origin well behind anal fin origin (above or slightly posterior) to anal fin mid-base. Further, the species differs from *C. macloti* in having a much shorter D1 inner margin, the absence of a hypercalcified rostrum, and in the presence of a strong serration on teeth, which is absent in *C. macloti*. Moreover, Borneo sharks differ from *C. porosus* and *C.* sp. A in having a row of 5–12 enlarged hyomandibular pores alongside mouth corners, in D2 height and in tooth formula. At least, *C. borneensis* differ from all mentioned species of the genus *Carcharhinus* in the average number of caudal vertebrae (58 in *C. borneensis* vs. 81 in *C. macloti*, 65 in *C. porosus* and 62 in *C.* sp. A).

Moreover, *C. borneensis* is frequently confused with *Rhizoprionodon* sp. because of its similar body size. However, *C. borneensis* is separable from members of the mentioned genus due to having a discrete row of enlarged hyomandibular pores alongside of mouth and by having the origin of D2 over or slightly behind mid-anal fin origin. Further, it differs from *Rhizoprionodon* sp. in having stronger serrated teeth, larger lateral cusplets on distal margin of upper teeth, and much shorter lower labial furrows (not more than 0.6 % TL, they are not visible when mouth is closed, whereas in *Rhizoprionodon* sp. they are not less than 1.0 % TL; GARRICK 1982, WHITE et al. 2010).

Geographical Distribution. Indo-West Pacific: Borneo, China, probably at the Philippines and around Java but validated records from these regions does not exist. – (GARRICK 1982, COMPAGNO 1984, COMPAGNO et al. 2005, WHITE et al. 2010).

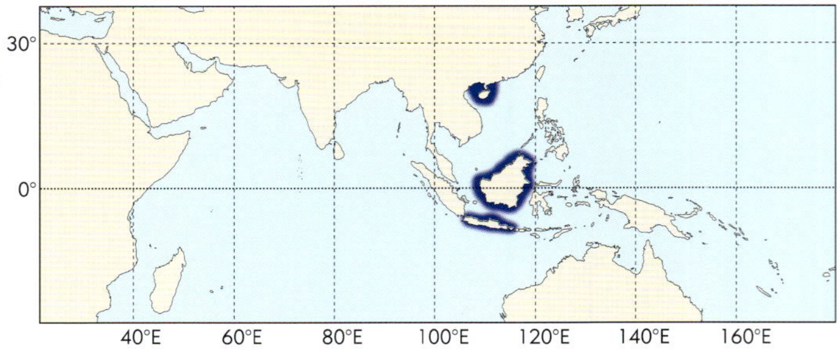

Habitat. A rare, coastal, inshore, in tropical waters living species but little is known about (COMPAGNO 1984).

Biology. No relevant data available.

Reproduction. Males seems to reach maturity with 54–58 cm TL, females at 60–65 cm TL. Reproductive mode is placental viviparous, (DULVY & REYNOLDS 1997). Normal litter size of 6 young, and at birth young have 23–27 cm TL (GARRICK 1982, WHITE et al. 2010).

Food. No data available but supposable the species feed mainly on bony fishes.

Red List status: EN C2a (ii) (IUCN 2010).

Carcharhinus brachyurus (GÜNTHER 1870)

En: Copper shark, New Zealand whaler, Cocktail shark; Fr: Requin cuivre; Sp: Tiburon cobrizo; Ge: Bronzehai

Carcharias brachyurus GÜNTHER (1870). [...] Four specimens listed under description (GARRICK 1982). Neotype: (DM) NMNZ 2262. Type material is a female of 242 cm TL, from Waganui, New Zealand, designated by GARRICK (1982).

Synonyms: *Carcharias lamiella* JORDAN & GILBERT (1883 a, b), *Eulamia ahenea* STEAD (1963), *Carcharhinus improvisus* SMITH (1952), *Carcharhinus rochensis* ABELLA (1972), *Carcharhinus remotoides* DENG et al. (1981), *Carcharhinus acarenatus* MORENO & HOYOS (1983a).

Synonyms in recent use: *Carcharhinus remotus* (not *Carcharias [Prionodon] remotus*, DUMERIL 1855 = *Carcharhinus acronotus*).

Authors note. In his extended discussion about the taxonomical status of the species GARRICK (1982) has grouped several different phenotypes (and species synonyms) under the name *Carcharhinus brachyurus* and designated them as variances. After studying many descriptions and reports of the "species" covered by this group name and by examination of a huge amount of type material we are convinced GARRICK was wrong in his argumentation to put them together because of distinct differences among the examined individuals. We recommend a differentiation and separation of one or even more species from *C. brachyurus*. The excluded species should become

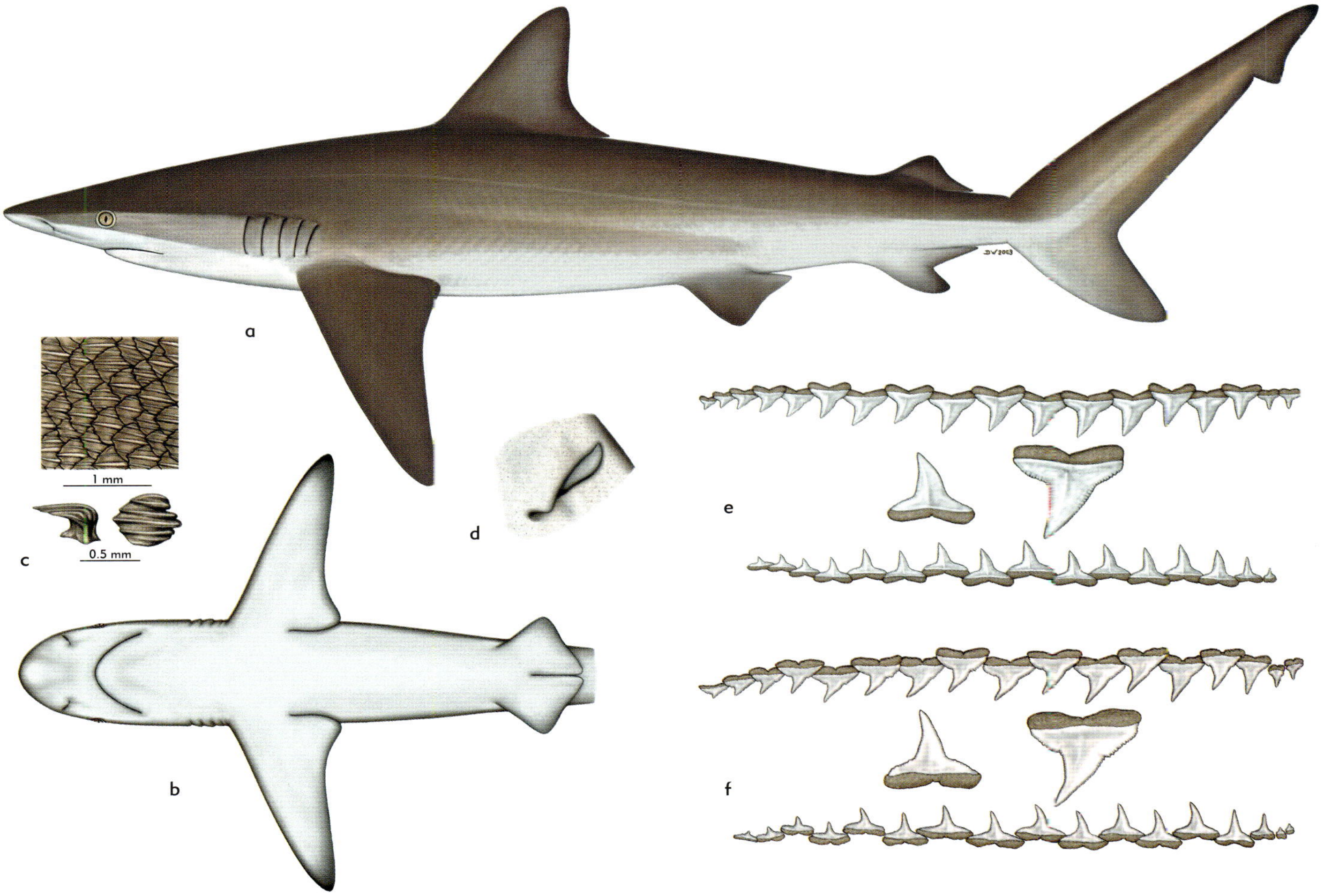

Figure 26. *Carcharhinus brachyurus*: 238 cm TL female from South Australia; **a**, left side; **b**, ventral view; **c**, dermal denticles; **d**, enlarged left nostril; **e**, left upper and lower teeth (symphysis to the right), inset are the enlarged fifth upper and lower teeth; **f**, left upper and lower teeth of a 224 cm TL male from New Zealand (symphysis to the right), inset are the enlarged fifth upper and lower tooth.

valid member(s) of the genus *Carcharhinus*. Hence, we favour the separation of *C. acarenatus* from the group synonym accordingly to the investigation results of MORENO & HOYOS (1983a).

Diagnostic features. Large, slender species, midline between dorsal fins smooth, without interdorsal ridge. Upper precaudal pit strongly developed, lower pit weak. Moderately long, narrowly rounded or pointed snout. Internarial width 0.9–1.6 times the preoral length. Eyes circular and small (diameter 1.8 %TL), eyes anterior margin above or slightly anterior to the front of mouth. Nostrils large and variations in shape from oblique, slit-like to round occur. Incurrent apertures small to large (in case of rounded nostrils). Anterior nasal flaps low and poorly developed. Upper labial furrows short and inconspicuous. Hyomandibular pores alongside corners of mouth inconspicuous. Gill slits large (2.4–3.3 % TL). First dorsal fin large (height 8.4 % TL), falcate, with a pointed or narrowly rounded apex. Its margin curving ventrally from fin apex, a short rear tip is present. D1 origin over or slightly anterior to pectoral rear tips. Small and low second dorsal fin (height 2.3 % TL), with a short rear tip and short inner margin. D2 origin over or slightly posterior to anal fin origin. Pectoral fins long (length of anterior margin 18.1 % TL), falcate, with narrowly rounded or pointed apices. Vertebrae: 103 precaudal centra and 92 caudal centra on average. Dermal denticles closed set, imbricate and sub-circular. With three low longitudinal ridges and a corresponding number of short posterior cusplets in smaller specimens, five posterior cusplets in larger ones.

Dental Formula:

$$\frac{15 - 2 - 15}{15 - 1 - 15} \quad \text{to} \quad \frac{14(16) - 1(3) - 14(16)}{14(15) - 1(3) - 14(15)}$$

Teeth of adult males are longer, narrower and more oblique or curved than in adult females. Furthermore, adult males have finer serrated teeth than juveniles. The variation in dentition seems to depend on age and specific habitat of the examined individuals were living in.

Upper teeth are narrow, crowns triangular to scythe-shaped, high and bent. Distal median edges are concave on two thirds of crowns apices, convex basally, deeply notched and with a concavity near the base. Both median edges finely serrated, basally slightly coarser, sometimes even with irregular serration. Roots are arched, slightly notched. One to three rows of symphysial teeth are present. Lower teeth are narrow, crowns erect to slightly oblique. Both median edges are concave, notched basally. Mesial edges convex apically, bent, fine and uniformly serrated, distal edges smooth-edged. Roots arched, slightly notched. One to three rows of symphysial teeth are present. Files in upper and lower jaw are arranged as alternate overlapping dentition.

Colour. Bronze-grey above (from other sources described as being bronze to pale-golden like a new penny but sometimes to olive-grey, white to creamy-white ventrally. A creamy grey-white strip extends along the body beginning from eyes to pelvic fins. Often fins have inconspicuous darker markings (edges). Dusky to black tips on fins can be present which are more distinct in juveniles than in adults.

Size. Maximum of 325 cm TL for males, and 288 cm TL for females (what correspond to an age of 20–22 years) but with an average for both sexes of 230–275 cm TL (SADOWSKY 1967, RANDALL et al. 1990). Maximum reported body mass of 304.6 kg (IGFA 2001).

Misidentifications. *Carcharhinus brachyurus* is characterised by having dusky tipped fins, coupled with the shape of its moderately long snout, a prominent tooth-morphology (oblique or "hooked" upper teeth). Because of its unique dental formula the species is almost without any other species to be confused in identification. Nevertheless, the copper shark could be confused with *C. amblyrhynchoides*, *C. limbatus* and *C. brevipinna* but differs due to the already mentioned features and in the average number of precaudal vertebrae from the other mentioned species (103 in *C. brachyurus* versus 91 in *C. amblyrhynchoides*, 82 in *C. brevipinna* and 96 in *C. limbatus*).

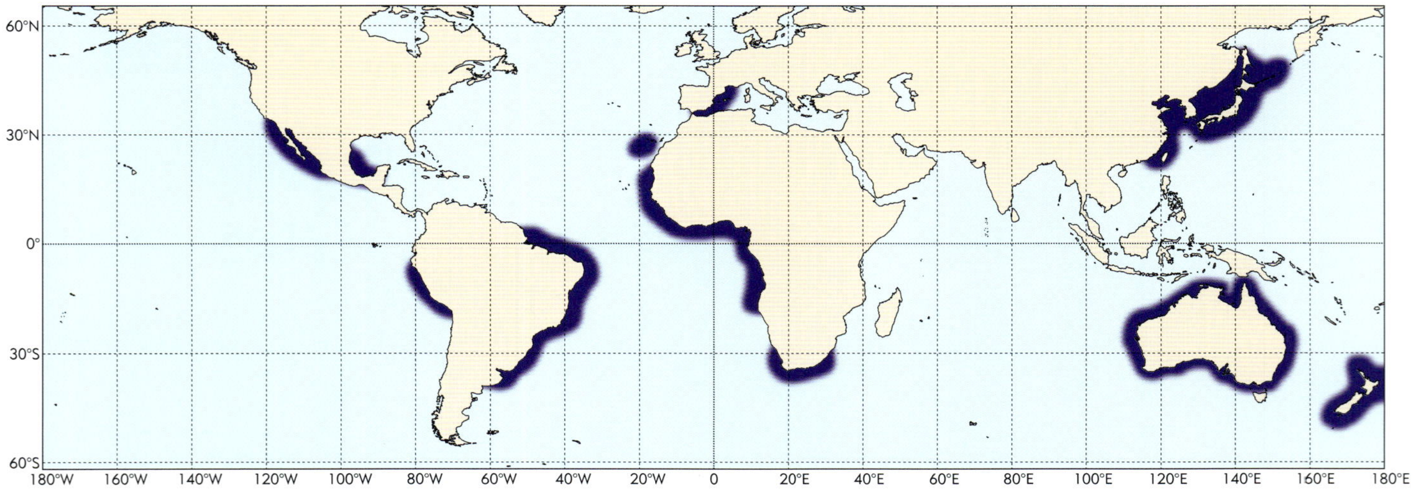

Geographical Distribution. Western Atlantic: Southern Gulf of Mexico, Brazil to Argentine. – Eastern Atlantic and Mediterranean: Balearic Islands, along French and Spanish Mediterranean coast to Algeria and Morocco, Canary Islands, Mauritania, Guinea to Namibia, South Africa. – Western Indian Ocean: South Africa. – Western Pacific: Japan, Korean Peninsula, China, southern Siberia, along the entire Australian coast, New Zealand. – Eastern Pacific: Southern California to the Gulf of California, Peru. – (ROSENBLATT & Baldwin 1958; BASS et al. 1973, 1986; GARRICK 1982; COMPAGNO 1984; SMALE 1991; COPPO 1992; VACCHI et al. 1996; CHIARAMONTE 1998; HEMIDA & LABIDI 2002; HOLTZHAUSEN 2002; MOREY & MASSUTI 2003; LUCIFORA 2003; COMPAGNO et al. 2005; LAST & STEVENS 2009).

Habitat. This species is living in tropical and sub-tropical regions between 40° N and 40° S, and occur offshore and inshore from surface to 100 m water depth. Occasionally, single specimens enter large coastal bays and inshore areas typically found swimming near the bottom (BRITO 1991, KAILOLA et al. 1993).

Biology. *Carcharhinus brachyurus* is a relatively abundant species. However, its biology is fairly unknown because this species is frequently mistaken with other shark species. Hence, statements about species distribution or behaviour are particularly doubtful. Copper sharks form large schools and migrate over long distances. They move northwards in the Gulf of Mexico inshore to the Florida and Louisiana coasts in spring and summer to reproduce and to feed. In autumn and winter the migration of the species is directed southwards presumably to follow its prey. Along the South African coast, the species moves southwards into the waters off southern Natal, and most sharks were caught at the southern beaches during the annual "Sardine run" in June and July when the sharks feed on shoals of pilchard *(Sardinops ocellatus)* (CLIFF & DUDLEY 1991, WALTER & EBERT 1991).

For the species a maximum age of 30 years was reported for males, and of 25 years for females (IGFA 2001).

Reproduction. Males reach maturity at 210–230 cm TL. Females reach maturity at 220–240 cm TL. Both sexes have matured before they are five years old. Reproductive mode is placental viviparous (DULVY & REYNOLDS 1997). Normal are 13–20 young per litter, and at birth young have 17–54 cm i. e. to 60–70 cm TL (BASS et al. 1973, 1986; CLIFF & DUDLEY 1991). Mature males are present at US and African coast for much of the year but mature females and immature individuals were only abundant in June and July (gonad indices in males are highest in March and April, whereas large ova are recorded in females in June and July). Mating occurs after July, and parturition is about 12 months later (MUNOZ-CHAPULI 1984, WALTER & EBERT 1991, LUCIFORA et al. 2003).

Food. The species feed mainly on bony fishes, e.g. sea catfish, jacks, gurnards, hakes and soles) but also smaller sharks (Squalidae), skates and cephalopods.

Red List status: NT (IUCN 2010).

Carcharhinus brevipinna (MÜLLER & HENLE 1839)

En: Spinner shark, Longnose grey shark, Inkytail shark, Smoothfang shark; Fr: Requin tisserand; Sp: Tiburon aleta negra; Ge: Spinnerhai, Langnasenhai

Carcharias (Aprion) brevipinna MÜLLER & HENLE (1839). Holotype: RMNH 2525. Type material comes from a 78.5 cm TL specimen, caught near Java, but only the mounted skin is present.

Synonyms: *Carcharias brevipinna* BLEEKER (1853), *Prionodon brevipinna* BLEEKER (1853, 1854), *Isogomphodon maculipinnis* POEY (1865), *Longmania brevipinna* WHITLEY (1939), *Uranga nasuta* WHITLEY (1943 a, b, c), *Longmania calamaria* WHITLEY (1944 a, b), *Aprionodon caparti* POLL (1951), *Carcharhinus johnsoni* SMITH (1952).

Synonyms in recent use. *Carcharhinus maculipinnis* POEY (1865), *Aprionodon brevipinna* MÜLLER & HENLE (1839).

Authors note. A sub-division of the spinner shark, *Carcharhinus brevipinna*, into two species (the first one is distributed in the Atlantic, the other one in the Pacific Ocean) is highly recommended. Individuals of this species distributed in the Atlantic ocean differ from the Pacific form in having a much smaller body size, teeth are much longer than in the Pacific variety, and even the cartilage in the Atlantic form differs in having less pronounced hyaline plates and Meckel's cartilage (MARK HARRIS, personal communication, 2005). Hence, we advocate further investigations to clear the taxonomic status of these variations.

Diagnostic features. Mid-sized to large species, without interdorsal ridge. Upper precaudal pit well developed, lower pit weak, internarial width 1.5–2.0 times the preoral length. Snout short and bluntly rounded. Eyes horizontally oval and small (diameter 1.5 % TL), eyes anterior margin above or slightly anterior to the front of mouth. Nostrils slit-shaped, strongly oblique, incurrent aperture oblique and long. Anterior nasal flaps low, nipple-shaped or pointed. Upper labial furrows short and inconspicuous. Hyomandibular pores alongside corners of mouth inconspicuous. Gill slits large (3.2–4.2 % TL). First dorsal fin large (height 7.3 % TL), falcate, with a narrowly rounded or pointed apex, and its posterior margin curving ventrally from fin apex. D1 origin slightly posterior of pectoral rear tips, its inner margin short, less than a third of dorsal base. Second dorsal fin large, high (height 2.2 % TL), the inner margin short. D2 origin over or slightly behind anal fin origin. Pectorals large (length of anterior margin 14.9 % TL), narrow and falcate. Fin apices narrowly rounded or pointed. Vertebrae: 82 precaudal centra and 89 caudal centra on average. Dermal denticles closed set, imbricate and sub-circular in smaller specimens, with three to five deep longitudinal ridges and a corresponding number of sharp-pointed, short posterior cusplets. In larger sharks shape of dermal denticles more rhomboid/ellipsoid, with seven, occasionally up to nine ridges and weak distinct cusplets.

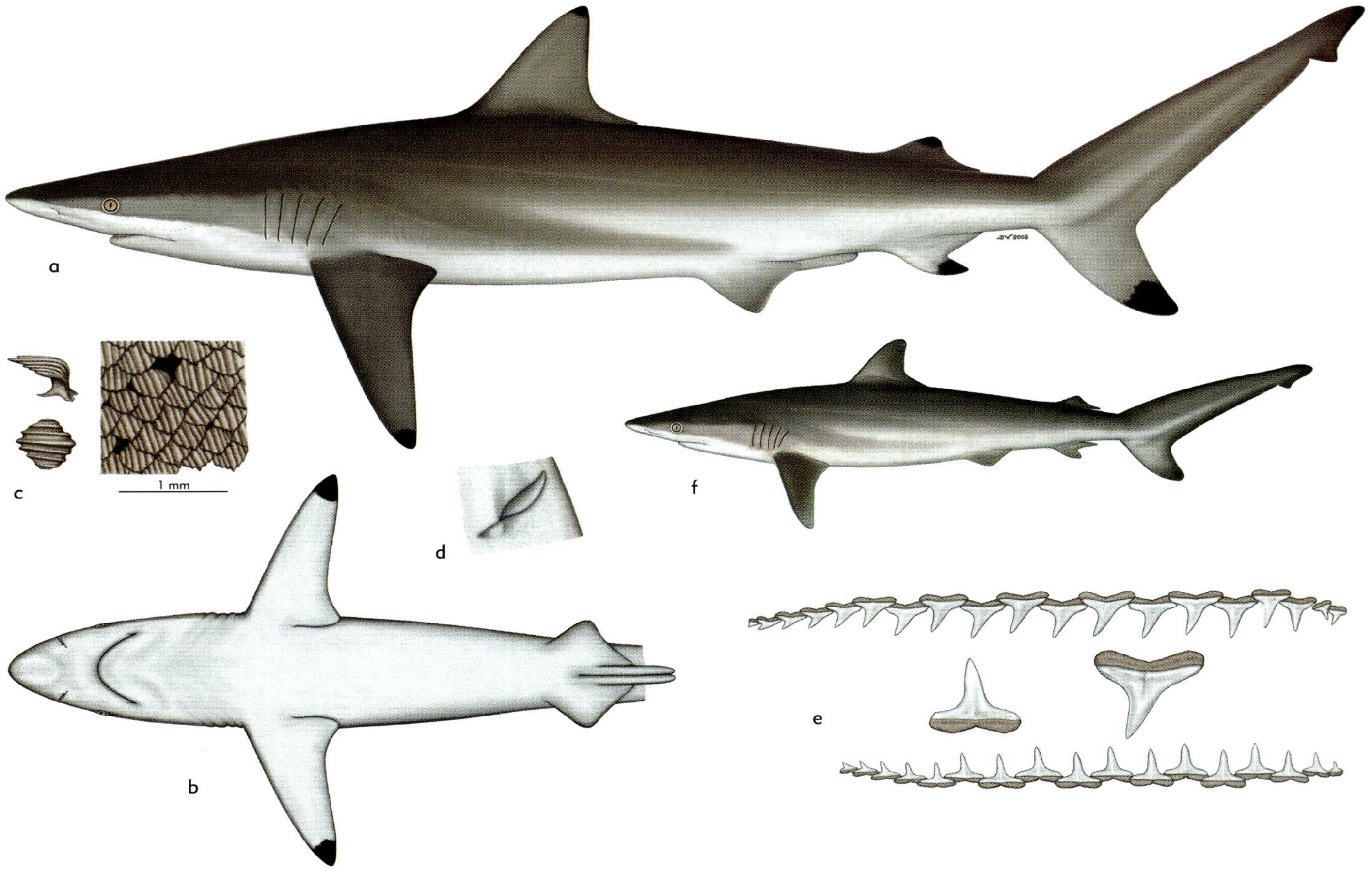

Figure 27. *Carcharhinus brevipinna*: 249 cm TL male from Natal, South Africa; **a**, left side; **b**, ventral view; **c**, dermal denticle; **d**, enlarged left nostril; **e**, left upper and lower teeth (symphysis to the right), and inset are the enlarged sixth upper and lower teeth; **f**, juvenile female of 82 cm TL from the Red Sea.

Dental Formula:

$$\frac{16 - 2 - 16}{16 - 1 - 16} \quad \text{to} \quad \frac{15(18) - 2(3) - 15(18)}{14(17) - 1(3) - 14(17)}$$

Upper teeth are narrow, crowns high, erect near the symphysis to slightly oblique (L to C rows). Distal median edges straight apically to convex on basal part of crown, deeply notched, coarsely serrated. Mesial edges straight to slightly concave, fine and uniformly serrated, except for juveniles of 80 cm TL (and smaller) where the median edges are smooth-edged. Serration in adults is slightly coarser basally, the roots are arched. Two or three rows of symphysial teeth are present. Lower teeth are narrow, crowns erect to slightly oblique, except the last three rows towards angles of jaw (P to C row), which are strongly oblique. Both median edges concave, slightly notched basally, mostly smooth-edged. Occasionally, slight irregular serration is distinct in adult specimens (especially in females) but normally teeth edges are smooth. Roots are arched, slightly notched. One to three rows of symphysial teeth are present. Files in upper and lower jaw are arranged as alternate overlapping dentition.

Colour. Grey or light brown above, ventrally pale-white. Black margins on D1 and D2, on ventral lobe of caudal fin and on pectoral fins are present, expanded to black tips on the apices of caudal lobe and pectorals. A black tip or a narrow black margin on anal fin, and sometimes a conspicuous white band on flanks is distinct. Mostly, juveniles are lacking dusky margins on fins.

Size. Maximum of 300 cm TL and a body mass of 90 kg (SANCHES 1991). An average of 195 cm TL and a body mass of 56 kg was reported. The maximum reported body mass for this species was 89.7 kg (IGFA 2001)

Misidentifications. *Carcharhinus brevipinna* is occasionally confused with its relatives, *C. amblyrhynchoides*, *C. limbatus* and *C. tilstoni*, but the latter two species (both forms of the blacktip shark) are slightly larger in body size, are more robust in body-contour, and their body-colouration is different. All mentioned three species of confusion have larger fins than the spinner shark, the tooth-shape of *C. brevipinna* is different to them (serration is normally absent in lower jaw teeth) and the dental formula doesn't match. Moreover, *C. limbatus* and

C. tilstoni have a shorter snout. An easy way to separate the mentioned species from spinner sharks is to check for the presence of a black margin colouration on pectorals and anal fin. This colouration pattern is usually absent in the three species of confusion but on the other hand juvenile spinner sharks are lacking often distinct markings and thus, identification is more complicated.

Carcharhinus brevipinna differs further in the average number of the precaudal vertebrae from the other species (82 in *C. brevipinna versus* 91 in *C. amblyrhynchoides*, 96 in *C. limbatus* and 88 in *C. tilstoni*).

Geographical Distribution. Western Atlantic: from North Carolina to Florida, Bahamas and Cuba, Gulf of Mexico to southern Brazil. – Southern Mediterranean: along the coast of Morocco and Algeria to Israel. – Eastern Atlantic: Cape Verde Islands, Senegal, Guinea, Sierra Leone to the coast of Angola, South Africa. – Indian Ocean: Red Sea, Gulf of Aden, Oman, south-eastern Africa, Madagascar, Mauritius, Seychelles, India, Singapore, Indonesia (Java and Sumatra), Vietnam, Japan (possibly at the Philippines). – Eastern Indian Ocean and western South Pacific: Australia (Queensland, western and northern Australian coast), New South Wales, New Guinea, Solomon Islands. – (BASS et al. 1973, 1986; GARRICK 1982; COMPAGNO 1984; ALLEN & CLIFF 2000; HEMIDA & LABIDI 2002; COMPAGNO et al. 2005).

Habitat. Commonly coastal-pelagic distributed species, in tropical waters over continental and insular shelves living, reported from inshore and offshore, with an affinity to shallower water (30 m) and a range from surface to 100 m depth. Juvenile spinner sharks move into lower portions of bays with the tides but normally they avoid areas of low salinity (COMPAGNO 1984, COMPAGNO et al. 1989, COMPAGNO & NIEM 1998).

Biology. Spinner sharks are very fast-swimming individuals and live quite often together in particularly large schools. They have an unusually prey hunting strategy by swimming rapidly upwards through the fish-shoals with open mouth, spinning along its longitudinal axis, snapping in all directions, and then spring out of the water after its feeding run (COMPAGNO 1984).

The oldest caught specimen ever was a 208 cm TL female, 11.3 years old. Growth rates are in excess of 20 cm per year for the first six months to two years, about 10 cm per year through adolescence, and 5 cm per year or less for adults (BRANSTETTER 1981, 1982, 1987).

Reproduction. Males reach maturity at approximately 150–190 cm TL, corresponding to an age of 6–7 years. Females reach maturity at 150–200 cm TL, at an age of 7–8 years. Reproductive mode is placental viviparous (DULVY & REYNOLDS 1997). 3–15 young per litter are usual, and with a mean of nine. At birth, young have 50–70 cm TL (JOUNG et al. 2005).

Both, large and small pregnant females produce pups of the same length but larger females generally have larger litters. Gonad indices in mature males were highest during December and January and in females during February and March. Mating occur between January and March. The hepatosomic index of pregnant females is highest at the onset of pregnancy and lowest at parturition. Reproductive cycle seems to be biennial with a gestation of 13–18 months. Other authors give duration of gestation as 12–15 months, e.g. ALLEN & CLIFF (2000). Different points of time for giving birth are known for this species. The young off Senegal are born in summer, whereas young of the population(s) in the Gulf of Mexico and off Florida are born in spring and early summer.

In the Gulf of Mexico, adult individuals remain in shallow water during the summer but retreat possibly southwards or into deeper water in autumn. The Natal coast serves as a nursery ground for a population of this shark species. Adult females occur there throughout the year while males seasonally occur during the summer. Older juveniles occur also off the eastern Cape of South Africa (BASS et al. 1973, 1986). Studies suggest young spinner sharks prefer slightly lower water temperatures than adults, and tend to move southwards and Cape-wards from Natal when temperatures increase (COMPAGNO 1984).

Food. The species take mainly bony fishes, often shoal-fishes, e.g. sardines, herring and anchovies, tunas, bonitos, croakers, mullets, grunts, tongue-soles, but feeds also on stingrays, sea catfish, squids and other sharks. Larger shark species are potential predators of spinner sharks, especially on juvenile and sub-adult individuals.

Red List status: NT (IUCN 2010).

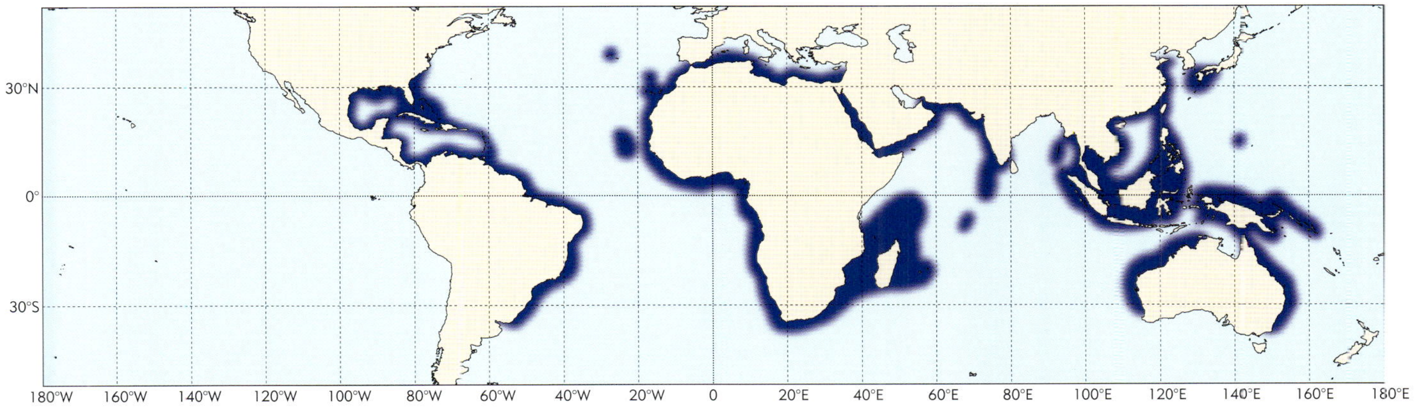

Carcharhinus cautus (WHITLEY 1945)

En: Nervous shark; Fr: Requin nerveux; Sp: Tiburon nervioso; Ge: Nervöser Hai

Galeolamna greyi cauta WHITLEY (1945). Holotype: AMS. Type material is a female of 162 cm TL (skin and teeth only) from Shark Bay, Herald Bight, Western Australia.

Diagnostic features. Small, slender species, without interdorsal ridge. Upper precaudal pit well developed, lower pit weak, internarial width 1.0–1.3 times the preoral length. Snout narrow, long and rounded or slightly pointed. Eyes circular and large (diameter 2.4 % TL), eyes anterior margin anterior to the front of mouth, by a distance equal to one-third to half of eye-diameter (GARRICK 1985). Anterior nasal flaps expanded as a narrow, nipple-shaped lobe, incurrent apertures oblique, slit-shaped and large. Upper labial furrows short and inconspicuous. Hyomandibular pores alongside corners of mouth inconspicuous. Gill slits short (2.0–2.8 % TL). First dorsal fin small (height 7.5 % TL), falcate, apex pointed or sharply rounded. D1 origin above posterior one-third of inner pectoral margin, and the rear tip 5.3–7.1 % TL. D2 origin above the middle of anal fin base (height of second dorsal fin 3.5 % TL). Pectorals large (length of anterior margin 16.0 % TL). Vertebrae: 89 precaudal centra and 78 caudal centra on average. Dermal denticles imbricate and sub-circular to ovoid, slightly wider than long, with three strong longitudinal ridges and a corresponding number of short posterior cusplets in small specimens. Adult specimens have five posterior cusplets.

Dental Formula:

$$\frac{12 - 2 - 12}{12 - 1 - 12} \quad \text{to} \quad \frac{12(14) - 1(2) - 12(14)}{11(13) - 1(2) - 11(13)}$$

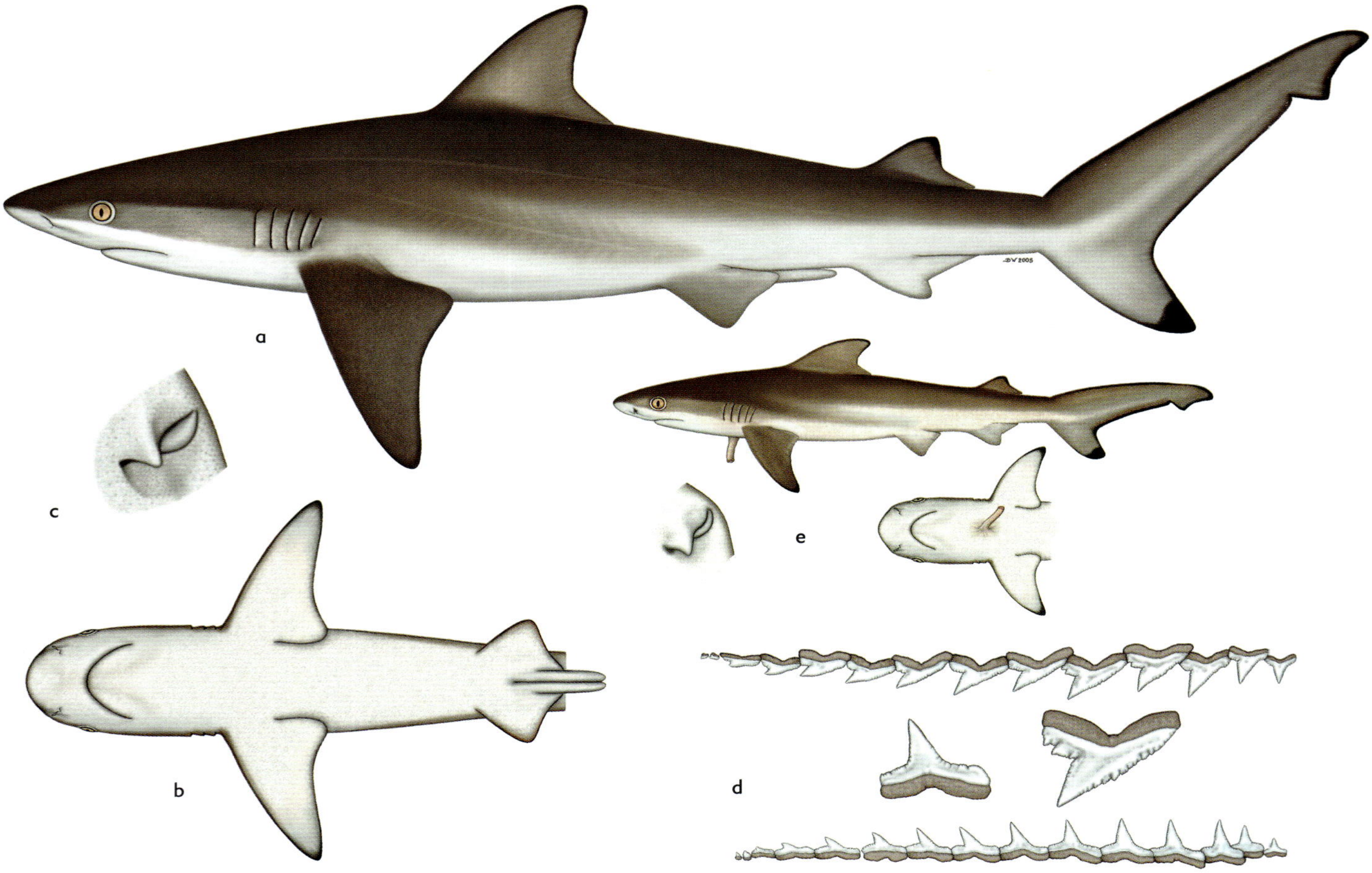

Figure 28. *Carcharhinus cautus*: 98 cm TL male from Queensland, Australia; **a**, left side; **b**, ventral view; **c**, enlarged left nostril; **d**, left upper and lower teeth (symphysis to the right), inset are the enlarged sixth upper and lower teeth; **e**, left side, ventral view and enlarged left nostril of a late female embryo from western Australia.

Upper teeth are narrow, crowns oblique or semi-oblique. Distal median edges straight apically to convex on basal part of crown, deeply notched, strongly serrated. Mesial edges straight to convex, strongly and irregularly serrated basally, but smooth-edged apically. Roots are very arched. One or two rows of symphysial teeth are present and they are only slightly smaller than the adjacent rows. Lower teeth are narrow, crowns erect (S to L rows) to oblique (LP to C rows). Both median edges concave, slightly notched. Fine and uniformly serrated edges in larger specimens but irregularly serrated in smaller individuals, roots are arched. One or two rows of symphysial teeth are present. Files in upper jaw are arranged as alternate overlapping dentition, and in lower jaw as mixed alternate overlapping dentition.

Colour. Grey or grey-brown, with a tinge of bronze above, white ventrally. Colouration of D1 and D2 is dusky with blackish margins, distinct on pectoral fins and upper caudal fin also. The blackish margins extend to tips on the caudal and pectoral fins, and a conspicuous white band on flanks is present. Sometimes trailing pale-white margins on pectorals, pelvics, anal and lower caudal fin are present.

Size. Maximum for females 133 cm TL, corresponding to an age of 16 years. Maximum for males 111 cm TL, corresponding to the age of 12 years (WHITE et al. 2002). Other sources give a maximum of 150 cm TL, and an average of 100–130 cm TL (COMPAGNO 1984, SHARKINFO).

Misidentifications. *Carcharhinus cautus* can be confused with *C. melanopterus*, *C. wheeleri* and *C. amblyrhynchos*. All three species share many similar body size and body-contour features but *C. cautus* differs from the latter two mentioned species in having a narrow, dark edging on the trailing margin of the caudal fin, rather than the prominent broad edging which characterize them, and in ratio of D1 : D2 height. Nervous sharks differ from *C. melanopterus* in lacking a prominent black blotch on apex of the first dorsal fin. Further, *C. cautus* has a different average number of precaudal vertebrae which is much less than those of the other three mentioned species (89 in *C. cautus versus* 115 in *C. amblyrhynchos*, 116 in *C. melanopterus* and 114 in *C. wheeleri*).

Geographical Distribution. Eastern Indian Ocean and western South Pacific: Australia (Queensland, northern and western Australian coast), off Ugi and Solomon Islands, New Guinea. – (COMPAGNO 1984, GARRICK 1985, LYLE & TIMMS 1987, COMPAGNO et al. 2005).

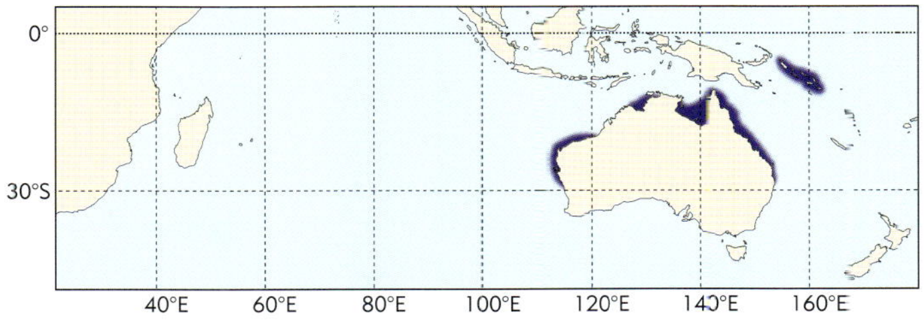

Habitat. It is a common species but little is known about. The species is living inshore, near continental and insular shelves. Nervous sharks occur in the offshore prawn trawling grounds in Albatross Bay in 7–45 m depth. In other locations the species occur in shallow nearshore waters (< 5 m), and in adjacent inshore waters of the Embley River estuary (SALINI et al. 1991).

Biology. No relevant data available.

Reproduction. Females and males reach maturity at 61–69 cm TL (WHITE et al. 2002). From habitats of northern Australia males and females are described to mature usually at 84–91 cm (LYLE 1987). Reproductive mode is placental viviparous (DULVY & REYNOLDS 1997).

Usually, 1–5 young per litter and at birth young have 35–40 cm TL. *C. cautus* copulates in late October to early November. Ovulation and conception occurs in late November to early December and parturition takes place approximately 8–11 months later (WHITE et al. 2002).

[...] Since mature non-pregnant females contain vitellogenic ova for 12–13 months, i. e. from November or December to the following December, and mature pregnant females contain embryos for 11 months, i. e. from December to October, *C. cautus* has a biennial reproductive cycle. By parturition, the females and males had reached 28 % and 32 % of their lengths at their maximum observed ages, respectively (LYLE 1987, WHITE et al. 2002).

Food: The species feed mainly on bony fishes, sometimes also molluscs and crustaceans, e. g. Penaeidae, Brachyura, Stomatopoda (SALINI et al. 1991).

Red List status: DD (IUCN 2010).

Carcharhinus dussumieri (VALENCIENNES 1839)

En: Whitecheek shark, White cheeked whaler shark; Fr: Requin a joues blanches; Sp: Tiburon cariblanco; Ge: Weißwangenhai

Carcharias (Prionodon) dussumieri VALENCIENNES (in MÜLLER & HENLE 1839). Lectotype: MNHN 1135. Type material is an immature male of 37 cm TL, (designated by GARRICK 1982), from Pondicherry, India.

Synonyms: *Carcharias (Prionodon) tjutjot* BLEEKER (1852), *Carcharias (Prionodon) javanicus* BLEEKER (1852), *Carcharias malabaricus* DAY (1873–78).

Synonyms in recent use: *Carcharhinus menisorrah* VALENCIENNES (in MÜLLER & HENLE 1839 = *C. falciformes* BIBRON (in MÜLLER & HENLE) 1839, as restricted by GARRICK [1982]).

Diagnostic features. Small, slender species, slightly stout in body contour. Normally, a low interdorsal ridge is present, internarial width 1.1–1.9 times the preoral length. Upper precaudal pit strongly developed, lower pit weak. Snout long and pointed or obtusely rounded but wide variations in shape and proportion of the snout are known. Eyes large and oval horizontally (diameter 2.4 % TL), eyes anterior margin above or slightly anterior to the front of mouth. Nostrils oblique, incurrent apertures broad, ovate, excurrent apertures small and slightly ovate. Anterior nasal flaps elongated and triangular, with sharply pointed lobes. Upper labial furrows short and inconspicuous. Hyomandibular pores alongside corners of mouth inconspicuous. First dorsal fin small (9.6 % TL), erect and broadly, triangular in shape, slightly falcate, with pointed or narrowly rounded apex. The posterior margin slopes backwards from apex, rear tip short. D1 origin over the posterior half of pectoral inner margin. Second dorsal fin large (3.2 % TL), its inner margin short. D2 origin over or slightly posterior to anal fin origin. Pectorals small (length of anterior margin 15.7 % TL), semi-falcate, narrow, with angular apices. Vertebrae: 62 pre-

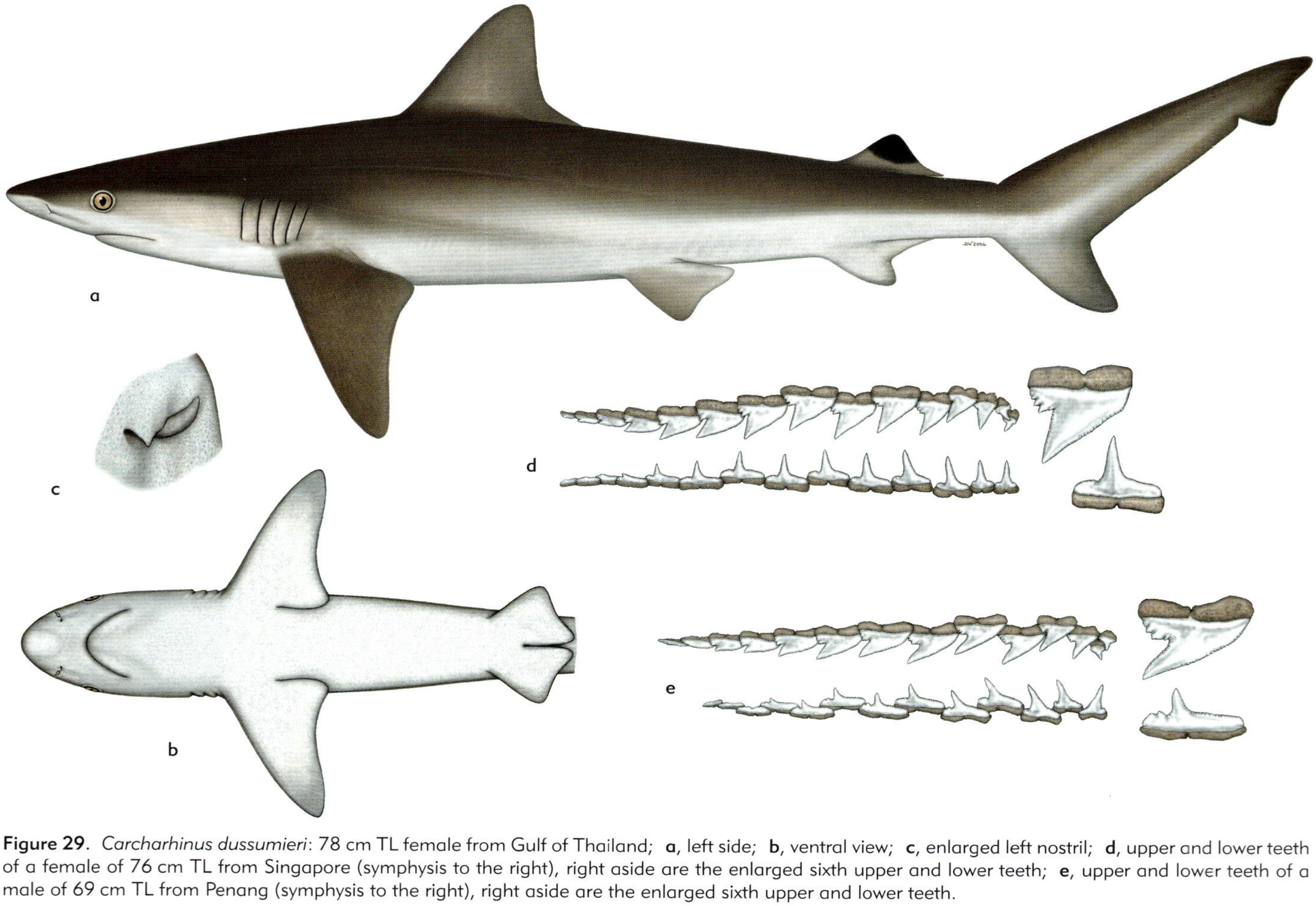

Figure 29. *Carcharhinus dussumieri*: 78 cm TL female from Gulf of Thailand; **a**, left side; **b**, ventral view; **c**, enlarged left nostril; **d**, upper and lower teeth of a female of 76 cm TL from Singapore (symphysis to the right), right aside are the enlarged sixth upper and lower teeth; **e**, upper and lower teeth of a male of 69 cm TL from Penang (symphysis to the right), right aside are the enlarged sixth upper and lower teeth.

caudal centra and 66 caudal centra on average. Dermal denticles closed set or only slightly imbricate in smaller individuals, up to totally imbricate in larger ones. Sub-circular to ovoid in shape, for smaller specimens with three longitudinal ridges and a corresponding number of posterior cusplets, five ridges and a corresponding number of cusplets for larger specimens.

Dental Formula:

$$\frac{13-2-13}{13-1-13} \quad \text{to} \quad \frac{12(14)-0(3)-12(14)}{11(15)-0(2)-11(15)}$$

Sexual dimorphisms in tooth-shape of half-grown and adult specimens, and morphological variations in dentition i.e. dental formula are known within this species. The dimorphism seems to occur in dependence of the food source of the specific habitat inhabited. Upper teeth of females and immature males are narrow to broad, crowns strongly oblique. Distal median edges straight to weak convex apically, slightly concave, deeply notched, and strongly serrated basally. Mesial edges convex, more finely and uniformly serrated. Upper teeth of males much narrower than those of females, crowns more oblique, finer serrated on median edges. Roots are almost straight to slightly oblique, deeply notched, dividing the roots almost into two root lobes. Usually two, seldom three or even none row(s) of symphysial teeth are present. Lower teeth are narrow, crowns oblique to slightly oblique, more oblique in mature males than in females and immature males. Distal median edges concave, strongly serrated, deeply notched. Mesial edges concave, slightly notched, finely serrated. Roots are arched (S to L rows) to almost straight (LP to C rows), slightly notched. One, even two or none row(s) of symphysial teeth is present. Files in upper and lower jaw are arranged as mixed alternate i.e. imbricate overlapping dentition.

Colour. Grey to grey-brown above, white ventrally. Often, a prominent dusky or black tip on the apex of D2 is present, other fins with light margins. Sometimes with a white band on flanks but this can be inconspicuous.

Size. Maximum of about 100 cm TL, and an average of 70–90 cm TL (BYKOV 1983).

Misidentifications. The species is frequently confused with *C. sealei* because of a similar body size and body-contour. Further, both species share a similar colouration and tooth-morphology but *C. dussumieri* differs in the number of precaudal vertebrae (62 in *C. dussumieri versus* 78 in *C. sealei*), the presence or absence of irregularities in vertebrae centra, and the shape of first dorsal fin. The whitecheek shark can also be confused with *C. acronotus* and *C. amblyrhynchos* but *C. acronotus* is easily identified by the prominent black tip on his snout, further Gray reef sharks differ in maximum body size and other body dimensions, and furthermore in the average number of precaudal vertebrae (62 precaudal vertebrae in *C. dussumieri versus* 84 in *C. acronotus* and 115 in *C. amblyrhynchos*).

Carcharhinus dussumieri can develop local populations which differ one from another in a few features presumably due to partial isolation and different nutrition (GARRICK 1982).

Geographical Distribution. Indo-West Pacific: Persian Gulf, Oman, Pakistan, India, Sri Lanka, Thailand, Singapore, Malaysia, Philippines, Java, Borneo, Vietnam, China, southern Japan, along coastline from north-western to north-eastern Australia, on south-eastern part of Papua New Guinea. – (GARRICK 1982; COMPAGNO 1984; SALINI et al. 1991, 1999; STOBUTZKI et al. 2002; COMPAGNO et al. 2005; LAST & STEVENS 2009).

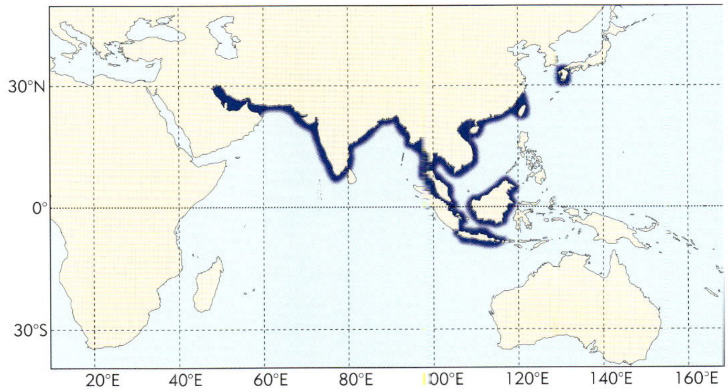

Habitat. It is a common species, inshore over continental and insular shelves, in shallow waters living but very little is known about (SIMPFENDORFER & MILWARD 1993).

Biology. No relevant data available.

Reproduction. Males reach maturity at 65–70 cm TL. Females reach maturity at 70–75 cm TL. Reproductive mode is placental viviparous (GILBERT & SCHLERNITZAUER 1966, DULVY & REYNOLDS 1997). Normal are 2–4 young per litter, and at birth young have 38 cm TL. *Carcharhinus dussumieri* has no definite reproductive season. The placenta is established when the embryo attains 5 cm TL. Off north-western Borneo and off Taiwan Islands gravid females with fully-termed young occur all year but with a peak in July and August (TESHIMA & MIZUE 1972, TESHIMA 1973).

Food. Mainly small bony fishes, crustaceans and cephalopods (SALINI et al. 1991, 1999; STOBUTZKI et al. 2002).

Red List status: NT (IUCN 2010).

Carcharhinus falciformes (Bibron 1839)

En: Silky shark; Fr: Requin soyeux; Sp: Tiburon jaqueton; Ge: Seidenhai

Carcharias (Prionodon) falciformis Bibron (in Müller & Henle 1839). Holotype: MNHN 1134. Type material is a female foetus of 53 cm TL, from Cuba.

Synonyms: *Carcharias menisorrah* Valenciennes (in Müller & Henle 1839), *Carcharias falcipinnis* Lowe (1839, 1840, 1843, 1860), *Prionodon menisorrah* Valenciennes (in Müller & Henle 1839), *Squalus tiburo* Poey (1860), *Gymnorhinus (Carcharias) pharaonis* Hemprich & Ehrenberg (1899), *Gymnorhinus pharaonis* Hilgendorf (1899), *Aprionodon sitankaiensis* Herre (1934), *Carcharhinus floridanus* Bigelow et al. (1943), *Eulamia malpeloensis* Fowler (1944), *Carcharhinus atrodorsus* Deng et al. (1981).

Synonyms in recent use: *Carcharhinus floridanus* Bigelow et al. (1943), Garrick et al. (1962a, b, 1964).

Diagnostic features: Large, slender species, a low interdorsal ridge is present. Upper precaudal pit well developed, lower pit weak, internarial width 1.1–1.7 times the preoral length. Snout long and bluntly pointed, eyes circular and large (diameter 1.8 % TL), eyes anterior margin slightly anterior to or above the front of mouth in small and sub-adult, and slightly behind front of mouth in larger specimens. Nostrils oblique to strongly oblique, slit-shaped. Anterior nasal flaps small, straight, with tiny and weakly developed lobes. Incurrent apertures ovate, upper labial furrows short and inconspicuous. Hyomandibular pores alongside corners of mouth inconspicuous. Gill slits short (2.2–3.3 % TL). First dorsal fin large, falcate (height 7.0 % TL), its apex bluntly pointed to sharply rounded, the rear tip relatively long. D1 origin behind pectoral free rear tips.

Figure 30. *Carcharhinus falciformes*: 287 cm TL female from Cocos Island; **a**, left side; **b**, ventral view; **c**, dermal denticles; **d**, enlarged left nostril; **e**, left upper and lower teeth (symphysis to the right), and inset are the enlarged fifth upper and lower teeth; **f**, juvenile female of 99 cm TL from Cuba.

Second dorsal fin long and low (height 1.7 % TL), its inner margin elongated. D2 origin over or slightly behind the origin of anal fin. Pectorals long and narrow, with rounded tips (length of anterior margin 16.8 % TL), anal fin long and low. Vertebrae: 102 precaudal centra and 104 caudal centra on average. Dermal denticles closed set, imbricate and sub-circular, a little bit wider than long, with three deep longitudinal ridges and three short, sharp-pointed posterior cusplets in embryos and juveniles. Sub-adults have five, adults seven ridges and cusplets.

Dental Formula:

$$\frac{15-2-15}{15-1-15} \quad \text{to} \quad \frac{14(16)-1(3)-14(16)}{14(17)-1(3)-14(17)}$$

Upper teeth are broad, crowns oblique, except the S and A rows. Both median edges notched halfway longitudinally. Distal edges straight apically, convex basally, deeply notched. Mesial edges are straight apically, to slightly convex basally, slightly notched. Both median edges coarsely serrated on basal part of crowns, roots are arched. In larger specimens the teeth of the ultimate row(s) are noticeable smaller in proportion than the penultimate row and could easily be overlooked (GARRICK 1982). One to three rows of symphysial teeth are present. Lower teeth are narrow, crowns erect. Both median edges concave basally, apically slightly convex, smooth-edged to fine and uniformly serrated. Roots are arched, only slightly notched. One to three rows of symphysial teeth present. Files in upper and lower jaw are arranged as alternate overlapping dentition.

Colour. Dark grey to brown-grey dorsally, and sometimes with an almost entire black colouration. Sides are metallic-grey, with a tinge of green, ventrally white. Tips on fins dusky but not black, except for D1.

Size. Maximum of about 350 cm TL, and an average of 200–240 cm TL (COMPAGNO & NIEM 1998). Maximum reported body mass of 346 kg (IGFA 2001).

Misidentifications. Silky sharks are frequently mistaken for *C. amblyrhynchos*, *C. sorrah*, *C. obscurus*, and *C. plumbeus*. The best way to identify and differentiate *C. falciformis* from the other mentioned shark species is to check for D1 origin which is behind inner pectoral free rear tips and therewith in contrast to the other species (ROSENBLATT & BALDWIN 1958). Furthermore, *C. falciformes* differs in the average number of precaudal vertebrae (102 in *C. falciformes versus* 115 in *C. amblyrhynchos*, 91 in *C. obscurus*, 90 in *C. plumbeus* and 70 in *C. sorrah*) and in dental formula.

Geographical Distribution. Western Atlantic: Massachusetts to southern Brazil (including Gulf of Mexico and the Caribbean Sea). – Central Atlantic: St. Paul's Rocks. – Eastern Atlantic: Madeira, along the Spanish coast, northern Africa to Senegal, northern Angola. – Indian Ocean: Red Sea, Oman, Mozambique, Tanzania, Madagascar, Comoros and Aldabra Island, Maldives Islands, India and around Sri Lanka. – Western Pacific: Thailand, China, Philippines, northern and western Australia (without southern Australian regions), New Caledonia, New Zealand. – Central and eastern Pacific: Hawaii, southern California to Peru. – Not distributed in the Mediterranean. – (GARRICK 1962a, b, 1964; BANE 1966; BASS et al. 1973, 1986; FOURMANOIR 1976a, b; GARRICK 1982; COMPAGNO 1984; HEMIDA & LABIDI 2002; COMPAGNO et al. 2005; JOUNG SHOOU-JENG et al. 2008).

Habitat: This is an abundant species circum-tropical, nearshore over continental and insular shelves living but also oceanic-epipelagic distributed. A distinct affinity of the species to a water of temperature of 20–24 °C was reported. Often, silky sharks are found over deepwater reefs, and near insular slopes from surface to 500 m depth (COMPAGNO 1984).

Biology. In the eastern and central Pacific, *C. falciformes* seems to be oceanic-pelagic but is more often seen and caught near land than in the open ocean compared with other oceanic living species, e.g. *Prionace glauca* and *C. longimanus*. Silky sharks are solitary living and often associated with shoals of tuna fish (KATO 1964). It

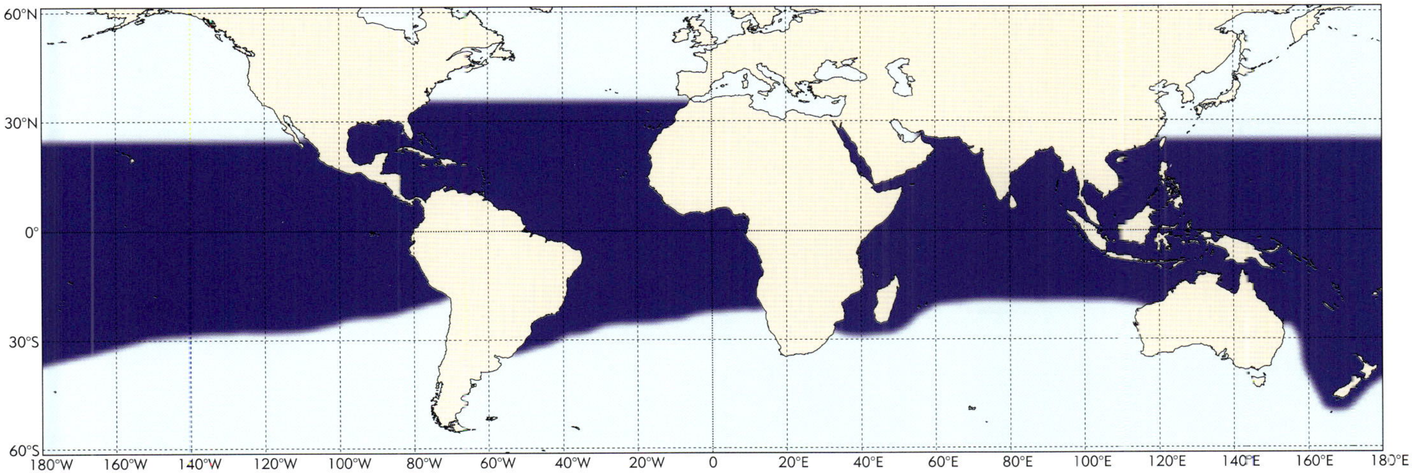

is a quick moving and fast swimming species, and is known to be dangerous to humans because of its size and abundance offshore. The species performs an agonistic or hunching display against divers and unknown things. Size segregation is known within the species but exact and relevant data are fairly unavailable.

Reproduction. Males reach maturity at 215–225 cm TL, corresponding to an age of approximately 10 years. Females reach maturity at 232–245 cm TL, and with an age of > 12 years. Reproductive mode is placental viviparous (GILBERT & SCHLERNITZAUER 1966, DULVY & REYNOLDS 1997). Normal are 2–14 young per litter, and at birth young have 70–85 cm TL. Silky sharks off Yucatan are born in early summer after a 12 month gestation period, with 70–76 cm TL. Young are born near deepwater reefs and are fast growing. After the juvenile have grown to a definite length they leave the reefs into blue water and live in schools or bigger aggregations. No differences in growth between the sexes were detected (GILBERT & SCHLERNITZAUER 1966, BONFIL 1993). Determination of the maximum age estimated by analysis of Alizarin-red-S-stained thin vertebral sections show a value of 22 years for females and 20 years for males. Other sources give a maximum age of 25 years (IGFA 2001).

Food: Mainly bony fishes, e. g. sea catfish, mullets, mackerel, yellowfin tuna, albacore, porcupine fish, but also cephalopods (squids, paper nautiluses) and pelagic crabs (CABRERA CHAVEZ-COSTA 2003).

Red List status: NT (IUCN 2010).

Carcharhinus fitzroyensis (WHITLEY 1943 a)

En: Creek whaler shark; Fr: Requin baleinier; Sp: Tiburon ballenero; Ge: Walbuchthai

Galeolamna (Uranganops) fitzroyensis WHITLEY (1943 a). Holotype: QM, I. 1229. Type material is a female of 117.5 cm TL, from Fitzroy River estuary, Connor's Creek, Queensland, Australia.

Diagnostic features. Mid-sized, robust species, without interdorsal ridge. Upper precaudal pit well developed, lower pit weak, internarial width 1.7–1.8 times the preoral length. Snout long, pointed to parabolic in shape. Eyes circular, fairly small (diameter 1.4 % TL), eyes anterior margin slightly anterior to the front of mouth in small specimens. In contrast to this description the margin of the holotype-specimen was described by GARRICK (1982) as slightly behind to front of mouth. Nostrils large, oblique, incurrent apertures big and ovate. Anterior nasal flaps small, nipple-shape, pointed lobes. Upper labial furrows short and inconspicuous. Hyomandibular pores alongside corners of mouth inconspicuous. Medium sized gill slits (2.4–3.0 % TL). First dorsal fin large and semi-falcate (height 9.4 % TL), with pointed or narrowly rounded apex. The posterior margin curving ventrally from fin apex. D1 origin over or slightly anterior to pectoral fin insertion, its inner margin short. Second dorsal fin large and high (height 3.2 % TL) but smaller than anal fin, with a short inner margin. D2 origin over or slightly posterior to anal fin origin. Pectorals large (length of anterior margin 18.1 % TL), triangular, with rounded apices. Vertebrae: 58 precaudal centra and 67 caudal centra on average. Dermal denticles closed set, imbricate, sub-circular, a little bit wider than long, and with three to five longitudinal ridges and three posterior cusplets.

Dental Formula:

$$\frac{14 - 2 - 14}{13 - 2 - 13} \quad \text{to} \quad \frac{14 - 2 - 14}{13(14) - 2 - 13(14)}$$

Upper teeth are narrow, with high crowns, semi-erect to oblique. Both median edges deeply notched on basal third. Distal median edges straight, mesial edges straight to slightly concave. Both edges are strongly serrated basally, fine serrated apically to even smooth-edged. Roots are arched, deeply notched. Two rows of symphysial teeth are present. Lower teeth are narrow, with erect crowns. Both median edges concave, slightly notched, partly very fine serrated. Roots are arched, deeply notched. Two rows of large symphysial teeth are present. Files in upper and lower jaw are arranged as alternate overlapping dentition.

Colour. Bronze-grey to brownish-grey (seldom pale, grey-blue) above, ventrally whitish. Without prominent markings on the fins, and without a conspicuous white or black band on flanks.

Size. Maximum of 150 cm TL, and an average of 110–130 cm TL (SHARKINFO, FROESE & PAULY 2010).

Misidentifications. Based on the combination of a smooth back, combined with lacking conspicuous markings on fins, the unique tooth-morphology, the body-dimensions and the body-shape, *C. fitzroyensis* differs from all other Australian shark species of the genus.

Nevertheless, the creek whaler shark could be sometimes confused with *C. porosus* and *C. borneensis* but both species have their second dorsal fin origin over or behind the middle of anal fin base. Furthermore, the mentioned species have a different number of precaudal vertebrae (58 in *C. fitzroyensis versus* 62 in *C. borneensis* and 59 in *C. porosus*).

Geographical Distribution. Western Pacific: Australia (Queensland, northern and north-western Australia). – (WHITLEY 1943 a, b, c, GARRICK 1982, COMPAGNO 1984, COMPAGNO et al. 2005).

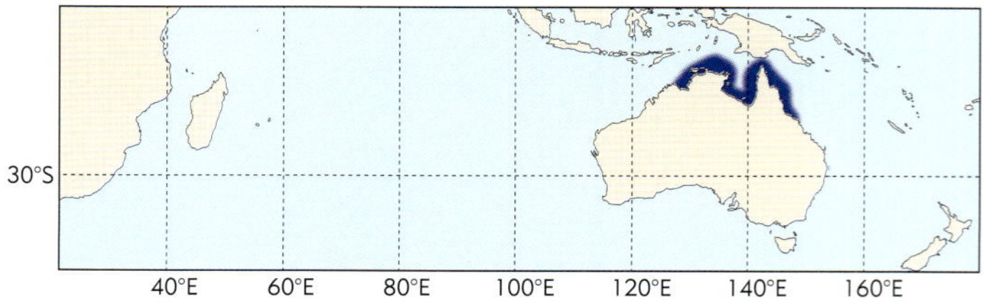

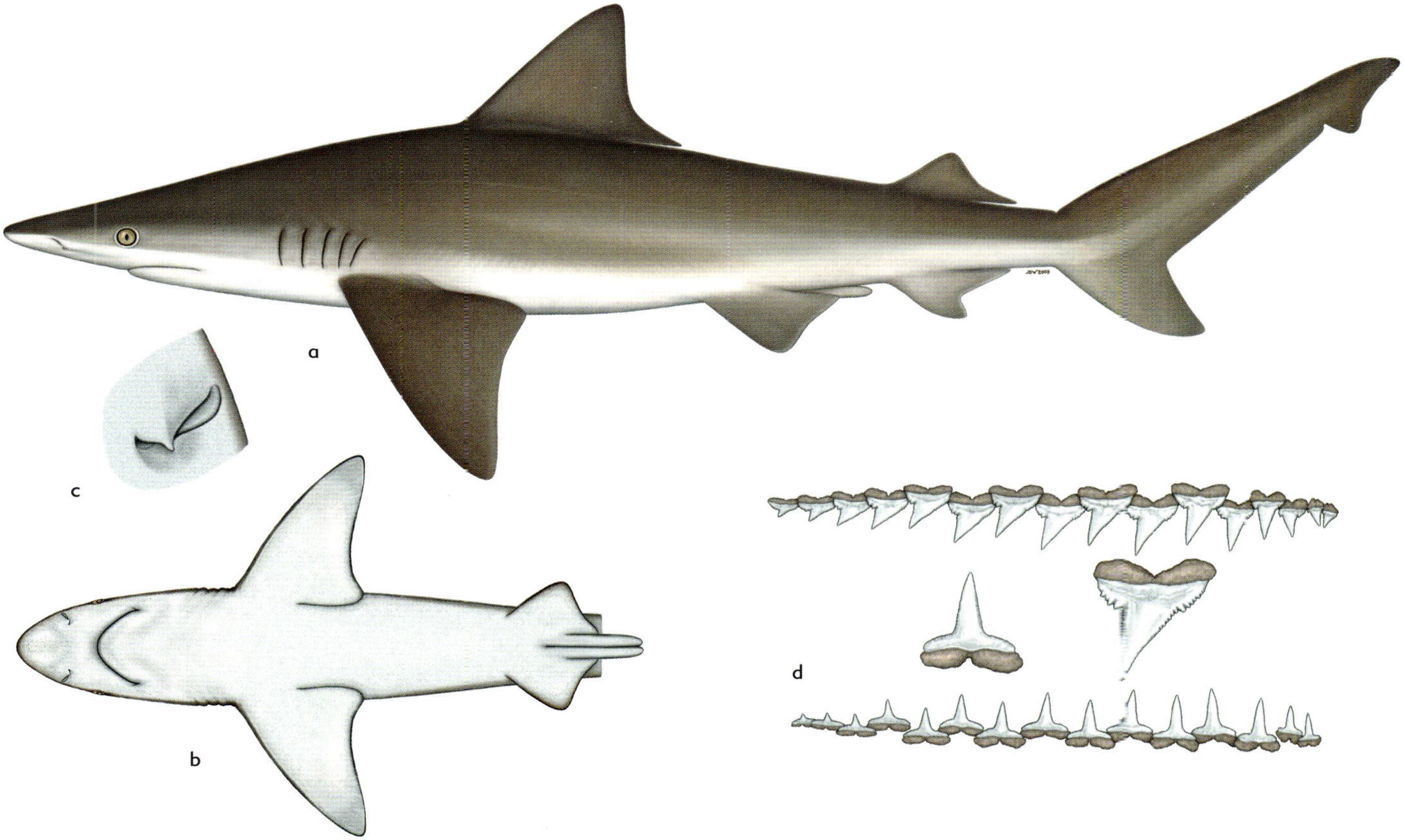

Figure 31. *Carcharhinus fitzroyensis*: 121 cm TL male from Queensland, Australia; **a**, left side; **b**, ventral view; **c**, enlarged left nostril; **d**, left upper and lower teeth (symphysis to the right), and inset are the enlarged sixth upper and lower teeth.

Habitat. In tropical and sub-tropical waters, over continental shelf, and in shallower waters living species. *Carcharhinus fitzroyensis* occur from intertidal up to 40 m depth (SIMPFENDORFER & MILWARD 1993).

Biology. It is a common and abundant species in the estuarine and inshore waters of northern Australia but there is little known about it (SIMPFENDORFER & MILWARD 1993).

Reproduction. Males reach maturity at 80–90 cm TL. Females reach maturity at 88–100 cm TL (LYLE 1984, 1987). Maturity is reached between May and July, ovulation is between July and September, and time for giving birth is in the following February to May. A distinct seasonal reproduction cycle is known within the species. Reproductive mode is placental viviparous (DULVY & REYNOLDS 1997). Normal are 1–7 young per litter, and at birth young have 35–50 cm TL (LYLE 1984, 1987).

Food. Mainly small bony fishes (pelagic and demersal), molluscs (mainly cephalopods) and crustaceans.

Red List status: LC (IUCN 2010).

Carcharhinus galapagensis (Snodgrass & Heller 1905)

En: Galapagos shark; Fr: Requin des Galapagos; Sp: Tiburon de Galapagos; Ge: Galapagoshai

Carcharias galapagensis Snodgrass & Heller (1905). Holotype: SU 12324. Type material is a fetus of 65 cm TL, but it is apparently lost. Type locality: Galapagos Islands.

Diagnostic features. Large, robust species, infrequently with a very low interdorsal ridge (which is usually absent). Upper precaudal pit well developed, lower pit weak, internarial width 0.9–1.4 times the preoral length. Snout long and bluntly rounded or pointed in smaller specimens Eyes circular and fairly large (diameter 2.0 % TL), eyes anterior margins slightly anterior to the front of mouth. Nostrils strongly oblique, ellipsoid to ovate-shaped. Anterior nasal flaps low, with tiny pointed lobes, incurrent apertures small. Upper labial furrows short and inconspicuous. Hyomandibular pores alongside corners of mouth inconspicuous. Gill slits relatively short (2.8–3.2 % TL). First dorsal fin triangular, large (height 10.5 % TL), its apex acute to sharply rounded. D1 origin over the mid-length or slightly posterior of inner pectoral margin. The D1 anterior margin short and weakly convex in smaller specimens, strongly convex in larger sharks. Second dorsal fin large and long (height 2.7 % TL), its distal margin concave (the same in anal fin). D2 origin over anal fin origin. Pectorals long, falcate to semi-falcate, apices pointed (length of anterior margin 20.0 % TL). Pectoral fin origin below the level of third or fourth gill opening. Vertebrae: 107 precaudal centra and 101 caudal centra on average. Dermal denticles closed set, imbricate, sub-circular in small specimens, with three longitudinal ridges and three strongly developed posterior cusplets. In larger specimens denticles are more rhomboid, arranged in almost regular, diagonal rows, and with five to seven ridges and up to five posterior cusplets.

Dental Formula:

$$\frac{14 - 1 - 14}{14 - 1 - 14} \quad \text{to} \quad \frac{13(15) - 1(3) - 13(15)}{13(15) - 1(3) - 13(15)}$$

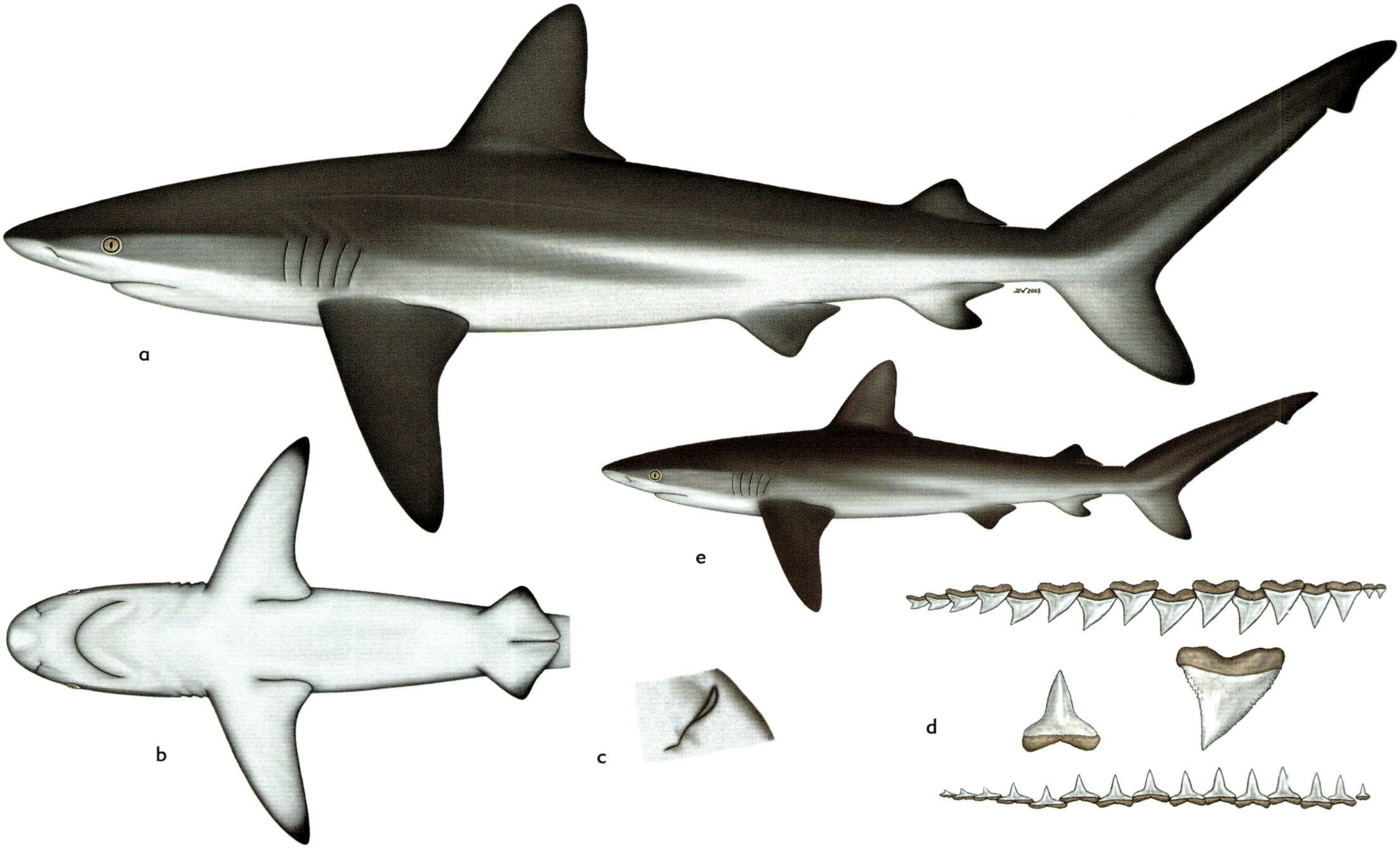

Figure 32. *Carcharhinus galapagensis*: 283 cm TL female from Galapagos Islands; **a**, left side; **b**, ventral view; **c**, enlarged left nostril; **d**, left upper and lower teeth (symphysis to the right), and inset are the enlarged fifth upper and lower teeth; **e**, juvenile female of 106 cm TL from St. Paul's Rock, Central Atlantic.

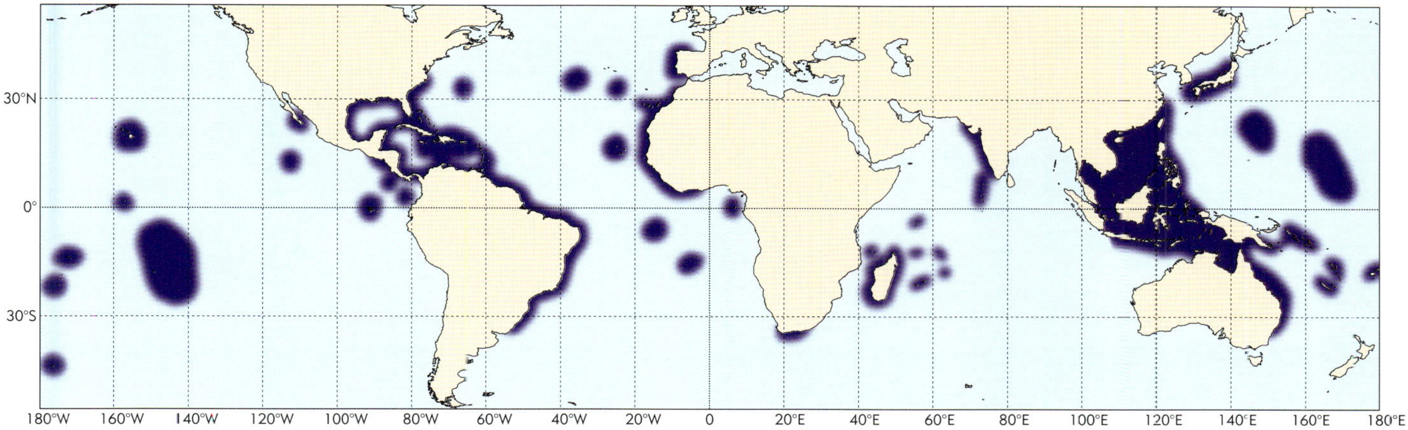

Upper teeth are broad, crowns triangular, erect (S to A rows) to oblique (for all other tooth rows). Distal median edges in adults and sub-adults concave, straight or slightly concave and weakly notched in juveniles. Mesial edges are straight (S to L rows) to convex (LP to C rows). Both edges strong and uniformly serrated, slightly coarser serration basally, roots arched. One, two, or seldom three rows of symphysial teeth are present. Lower teeth are narrow, crowns erect, triangular. Distal and mesial edges concave, very fine and uniformly serrated. Roots are arched, slightly notched. One, two or even three rows of symphysial teeth are present. Files in upper jaw are arranged as alternate overlapping dentition, and in lower jaw as mixed alternate i. e. imbricate overlapping dentition.

Colour. Dusky-grey or brownish-grey, with a tinge of green dorsally. The sides light grey with a tinge of metallic-green, yellow-whitish ventrally. Most tips on fins are dusky but never truly black or white (particularly apices of pectorals and underside can be dark), and with an inconspicuous white band on flanks.

Size. Maximum up to 370 cm TL, and an average of 300 cm TL (MYERS 1991). Maximum reported body mass of 85.5 kg (IGFA 2001).

Misidentifications. This species could be confused with *C. obscurus*, *C. plumbeus* and *C. altimus* but *C. galapagensis* differs from all other mentioned species in having a different colouration of fin apices, in dental formula, tooth-shape, and in having an interdorsal ridge. In contrast to *C. obscurus*, Galapagos sharks have a larger first dorsal fin and a higher average number of precaudal vertebrae (107 in *C. galapagensis versus* 109 in *C. altimus*, 91 in *C. obscurus*, and 90 in *C. plumbeus*). Moreover, Galapagos sharks differ from *C. amblyrhynchos*, *C. wheeleri*, *C. albimarginatus*, *C. brachyurus* and *C. falciformes* in tooth-shape and dental formula, in body size, colouration and in snout length and shape.

Geographical Distribution. Western Atlantic: North Carolina to Florida, Cuba, Bermudas, Virgin Islands and around off Caribbean Islands, Gulf of Mexico, Latin America to southern Brazil. – Central and eastern Atlantic: Azores, along Spanish and Portuguese Atlantic coast, Madeira, Cape Verde, São Tomé, Ascension and St. Helena to South Africa. – Indian Ocean: Red Sea to South Africa, and off Seychelles, Mauritius and Madagascar, India, Maldives. – Western and Central Pacific: Gulf of Thailand, South China Sea, China to southern Japan, Borneo, Philippines, Celebes Sea, northern to eastern coast of Australia, Papua New Guinea, Middleton, Marianas, Marshall Islands, Tuamoto Archipelago, Hawaiian Islands. – Eastern Pacific: Off some small Pacific islands and island-groups, Galapagos, Cocos, Malpelo and Clipperton Island, southern Baja California, to Guatemala, Colombia and Peru. – (RANDALL 1963a, b; BASS et al. 1973, 1986; EDWARDS & LUBBOCK 1982, GARRICK 1982; COMPAGNO 1984; TANIUCHI et al. 1985; BRUM & AZEVEDO 1993; FERGUSSON & COMPAGNO 1999; COMPAGNO et al. 2005).

Habitat. In contrast to the vernacular name the species is not limited to the Galapagos Islands than is rather circum-tropically and sub-tropically distributed. Generally, *C. galapagensis* is associated to oceanic islands and lives over insular and continental shelves (PAPASTAMATIOU et al. 2006). Nevertheless, the species occurs also inshore, near coral reefs, over grain-coarse rocky bottom. Galapagos sharks are common and abundant there and have a strong affinity to clear waters with strong currents. The species was reported from close to the surface (0–2 m) to 180 m depth.

Carcharhinus galapagensis is not truly oceanic living, but coastal-pelagic and capable of crossing considerable distances of the open ocean between islands, at least 50 km per day (COMPAGNO 1984).

Biology. This species can occasionally occur in larger but loose aggregations (not really schools). If a social behaviour is present in these aggregations is not known yet. Individuals of the species often swim just above the bottom substrate (COMPAGNO 1984). A sexual segregation is known within this species and hence, the average depth of captures is greater for juveniles (45.1 m) and mature males (60.2 m), than for sub-adults (38.8 m) and mature female individuals (34.2 m, WETHERBEE et al. 1996). The young are living near the shore, close to their birth places and nursery grounds, in shallower water of 25 m depth or less. Galapagos sharks are known to show an agonistic or hunching display to one another, and to divers and swimmers. Especially the young can be very aggressive in their behaviour.

The estimated maximum longevity is approximately 24 years (IGFA 2001.

Reproduction. Males reach maturity at 170–240 cm TL. Females reach maturity at 215–245 cm TL but this fact is not completely clear yet (WETHERBEE et al. 1996). The age of specimens of this species at first reproductive effort is approximately 10 years. Reproductive mode is placental viviparous (DULVY & REYNOLDS 1997). The litter size has a range of 4–16, with an average of 10 young per litter. In Hawaiian waters, at birth Galapagos sharks have 60–80 cm TL (WETHERBEE et al. 1996). For other regions length at birth seems to be similar but this is not clear yet. Mating and parturition apparently occur early in the year, and duration of the gestation is estimated to be about 12 months (DECROSTA 1981, WETHERBEE et al. 1996).

Food. The species feed mainly benthic living bony fishes, e.g. eel, flatfish and triggerfish, also flying fishes, other not further specified shark species, molluscs, cephalopods, birds and garbage. A reduction in the consumption of teleosts and a rising part of elasmobranches in the diet with increasing body size is proven for this species. In addition, at the Galapagos Island and off the coast of Ecuador the species feed on sea lions and on marine iguanas as well (WETHERBEE et al. 1996).

Red List status: NT (IUCN 2009).

Carcharhinus hemiodon (VALENCIENNES 1839)

En: Pondicherry shark; Fr: Requin baliai; Sp: Tiburon de Pondicherry; Ge: Pondicherryhai

Carcharias (Hypoprion) hemiodon VALENCIENNES (in MÜLLER & HENLE 1839). Holotype: MNHN 1040. Type material is an immature male of 47 cm TL, from Pondicherry, India.

Synonyms: *Carcharias watu* SETNA & SARANGDHAR (1946), *Hypoprion atripinna* CHU (1960).

Diagnostic features. Small, sturdy species, normally without an interdorsal ridge (seldom a low ridge can be present). Upper precaudal pit well developed, lower pit weak. Internarial width 1.1–1.3 times the preoral length. Snout long and narrowly rounded or pointed. Eyes circular and large (diameter 2.3 % TL), eyes anterior margin above or slightly anterior to the front of mouth. Nostrils strongly oblique, incurrent apertures large and ovate. Anterior nasal flaps expanded as narrow, pointed lobes. Upper labial furrows short and inconspicuous. Hyomandibular pores alongside corners of mouth inconspicuous. Gill slits long (2.9–3.5 % TL). First dorsal fin large (height 10.5 % TL), falcate, its apex pointed or narrowly rounded. The posterior margin curving ventrally from apex, with a long free rear tip. D1 origin above or slightly anterior to pectoral inner margin, just behind insertion. Second dorsal fin large and high (height 2.3 % TL), its inner margin short. D2 origin over or slightly posterior to anal fin origin. Pectorals large, broad and falcate, apices narrowly rounded or pointed (length of anterior margin 17.6 % TL). Vertebrae: 70 precaudal centra and 82 caudal centra on average. Dermal denticles in small specimens imbricate, ovoid, wider than long, with three longitudinal ridges and three, seldom five relatively short, posterior cusplets.

Dental Formula:

$$\frac{14-1-14}{13-1-13} \quad \text{to} \quad \frac{14(15)-1(2)-14(15)}{12(14)-1(2)-12(14)}$$

Sexual dimorphism in tooth-shape is known within this species. Upper teeth are narrow, crowns high, erect to oblique. Distal median edges oblique basally, convex apically, deeply notched on basal third longitudinally. Mesial edges are straight or slightly concave, deeply notched on basal third longitudinally. Serration on both edges is present which is very large basally and more distinct on distal median edges but apically smooth-edged. Sometimes in smaller specimens both edges can be smooth-edged. Roots are arched and deeply notched. One or two rows of symphysial teeth are present. Lower teeth are narrow, with high semi-erect crowns. Distal median edges convex basally, straight apically, fine and uniformly serrated, deeply notched on basal third of crowns. Mesial edges are concave, fine and uniformly serrated. Roots are arched (S to L rows) to almost straight (LP to C rows), deeply notched. One or two rows of symphysial teeth are present. Files in upper and lower jaw are arranged as alternate overlapping dentition.

Colour. Grey dorsally and white ventrally, conspicuous black tips are present on the apices of the pectorals and D2. A small margin on the dorsal lobe and a big black tip on the ventral lobe of the caudal finis present, a narrow black margin is curving ventrally from D1 apex, and a conspicuous white band on flanks is distinct.

Size. Maximum of about 200 cm TL, and an average of 150 cm TL (COMPAGNO & NIEM 1998).

Misidentifications. *Carcharhinus hemiodon* could be confused with *C. amblyrhynchoides*, *C. brevipinna*, *C. limbatus*, *C. melanopterus*, *C. sorrah* and *C. leiodon* but the Pondicherry shark differs from the other mentioned species in body size, tooth-shape, dental formula, and in the average number of precaudal vertebrae (70 in *C. hemiodon versus* 91 in *C. amblyrhynchoides*, 82 in *C. brevipinna*, 96 in *C. limbatus*, 116 in *C. melanopterus* and 115 in *C. leiodon*). The spot tail shark has the same average number of precaudal vertebrae but differs in the average number of the caudal vertebrae (82 in *C. hemiodon versus* 88 in *C. sorrah*). Furthermore, the tooth-shape and the dental formula are different. Another sympatric species which is very similar in morphology to *C. hemiodon* is the blacktip reef shark, *C. melanopterus*, but this species is easily to identify by its prominent D1 colouration. The second dorsal fin is higher in *C. melanopterus* than in *C. hemiodon*, and the snout contour is much blunter. *Carcharhinus macloti* and the Pondicherry shark share many features which are similar or almost equal and this complicates a doubtless identification. Both species have a very similar body size, the same dental formulae, the same number of precaudal vertebrae

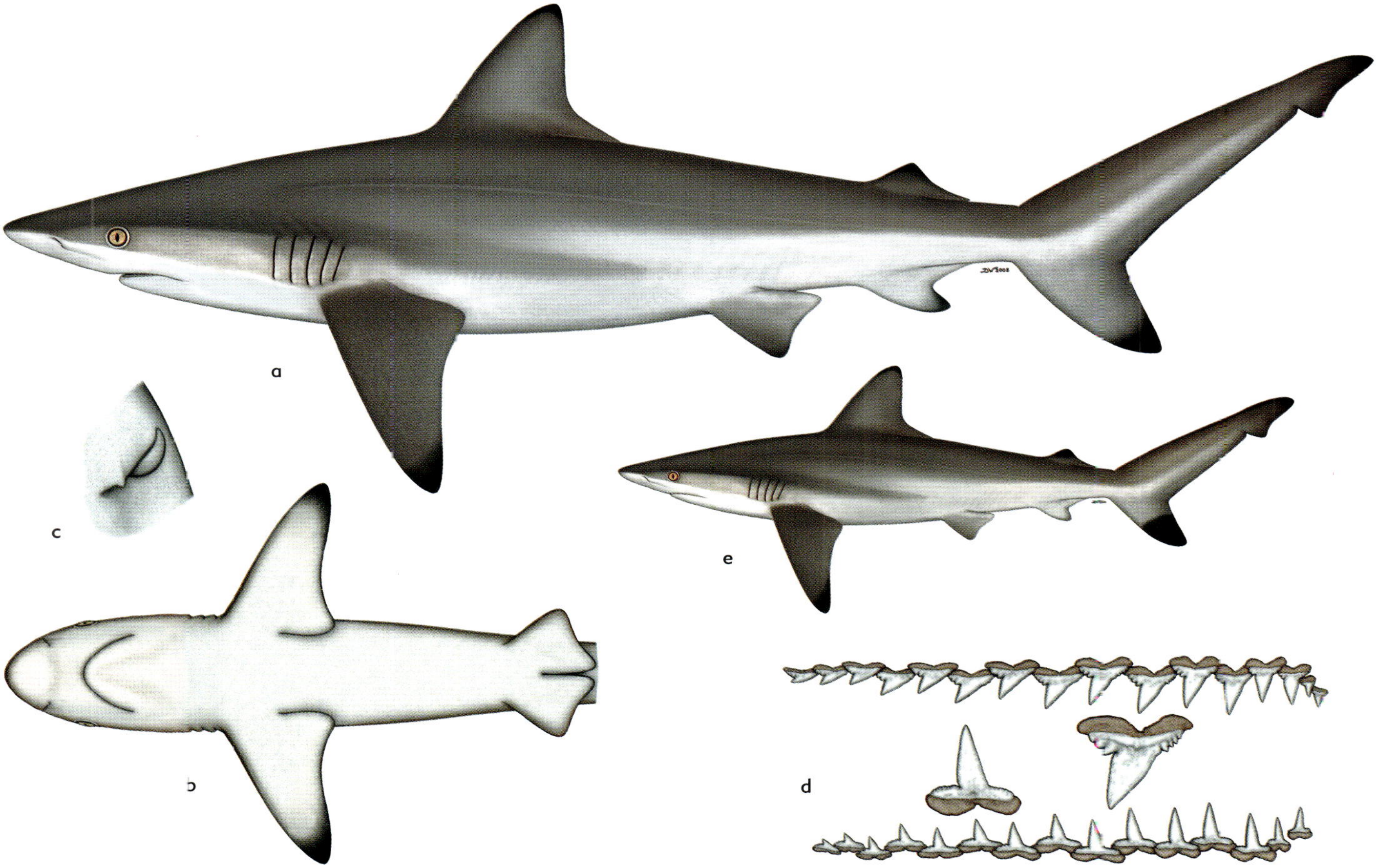

Figure 33. *Carcharhinus hemiodon*: 67 cm TL female from India; **a**, left side; **b**, ventral view; **c**, enlarged left nostril; **d**, left upper and lower teeth (symphysis to the left), and inset are the enlarged sixth upper and lower teeth; **e**, juvenile female of 25 cm TL from India.

and a similar geographical distribution. However, *C. hemiodon* differs from the hardnose shark in snout length and shape, and in fin colouration patterns.

Geographical Distribution. Indo-West Pacific: From Oman to Pakistan, India, Indonesia (Sulawesi, Java, Waigeo), Vietnam, southern coast of China, Philippines, Papua New Guinea, northern Australian coast. – (ZEITZ 1888, COMPAGNO 1984, GARRICK 1985, COMPAGNO et al. 2005).

Habitat. Supposable, it is a common and wide ranging species, typically over continental and insular shelves living but little is known about it (COMPAGNO 1984).

Biology. Reports of the occurrence of this species in river mouths and in freshwater exist but these reports have not been confirmed yet and seem to be doubtful. Presumably, the species was mistaken with other species e.g. *Glyphis gangeticus*, however if the species identification of these records were correct it would prove the ability of the species to tolerate low salinity (COMPAGNO 1984).

Reproduction. Reproduction mode is placental viviparous (DULVY & REYNOLDS 1997) but relevant data are not available.

Food. Poorly known, but probably the species feed on bony fishes.

Red List status: CR A2acd C2a (i), (IUCN 2010, SHARKINFO).

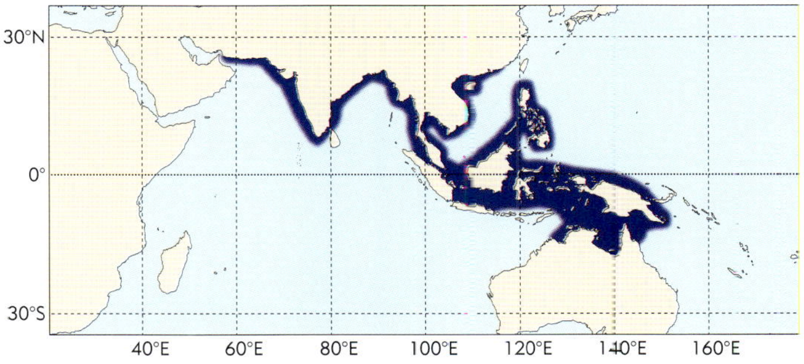

Carcharhinus isodon (Valenciennes 1839)

En: Finetooth shark; Fr: Requin à petites dents; Sp: Tiburon dentiliso; Ge: Feinzahnhai

Carcharias (Aprion) isodon Valenciennes (in Müller & Henle 1839). Holotype: MNHN 1037. Type material is an immature male of 65 cm TL, (probably) from the coast of New York, USA.

Synonyms: *Aprionodon punctatus* Gill (1861, 1862, 1864).

Diagnostic features. Small species, without interdorsal ridge. Upper precaudal pit well developed, lower pit weak, internarial width 1.1–1.5 times the preoral length. Mid-sized long and sharply pointed snout, eyes circular, fairly large (diameter 1.8 % TL), eyes anterior margin anterior to the front of mouth. Nostrils small, strongly oblique and slit-shaped. Anterior nasal flaps low and broadly angular but not expanded, with pointed lobes. Upper labial furrows short and inconspicuous. Hyomandibular pores alongside corners of mouth inconspicuous. Gill slits very long (4.4–5.5 % TL). First dorsal fin large (height 9.0 % TL), falcate to semi-falcate, with a sharply pointed or rounded apex, posterior margin curving ventrally from fin apex. D1 origin over or slightly posterior to pectoral free rear tips, the inner margin of first dorsal fairly short. Second dorsal fin large and high (height 2.7 % TL), its inner margin short (1.4 times the height). D2 origin over or slightly anterior or posterior to anal fin origin. Pectorals small (length of anterior margin 15.5 % TL), falcate, narrowly rounded or pointed apices. Vertebrae: 79 precaudal centra and 86 caudal on average. Dermal denticles closed set, imbricate, sub-circular, each with three longitudinal ridges and a corresponding number of posterior cusplets, sharp-pointed and fairly short in small specimens.

Dental Formula:

$$\frac{15 - 2 - 15}{14 - 2 - 14} \quad \text{to} \quad \frac{15 - 1(2) - 15}{14(15) - 2(3) - 14(15)}$$

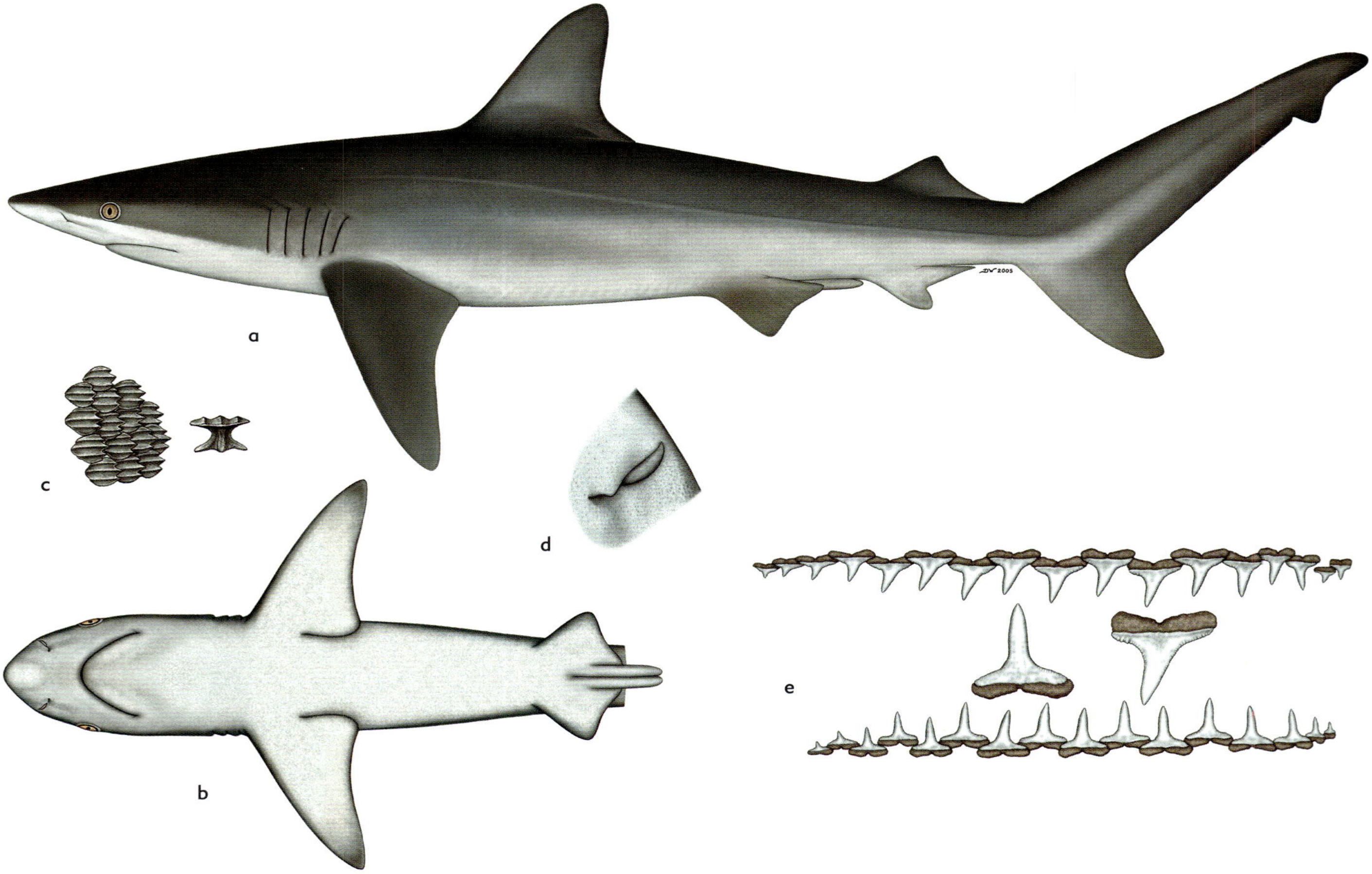

Figure 34. *Carcharhinus isodon*: 141 cm TL female from Florida; **a**, left side; **b**, ventral view; **c**, dermal denticles; **d**, enlarged left nostril; **e**, left upper and lower teeth (symphysis to the left), and inset are the enlarged sixth upper and lower teeth.

Upper teeth are narrow, crowns erect to slightly oblique. Distal median edges concave, mesial edges slightly concave. Both edges are smooth-edged or weakly serrated. Sometimes slightly irregular serration can be present on both basal parts of crowns. Roots are arched, deeply notched. One or two rows of symphysial teeth are present. Lower teeth are narrow, crowns erect. Both median edges concave and smooth-edged, roots arched, and deeply notched. Two or three rows of symphysial teeth are present. Files in upper and lower jaw are arranged as alternate overlapping dentition.

Colour. Dark grey dorsally, white ventrally, without distinct markings on fins. Both margins of caudal fin can be narrowly dark-edged, and sometimes an inconspicuous white band on flanks is present.

Size. Maximum of 190 cm TL, possibly up to 200 cm TL (SPRINGER 1990).

Misidentifications. The species could be confused with *C. amblyrhynchoides*, *C. brevipinna*, *C. limbatus* (and *C. leiodon*) which have a comparable tooth-morphology, habitus and body size. However, *C. isodon* differs in lacking black tips on fins, further in the position of D1 and D2 origin, in fins height, and in the average number of precaudal vertebrae (79 in *C. isodon versus* 91 in *C. amblyrhynchoides*, 82 in *C. brevipinna*, 96 in *C. limbatus* and 115 in *C. leiodon*).

Geographical Distribution. Western Atlantic: East coast of the US, from New York to Florida, Cuba, Gulf of Mexico and southern Brazil. – Eastern Atlantic: Senegal (very rare in this area) and Guinea-Bissau. – (BRANSTETTER & SHIPP 1980; COMPAGNO 1984; GARRICK 1985; CARLSON et al. 1999; CARLSON & CORTES 2003; COMPAGNO et al. 2005; DRIGGERS & HOFFMAYER 2009).

Habitat. *Carcharhinus isodon* is a common and abundant species of the south-eastern coast of the United States and inhabits shallow, coastal, inshore waters to depths of 25 m (CASTRO 1993).

Biology. The species spend the summer in the area off Georgia and the Carolinas, and the winter off Florida. *Carcharhinus isodon* appear in the nursery grounds and mating areas of South Carolina when the water surface temperature has risen above 20 °C in late April and early May. The finetooth sharks leave the Carolinas in early fall and migrate southward when the water surface temperature has decreased below 20 °C (CASTRO 1993). The species often forms large school during their migrational journeys.

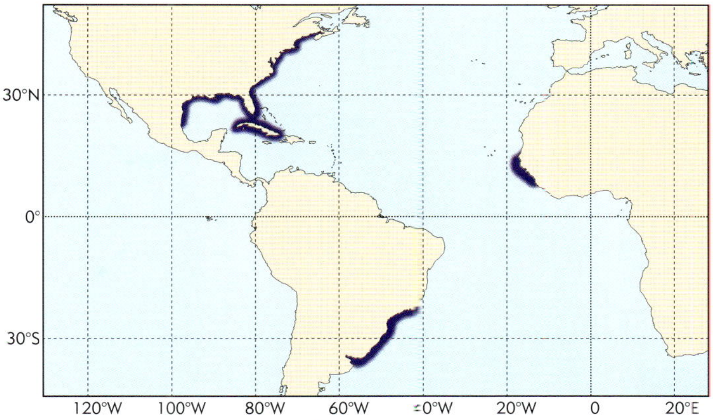

The oldest caught specimens (no data of the sex of the individuals available) were 8.0–8.1 years. A theoretical longevity of 14.4 years for females and of 8.5 years for males was calculated (CARLSON et al. 1999, 2003; CARLSON & CORTES 2003).

Reproduction. Females and males reach maturity at 130 cm TL and up. Median length at maturity is 119–123 cm TL, equivalent to 3.9–4.3 years for both sexes (CASTRO 1993). Reproductive mode is placental viviparous (DULVY & REYNOLDS 1997). Normal are 1–6 young per litter, and at birth young have 48–60 cm TL (CASTRO 1993). The young are born close to the shore and are rapid-growing. Embryos are lecithotrophic during their first fifteen weeks of development. Subsequently, the embryos establish a placental connection to the mother. Implantation occurs when the embryos measure about 13 cm or at about the fifteenth week of gestation. Gravid females enter the shallow water of nurseries off South Carolina in late April. Parturition occurs from late May to mid-June after a gestation of about twelve months, plus or minus two weeks. Mating occurs from early May to early June (CARLSON et al. 1999, CARLSON & CORTES 2003).

Food. Mainly small bony fishes and cephalopods. Finetooth sharks may occasionally fall prey to larger sharks, e.g. *C. obscurus*.

Red List status: LC (IUCN 2010).

Carcharhinus leiodon GARRICK (1985)

En: Smooth tooth blacktip shark; Fr: –; Sp: –; Ge: Glattzahnhai

Carcharhinus leiodon GARRICK (1985). Holotype: NMV. Type material is an immature male of 75 cm TL, collected by HEIN (1902). Evaluators: MUSICK & FOWLER (2000) (Shark Red List Authority), Shark Specialist Group (2000).

Authors note. After the studies of species description and argumentation of GARRICK (1985) doubts have appeared regarding this species as a valid member of the genus *Carcharhinus*. We assume the reported specimen could be a mutation, a variation or a local phenotype of another species, e.g. *C. brevipinna*, *C. limbatus* or *C. wheeleri* because all mentioned species co-occur in the same area where the holotype specimen was caught.

Our theory is based on the strong similarity of body size features and colouration patterns of the holotype specimen to *C. brevipinna* and *C. wheeleri*, further due to the very similar colouration, tooth-shape and dental formula of *C. leiodon* and *C. limbatus*.

Depending on the fact one specimen (the holotype) was hitherto caught and reported in a region with a fairly huge fishery impact more individuals of this "species" would have been caught until now. MOORE et al. (2010) announced at the shark congress 2010 in Australia a re-description of the species, including further details about geographical distribution, body colouration, size range and maturity but at time of going to press the source was still pending and not available.

Diagnostic features. A medium sized species, without interdorsal ridge. Upper precaudal pit well developed, lower pit weak. Internarial width 1.1 times the preoral length. Snout long and bluntly pointed, eyes small (diameter 1.9 % TL), eyes anterior margin above the front of mouth. Nostrils strongly oblique, incurrent apertures slit-shaped. Anterior nasal flaps low, with pointed lobes. Upper labial furrows short and inconspicuous. Hyomandibular pores alongside corners of mouth inconspicuous. Gill slits very large (3.2–4.5 % TL). First

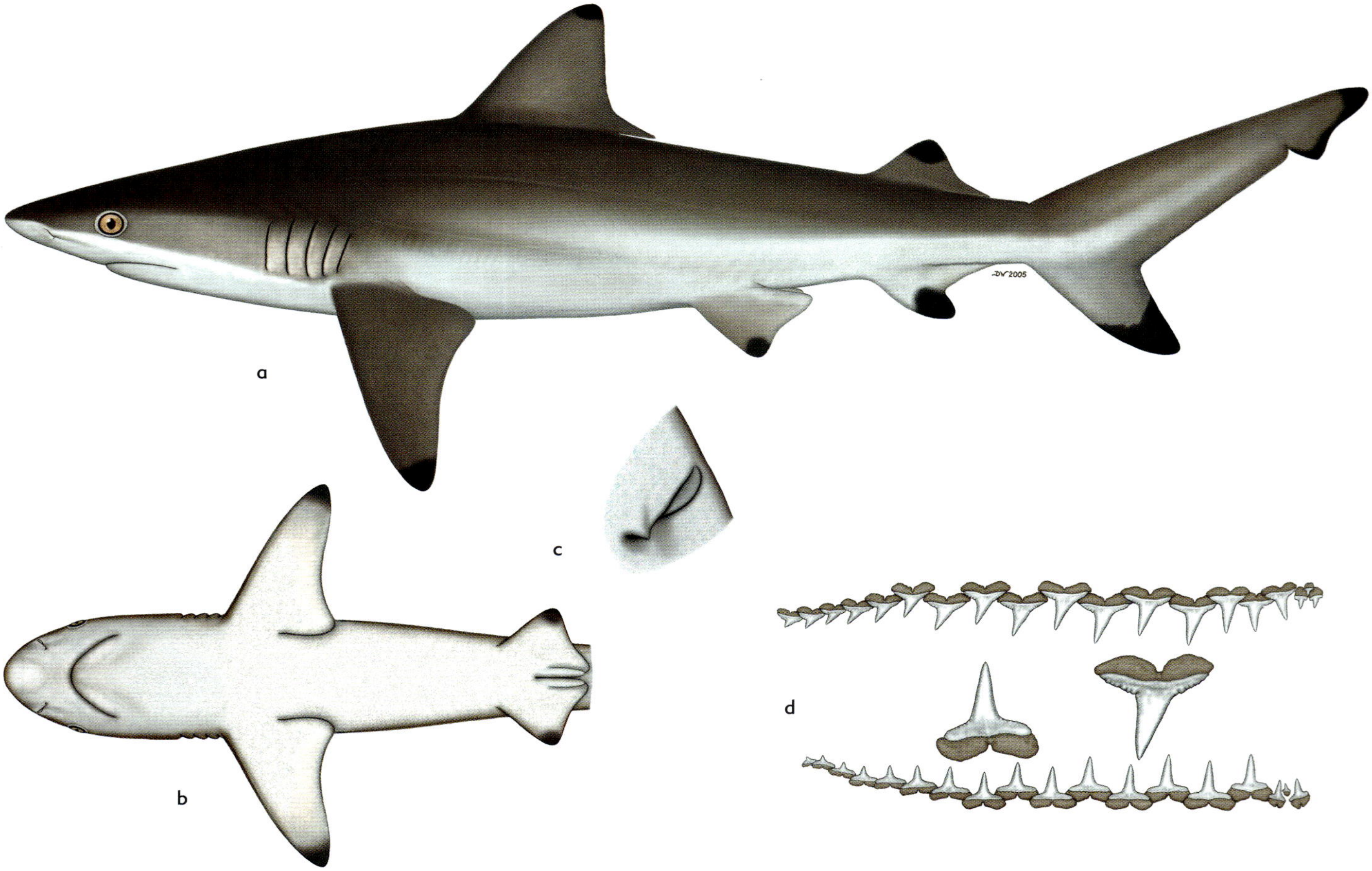

Figure 35. *Carcharhinus leiodon*: 75 cm TL immature male from the Gulf of Aden, Red Sea; **a**, left side; **b**, ventral view; **c**, enlarged left nostril; **d**, left upper and lower teeth (symphysis to the left), and inset are the enlarged fifth upper and lower teeth.

dorsal fin high (height 10.5 % TL), more erect than falcate, slightly triangular, with a moderately long inner margin, its apex pointed. D1 origin slightly behind middle of pectoral inner margin. Second dorsal fin large and high (height 4.1 % TL), its inner margin long, almost equal to anal fin height. D2 origin slightly behind anal fin origin. Pectorals large and narrow (length of anterior margin 17.4 % TL), weakly falcate, with pointed apices distally. Vertebrae: 115 precaudal centra and 83 caudal centra on average. Dermal denticles slightly imbricate-overlapping, ovoid, wider than long, with three strong longitudinal ridges and three, occasionally five sharp pointed posterior cusplets.

Dental Formula:

$$\frac{16 - 3 - 16}{15 - 3 - 15}$$

Upper teeth are narrow, crowns erect (S to AL rows) to slightly oblique (L to C rows). Distal median edges slightly concave apically, convex basally, deeply notched, mesial edges concave. Both edges are smooth-edged on apical half of crowns, more coarsely serrated basally. Sometimes slight or even coarse irregular serration on distal edges is distinct. Roots are arched, deeply notched. Three rows of symphysial teeth are present. Lower teeth are narrow, crowns high and erect. Both median edges concave, deeply notched and smooth-edged. Roots arched, deeply notched. Three rows of symphysial teeth are present.

Files in upper jaw are arranged as alternate overlapping dentition (A to LP rows), then as imbricate overlapping dentition type I (P to C rows). Files in lower jaw arranged as alternate overlapping dentition (S to P rows), then as imbricate overlapping dentition type I (PC to C rows).

Colour: Yellowish-brown above, yellowish ventrally and with a tongue of this paler underside colour extending forward along the side of the trunk from the pelvic region to below the first dorsal fin. Upper surface of caudal peduncle behind D2 is dusky. All fins with black tips, markings on apices of D1 and on pelvic fins are smaller than those of the other fins. The black tip on upper caudal lobe is extending forward.

However, the body colouration of the holotype individual appears to be faded i.e. is too much yellowish-brown distinct (as the result of preservation), and is probably more darker brownish on freshly caught specimen.

Size. The immature holotype male has 75 cm TL (GARRICK 1985).

Misidentifications. The narrow, erect and smooth-edged upper teeth of *C. leiodon* have very close counterparts in *C. isodon* and juvenile *C. brevipinna*, also in *C. amblyrhynchoides*, *C. limbatus* and in adult *C. brevipinna*. Other common diagnostic features which *C. leiodon* share with the mentioned species are the absence of an interdorsal ridge, a similar shape of snout, and the length of gill openings which are rather large. *Carcharhinus leiodon* differ from *C. isodon* in having black fin tips, a much higher second dorsal fin and in average more precaudal vertebrae (115 in *C. leiodon versus* 79 in *C. isodon*). The smoothtooth blacktip shark are separable from the other black-tipped species *C. amblyrhynchoides*, *C. limbatus*, *C. tilstoni* and *C. brevipinna* in having its D1 origin above the middle of the pectoral fin inner margin (*C. amblyrhynchoides*, *C. limbatus* and *C. tilstoni* have it over the pectoral axis, *C. brevipinna* over or behind the pectoral inner corner) by the higher average number of precaudal vertebrae, and to a less obvious extent by several differences in fin sizes and proportions (115 precaudal vertebrae in *C. leiodon versus* 91 in *C. amblyrhynchoides*, 96 in *C. limbatus*, 88 in *C. tilstoni* and 82 in *C. brevipinna*). In comparison of external features *C. leiodon* is most likely to be confused with *C. melanopterus*, and perhaps with *C. hemiodon*. The species differs from *C. melanopterus* in shape of snout, upper tooth-shape, and in dental formula. It is readily distinguished from *C. hemiodon* by its much shorter first dorsal rear tip, its higher second dorsal fin and upper tooth-shape.

In comparison of precaudal counts *C. leiodon* is encompassed only by *C. amblyrhynchos*, *C. wheeleri*, *C. melanopterus* and *C. leucas*, but these species differ from *C. leiodon* in numerous other features.

Geographical Distribution. Western Indian Ocean: known only from southern Arabia, Gulf of Aden (probably at Qishn in Yemen). Hitherto, the species is known from a single record from the Gulf of Aden only, where sharks are heavily fished. Presumed to have a very restricted distribution in the northern Indian Ocean it must be a small population. – (GARRICK 1985).

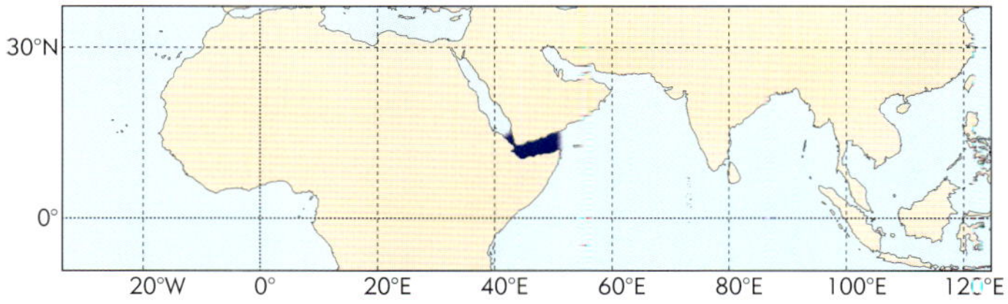

Habitat. No data available.

Biology. No data available.

Reproduction. Reproductive mode is (probably) placental viviparous but further relevant data are not available at the moment.

Food. No data available but probably the species feed on smaller bony fishes.

Red List status: VU B1ab (iii), C2a (ii) (IUCN 2010).

Carcharhinus leucas (VALENCIENNES 1839)

En: Bull shark, Zambezi shark, Lake Nicaragua shark, River Whaler, Freshwater Whaler, Swan River whaler; Fr: Requin bouldogue; Sp: Tiburon sarda; Ge: Bullenhai, Stierhai, Gemeiner Grundhai

Carcharias (Prionodon) leucas VALENCIENNES (in MÜLLER & HENLE 1839). Syntype: Two stuffed specimens are still existing, MNHN A 9650 (an adult male of 160 cm TL) and MNHN A 9652 (a female of 186 cm TL), from the Antillean Islands.

Synonyms: *Carcharias (Prionodon) zambezensis* PETERS (1852), *Prionodon platyodon* POEY (1860), *Squalus obtusus* POEY (1861), *Eulamia nicaraguensis* GILL & BRANSFORD (1877), *Carcharias azureus* GILBERT & STARKS (1904), *Carcharias spenceri* OGILBY (1910a, b), *Galeolamna (Bogimba) bogimba* WHITLEY (1943a, b, c), *Lamnarius spenceri* WHITLEY (1943a, b, c), *Galeolamna greyi mckaili* WHITLEY (1945), *Carcharhinus vanrooyeni* SMITH (1958).

Synonyms in recent use: *Carcharhinus nicaraquensis* GILL & BRANSFORD (1877), *Carcharhinus zambezensis* PETERS (1852), *Carcharhinus azureus* GILBERT & STARKS (1904).

Diagnostic features. Large, very brawny or heavy-bodied species, without interdorsal ridge. Upper precaudal pit well developed, lower pit weak. Internarial width 0.7–1.1 times the preoral length. Snout very short, broad and bluntly rounded. Eyes circular and fairly small (diameter 1.2 % TL), eyes anterior margin occasionally above but usually slightly anterior to the front of mouth. Nostrils strongly oblique, incurrent apertures large and ovate-shaped. Anterior nasal flaps expanded or broad-based to triangular lobes. Upper labial furrows short and inconspicuous. Hyomandibular pores alongside corners of mouth inconspicuous. Gill slits large (2.3–3.7 % TL). First dorsal fin large (height 9.2 % TL), and broad, rather long based, triangular to semi-falcate in shape. Its apex pointed or sharply rounded, the posterior margin curving ventrally from fin apex. D1 origin over to

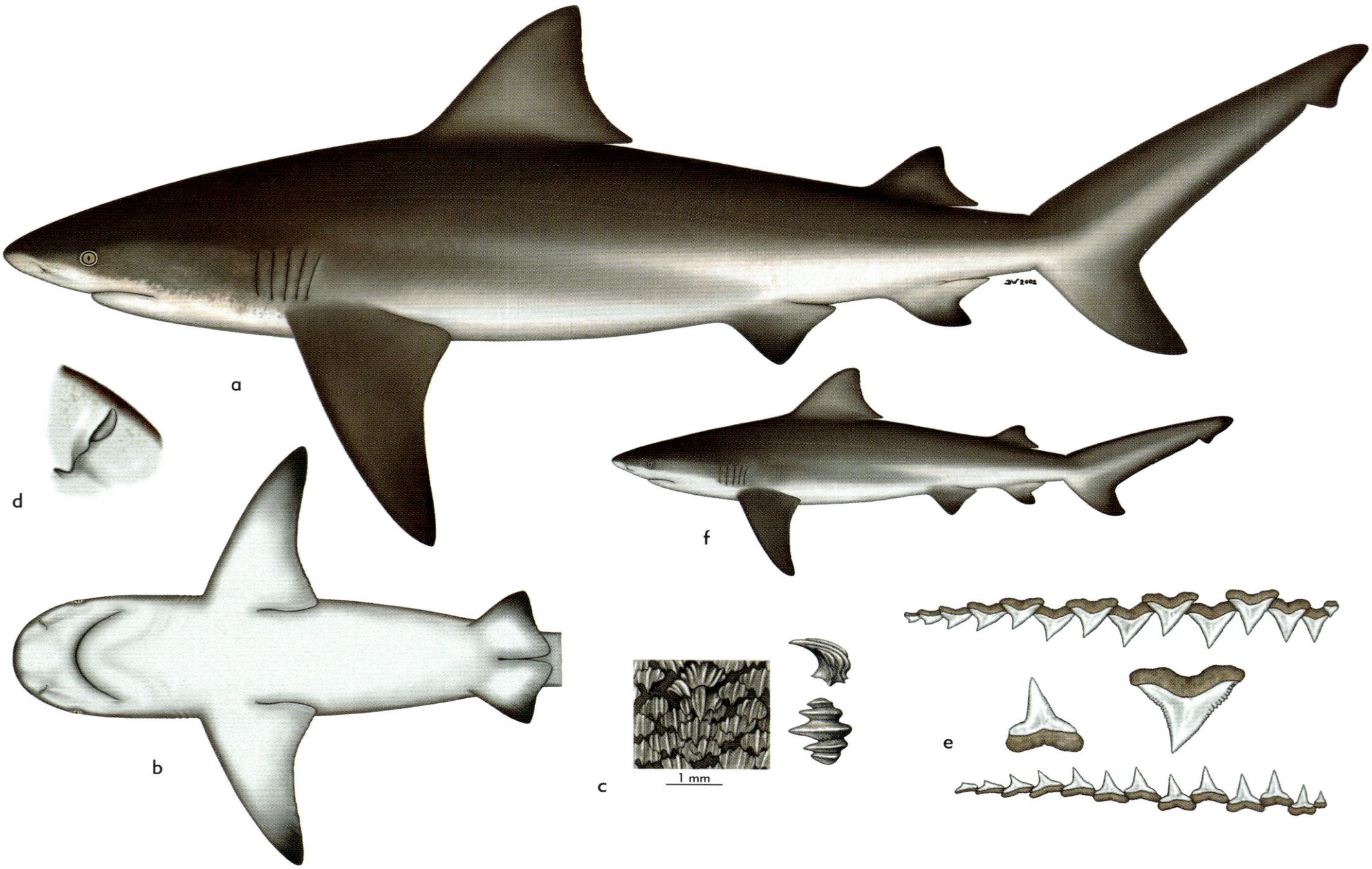

Figure 36. *Carcharhinus leucas*: 302 cm TL female from the Bahamas; **a**, left side; **b**, ventral view; **c**, dermal denticles; **d**, enlarged left nostril; **e**, left upper and lower teeth (symphysis to the left), and inset are the enlarged sixth upper and lower teeth; **f**, juvenile male of 94 cm TL from Senegal, West Africa.

halfway posterior of pectoral fin insertion, the inner margin short. Second dorsal fin large and high (height 3.8 % TL), with a short inner margin. D2 origin anterior to anal fin origin. Pectorals large and broad (length of anterior margin 19.4 % TL), triangular to semi-falcate, apices narrow and pointed. Vertebrae: 116 precaudal centra and 101 caudal centra on average. Dermal denticles relatively large, loose-spaced in smaller individuals but closed set and imbricate in larger specimens. Ovoid in shape, with three (in smaller specimens), and five to seven strong longitudinal ridges and a corresponding number of sharp-pointed posterior cusplets in larger individuals (BARGAR & THORSON 1992).

Dental Formula:

$$\frac{13-1-13}{12-1-12} \quad \text{to} \quad \frac{12(14)-1-12(14)}{12(13)-1(2)-12(13)}$$

Upper teeth are broad, crowns erect to slightly oblique. Distal median edges concave, mesial edges straight to convex. Both edges are fine and uniformly serrated on apical half, coarser serrated on basal half of crowns, roots are arched. One row of symphysial teeth is present. Lower teeth are broad, crowns erect to slightly semi-oblique. Both median edges are serrated but serration is a little bit coarser on halfway to apically. Roots are arched, only slightly notched. One or two rows of symphysial teeth are present. Files in upper jaw are arranged as alternate overlapping dentition, and in lower jaw as mixed alternate / imbricate overlapping dentition.

Colour. Brownish-grey to pale dusky-grey dorsally, sometimes with tiny blue spots (KATO 1964). Light grey laterally, white to yellow-white ventrally. White to dusky tips are present on the underside of pectorals, pelvics and on anal fin but they are never deep black. Dark tips are present on most or on all fins, especially in juveniles, being more distinct than in adults but younger individuals are very often not strikingly marked. Occasionally, juveniles and adults have an inconspicuous white band on flanks.

Size. Maximum of 350 cm TL and a body mass of 230 kg. The average total length for males is 225 cm TL, corresponding to a body mass of 95 kg, and of 240 cm TL, and a body mass of 130 kg for females (SOMMER et al. 1996). Maximum body mass for the species was determined to be of 316.5 kg (IGFA 2001).

Misidentifications. This species is often confused with *Glyphis gangeticus* because both sympatric species can occur temporarily in freshwater, riverine and lacustrine systems of the Indian subcontinent. Furthermore, the species could be mistaken for its relative *C. amboinensis*. All mentioned species share a similar body-contour and body size features, e. g. very short, bluntly rounded snout, the tooth-shape is very similar but only the mentioned *Carcharhinus* species have tips on their fins which are dusky but never deep black in bull shark. *Carcharhinus leucas* differs from *C. amboinensis* in the ratio of D1:D2 height, in dental formula, and in the average number of precaudal vertebrae (116 in *C. leucas versus* 92 in *C. amboinensis*). The Ganges shark, *Glyphis gangeticus*, differs from the bull shark mainly in tooth-shape and body size.

Geographical Distribution. Western Atlantic: Massachusetts (Mississippi River and Atchafalaya Rivers) to the Bahamas. Gulf of Mexico, Caribbean Sea, Lake Nicaragua and San Juan River (Nicaragua), Lake Ysabel (Guatemala), Patula River (Honduras), Panama Canal, Amazon River (to Peru) to southern Brazil and Argentina. – Eastern Atlantic: Cape Verde, Morocco, from Senegal to Angola, in Gambia River and Ogooue River (Gabon). – Western Indian Ocean: Southern Red Sea, Djibuti, selectively along coastline to South Africa (including Zambezi and Limpopo River, Ruenwa River (Zimbabwe), Kenya, Tigris River and Shad'al Arab River (Iraq), Seychelles, Mauritius and Madagascar, Hooghly River (India) and off Indian Peninsula, Maldives. – Western Pacific: Andaman Sea, Indonesia, Gulf of Thailand, Vietnam, South China Sea to southern Japan, Borneo, Philippines, Celebes Sea, Australia (western to eastern coast, including Lake Macquarie and Swan, E. Alligator Rivers), New Guinea (including Lake Jamoer), New Caledonia, Fiji, Tuamotu Atoll and Tonga, and off several other Micronesian Islands. – Eastern Pacific: Southern Baja California and Gulf of California to Ecuador and Peru.

The species is distributed in some smaller lakes and rivers along

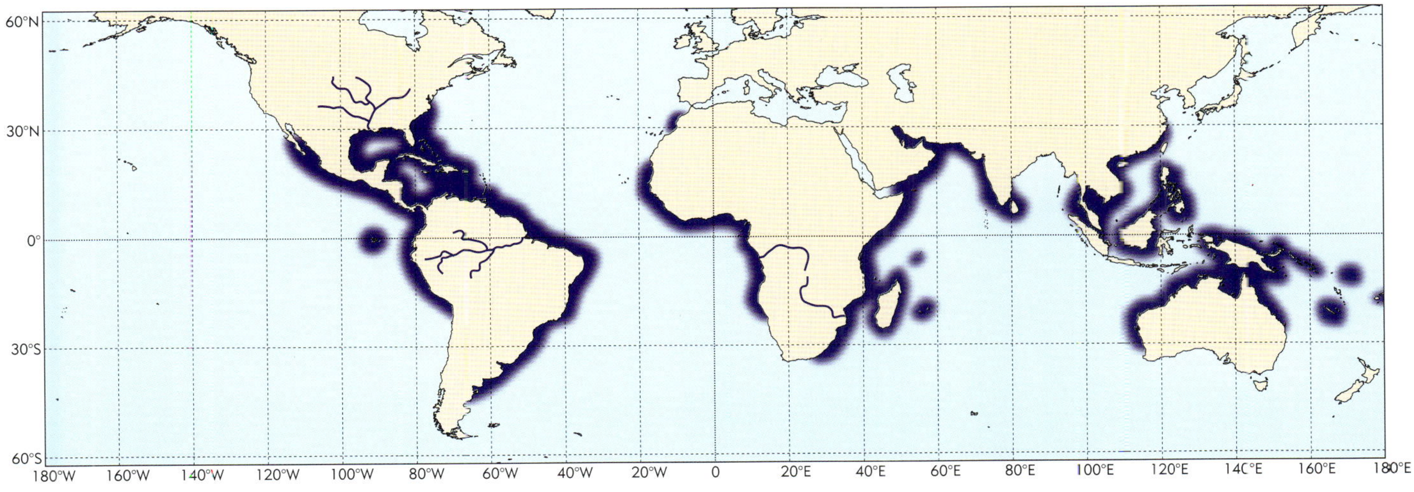

the mentioned areas too, but they are not mentioned in detail here.

Only the biggest rivers and river systems are shown in the map of geographical distribution. – (BIGELOW & SCHROEDER 1961; SADOWSKY 1971; THORSON et al. 1966a, b, 1971, 1976; BASS et al. 1973, 1986; FOURMANOIR 1976a, b; THOMERSON et al. 1977; GARRICK 1982; COMPAGNO 1984; CHIARAMONTE 1998; TAN & LIM 1998; TANIUCHI et al. 2003; COMPAGNO et al. 2005; BRUNNSCHWEILER & COMPAGNO 2008).

Habitat. The species inhabits coastal, estuarine, riverine and lacustrine tropical and sub-tropical habitats, and is usually wide-spread along coasts. Bull sharks occur close inshore, swimming directly under the water surface up to 30 m depth, up to 150 m and deeper (COMPAGNO 1984).

Biology. Commonly, *C. leucas* occur in hypo- and hypersaline habitats but is often found in muddy areas with low salinity also, living together with few other shark competitors, e.g. in Australian rivers. *Carcharhinus leucas* penetrate far into freshwater and is able to exist and to breed (in a minor extent compared to their use of brackish water areas close to the coast), over a longer time period in tropical lakes and rivers, e.g. up to 3700 km from the coast in the Peruvian Amazon River.

In South Africa (St. Lucia Lake) bull sharks can tolerate hypersaline water of 53 PSU (Part Salinity Unit) but normally they tend to move out when salinity exceed 50 PSU. Bull sharks caught from lakes and rivers are often in poor condition despite adequate food ability, and the hypersaline water conditions seems to be sub-optimal for their metabolism and ionic balance (PILLIANS et al. 2005, HEUPEL et al. 2010). Newborn bull sharks live in euryhaline water, and juveniles migrate commonly into freshwater.

Occasionally, juveniles are seen spinning out of water, possibly to dislodge parasites or for feeding, but usually they occur near the bottom, and are rarely seen near the surface (ECCLES 1992). Off Florida, young and juveniles can be found inshore in coastal lagoons, but sub-adults and adults live in deeper coastal waters. Adult females return to lagoons to drop their young.

It is a slow swimming species, especially when travelling but can be quick and agile when attacking its prey. In the western Atlantic Ocean, the sharks make a northward movement along the US-Atlantic coast during the summer away from areas of warmer water, back towards the south when water cools in winter (YEISER et al. 2008, ORTEGA et al. 2009).

The maximum reported age for this species was reported to be 32 years (COMPAGNO 1984, SNELSON & WILLIAMS 1981, SNELSON et al. 1984, HEITHAUS et al. 2001, IGFA 2001).

Reproduction. Males reach maturity at 210–220 cm TL, corresponding to an age of 14–15 years. Females reach maturity at > 225 cm TL, corresponding to an age of 18 or more years (BRANSTETTER & STILES 1987). In the western Atlantic mating occurs in the late spring and summer (May–July, COMPAGNO 1984). Reproductive mode is placental viviparous (DULVY & REYNOLDS 1997). The number of young per litter varies from 7–9 (other sources give number of young per litter with 1–13 (FROESE & PAULY 2010). BASS et al. (1973, 1986) gave the number of pups with 10–12 per litter. At birth, young have 60–85 cm TL. The juveniles are slow growing and the length of the smallest caught free living young was 49.7 cm, but young presumably 9 to 12 months old were 98–112 cm TL.

[...] Between 21 and 24 months they reached 124–128 cm TL, which is the same size as when they start migrating to the open sea (SADOWSKY 1971). Duration of the gestation was calculated to be 10–11 months. Birth place of the young and pupping grounds are near shore, in estuaries and river mouths, also in freshwater lakes, e.g. Lake Nicaragua. In the western North Atlantic, off Florida, in the Gulf of Mexico and off the South African coast, time for giving birth is in late spring to early summer (April–June). In the Lake Nicaragua young are born throughout the year. It was thought pregnant females are inhibited in their feeding shortly before giving birth but this issue is not clear yet (JENSEN 1976, THORSON & LACY 1982, COMPAGNO 1984).

Food. The species has a very broad food spectrum which is mainly dominated by bony fishes, e.g. garfish, tarpon, freshwater eels, snake eels, shad, sardines, anchovies, lizardfish, milkfish, gonorhynchids, characins, sea catfish, mullet, mackerel, tuna, sea bass, perch and striped bass, cichlids, snappers, bluefish, grunts, croakers, parrotfish, soles and flounders, gurnards, toadfish, boxfish and many other not in detail listed bony fishes.

Bull sharks are important predators of other elasmobranches, particularly on smaller, young sharks and stingrays in inshore nursery grounds. They take bramble sharks (not specified), spiny dogfish (*Squalus* sp.), requiem sharks (*Carcharhinus* sp.), sharpnose shark (*Rhizoprionodon* sp.), hammerhead sharks (*Sphyrna* sp.), skates (not specified), butterfly rays (*Gymnura* sp.), eagle rays (*Myliobatis* sp.), and devil rays (*Mobula* sp.) (COMPAGNO 1984). Adults capture more elasmobranches than juveniles due to their larger body size, and occasionally, young bull sharks display cannibal tendencies.

Juveniles of the Lake Nicaragua population have a more fish dominated diet than adults, which also feed on mammals. This species is known also to eat sea turtles, birds, dolphins, whale carcasses and terrestrial mammals, in whole and in parts. Bull sharks inhabiting the South African rivers are mostly immature individuals and they have a different food spectrum than adults, e.g. antelopes are included. The sharks are able to attack them, when the antelopes are on their annual migration during dry season to water wholes crossing rivers. Furthermore, the species feed also on cattle, dogs and rats when they are in the water, and is made responsible for fatal attacks on humans. Invertebrate prey includes crabs (marine, freshwater and land crabs), shrimp, hermit crabs, cephalopods (squids), sea snails, and sea urchins. Moreover, the species is known to use food sources like slaughterhouse offal, fish and animal garbage, especially in river systems near bigger cities (SPRINGER 1964, KATO 1964, TUMA 1976, SNELSON et al. 1984).

Red List status: NT (IUCN 2010).

Carcharhinus limbatus (VALENCIENNES 1839)

En: Blacktip shark, Blacktip whaler; Fr: Requin borde; Sp: Tiburon macuira; Ge: Großer Schwarzspitzenhai, Atlantischer Schwarzspitzenhai

Carcharias (Prionodon) limbatus VALENCIENNES (in MÜLLER & HENLE 1839). Syntype: MNHN 3468. Type material comes from a male of 72 cm TL, from Martinique, but only the mounted skin is present.

Synonyms: *Carcharias (Prionodon) pleurotaenia* BLEEKER (1852), JORDAN & SEALE (1906), *Carcharias (Prionodon) muelleri* STEINDACHNER (1867), *Carcharias abbreviatus* KLUNZINGER (1871), *Carcharias ehrenbergi* KLUNZINGER (1871), *Carcharias aethalorus* JORDAN & GILBERT (1882), *Carcharias phorcys* JORDAN & EVERMANN (1904, 1905), *Carcharhinus natator* MEEK & HILDEBRAND (1923), *Galeolamna pleurotaenia tilstoni* WHITLEY (1950).

Diagnostic features. Large, robust species, without interdorsal ridge. Upper precaudal pit well developed, lower pit weak. Internarial width 1.1–1.7 times the preoral length. Snout long and pointed, eyes circular and relatively small (diameter 1.7 % TL), eyes anterior margins slightly anterior to the front of mouth. Nostrils oblique, incurrent apertures large and ovate-shaped. Anterior nasal flaps low with triangular, pointed lobes. Upper labial furrows short and inconspicuous. Hyomandibular pores alongside corners of mouth inconspicuous. Gill slits large (3.2–4.4 % TL). First dorsal fin large and falcate (height 10.4 % TL), its apex narrowly rounded. The posterior margin curving ventrally from fin apex. D1 origin above or

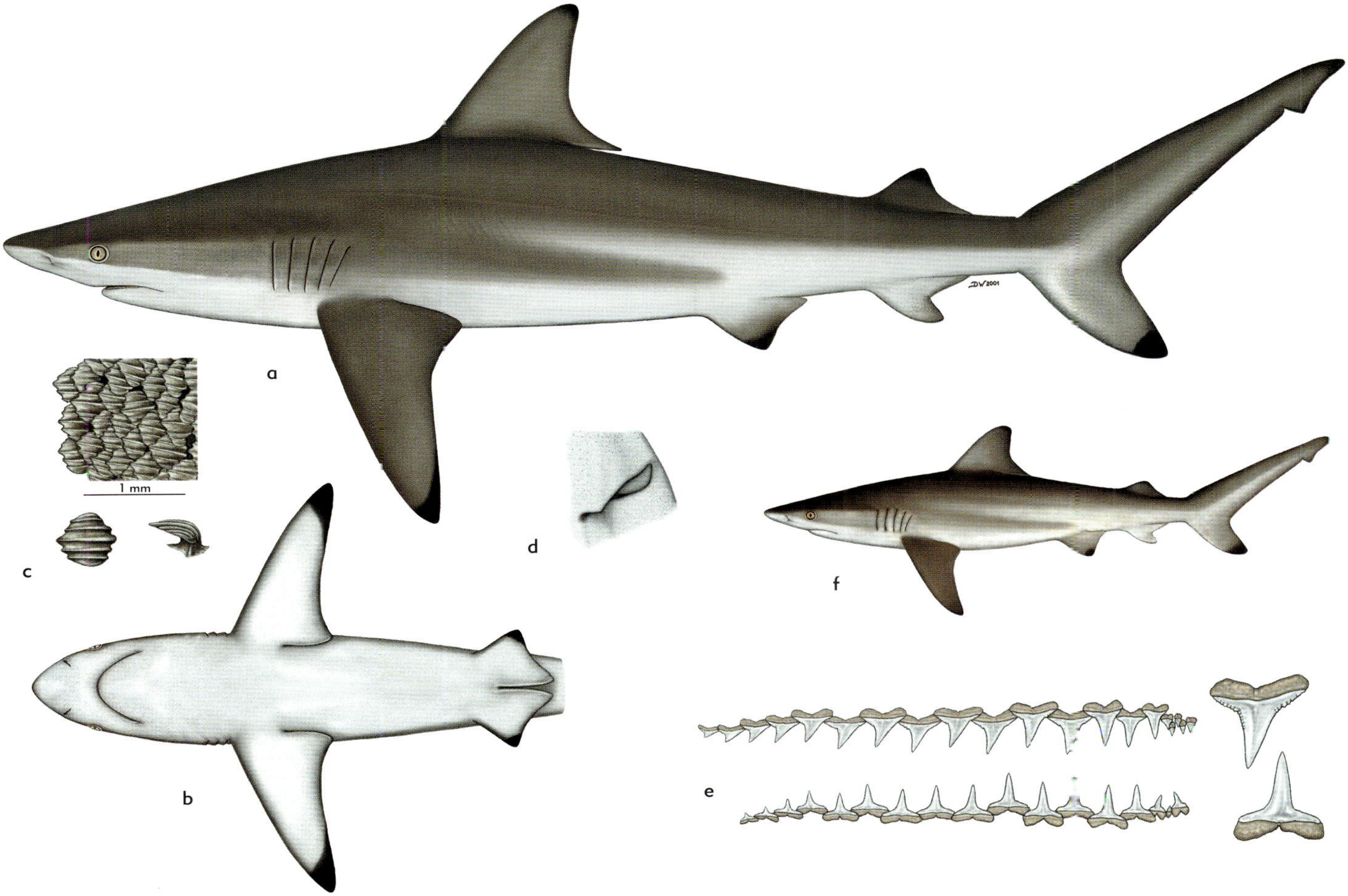

Figure 37. *Carcharhinus limbatus*: 219 cm TL female from Cuba; **a**, left side; **b**, ventral view; **c**, dermal denticles; **d**, enlarged left nostril; **e**, left upper and lower teeth (symphysis to the left), and teeth left aside are the enlarged sixth upper and fifth lower teeth; **f**, juvenile female of 46 cm TL from the Virgin Islands.

halfway posterior to pectoral fin insertion, but near the pectoral free rear tip, with a short inner margin. Second dorsal fin large and high (height 2.9 %TL), its inner margin short. D2 origin over or slightly anterior to anal fin origin. Pectorals large (length of anterior margin 18.1 % TL), falcate, their apices narrowly rounded or pointed. Vertebrae: 96 precaudal centra and 98 caudal centra on average. Dermal denticles closed set, imbricate and sub-circular, with three longitudinal ridges and three posterior cusplets in smaller specimens. Denticles of larger individuals more rhomboid in outline, with five to seven longitudinal ridges and a corresponding number of posterior cusplets. In addition, denticles weakly developed and more regularly arranged in diagonal rows in larger individuals.

Dental Formula:

$$\frac{15 - 2 - 15}{14 - 2 - 14} \quad \text{to} \quad \frac{14(16) - 1(3) - 14(16)}{13(16) - 1(3) - 13(16)}$$

Upper teeth are narrow, crowns erect to slightly oblique. Distal median edges concave, notched halfway longitudinally and straight apically, mesial edges weakly concave basally, to convex apically. Serration is fine and uniformly on the central part of crowns, slightly coarser and often somewhat irregular basally, roots arched. Two, one i. e. three rows of symphysial teeth are present. Lower teeth are narrow, high crowns. Both median edges concave, notched on mesial edges, very fine and uniformly serrated. Serration is absent in newborn, the margins of juveniles are smooth-edged. Roots are arched, deeply notched. Two, i. e. one or three rows of symphysial teeth are present. Files in upper jaw are arranged as alternate overlapping dentition, and in lower jaw as mixed alternate / imbricate overlapping dentition.

Colour. Grey to dusky grey-brown, with a tinge of bronze dorsally, white ventrally. With a conspicuous white band on flanks which begins on pectoral fin axis. Black tips are distinct on almost all fins and on the ventral caudal lobe, with exception of the anal fin which is usually plain. Black edges are usually on D1 and on dorsal caudal fin lobe present. In some areas adult specimens may have plain or virtually plain fins.

Size. Maximum of 275 cm TL, and an average of 150 cm TL. The average body mass of approx. 18 kg, and a maximum reported body mass of 122.8 kg was reported (IGFA 1991, LAFRANCE 1994).

Misidentifications. The blacktip shark is often confused with the smooth-backed species *C. amblyrhynchoides*, *C. brevipinna* and *C. tilstoni*. All mentioned species share a similar body-contours, similar fin colouration (on tips), and a comparable tooth-shape. *Carcharhinus limbatus* differs from *C. brevipinna* in having D1 origin over or slightly behind pectoral axis, in presence of serration on teeth of lower jaw and in having a shorter snout but identification of both species is very difficult, especially of juveniles (Branstetter 1982). *Carcharhinus amblyrhynchoides* has a different snout-shape and a different second dorsal fin proportion than the blacktip shark. The close related species, *C. tilstoni*, differs slightly from *C. limbatus* in presence or absence of dark or black fin colouration, in dental formula, and in the average number of the precaudal vertebrae (96 in *C. limbatus*, versus 82 in *C. brevipinna*, 91 in *C. amblyrhynchoides* and 88 in *C. tilstoni*, Lavery & Shacklee 1991). Both forms of blacktip shark co-occur in subtropical and tropical areas of Australia and were divided after genetic relationship investigations (Ovenden et al. 2009, 2010). *Carcharhinus limbatus* was believed to be relatively rare in the northern part of Australia, in an occurrence ratio of *C. limbatus* to *C. tilstoni* of 1 : 300 but recent studies have revealed *C. limbatus* is not rare around Australia, and a species abundance ratio of approximately 50 : 50 was estimated (Ovenden et al. 2010). Furthermore, the studies of Ovenden et al. 2010 (and others) provides genetically evidence that *C. limbatus* and *C. amblyrhynchoides* are more closely related to each other than to *C. tilstoni*, and *C. tilstoni* seems to be more endemic than known before as suspected by Compagno et al. (2005) and Last & Stevens (2009).

Geographical Distribution. Western Atlantic: from Massachusetts to Florida, Gulf of Mexico, Caribbean Sea, Bahamas and Martinique, to southern Brazil. – Eastern Atlantic and Mediterranean: Western Mediterranean, Madeira, Canary Islands, Senegal to Democratic Republic of Congo. – Indo-West Pacific: Red Sea, Somalia to Madagascar, South Africa, Pakistan, India, Sri Lanka, Thailand, China, Taiwan, Japan, Philippines, Java, Borneo, Australia (western Australia, Queensland and Northern Territory), New Guinea, New

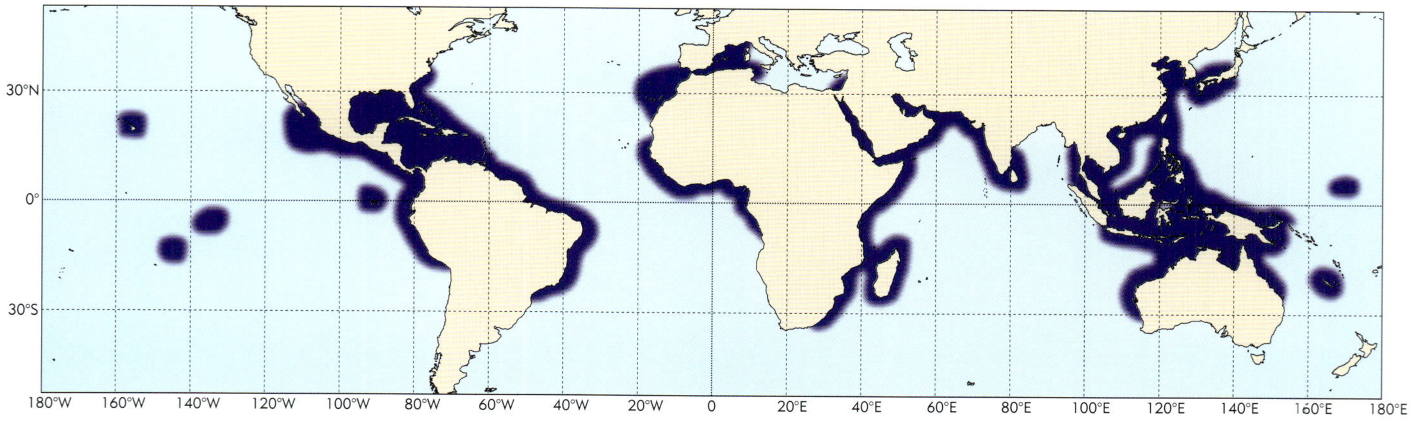

Caledonia. – Central Pacific: Tahiti, Marquesas, Hawaiian Islands. – Eastern Pacific: Southern Baja California (exceptionally to San Diego, California), to Peru, Revillagigedo and Galapagos Islands. – (BASS et al. 1973, 1986; FOURMANOIR 1976 a, b; BRANSTETTER 1981, 1982, 1987; GARRICK 1982; COMPAGNO 1984; DUDLEY & CLIFF 1993; CROW et al. 1996; TAVARES & PROVENZANO 2000; HEMIDA & LABIDI 2002; COMPAGNO et al. 2005).

Habitat. It is a common species of tropical and warm-temperate waters. The species is inshore and pelagic living, typically found on or adjacent to the continental and insular shelves but not truly oceanic (RIEDE 2004).

[...] Commonly occurs close inshore, off river mouths and in estuaries, in shallow muddy bays and in more saline parts of mangrove swamps, in island lagoons and along drop-offs on coral reefs as well far offshore, rarely found deeper than 30 m (COMPAGNO 1984).

Biology. The species can tolerate reduced salinity in estuaries and of river mouths but does not penetrate as far into freshwater as bull sharks (COMPAGNO 1984). Seasonal migrations are known for the species, and could be a reason for absence or presence of the species during few months of the year in specific habitats (e.g. the species appears only in few months off Florida's coast). Age and sexual segregation, especially when the females are pregnant was proven within the species (HEUPEL & SIMPFENDORFER 2002). Juveniles live inshore and in estuarine nursery grounds, along shoreline, where they remain for the first years of their lives (LIESKE & MYERS 1994, HEUPEL & SIMPFENDORFER 2002). Occasionally, blacktip sharks live together in aggregations or schools. The species is known to show an agonistic or hunching display when being aggressive one to another and to divers.

Carcharhinus limbatus is a fast swimming species, often hunting in midwater due to its unusual hunting strategy. Individuals of this species can leap out themselves of the water similar to the related spinner shark, *C. brevipinna* (BASS et al. 1986). The sharks launch themselves vertically through the schools of prey-fishes, spinning and snapping around in all directions, and then breach the surface after their feeding run (HEUPEL & HUETER 2002).

A theoretical longevity of 12 years was estimated for this species (WINTNER & CLIFF 1996, COMPAGNO 1984, KULKARNI et al. 1988, TAVARES & PROVENZANO 2000).

Reproduction. Males reach maturity at 140–145 cm TL, corresponding to an age of 4–5 years, and all males over 145 cm TL are mature. Females reach maturity at about 150–180 cm TL, corresponding to an age of 7–8 years (BASS et al. 1973, 1986; BRANSTETTER & STILES 1987; LESSA et al. 1999 a, b). Reproductive mode is placental viviparous (DULVY & REYNOLDS 1997) and parthenogenesis was reported to occur within the species (CHAPMANN et al. 2008). The litter size of this species has an average of six embryos, with a maximum of eleven young per litter. The mean embryo length within a litter is 13–49 cm TL. Thereby, larger females can have more and larger embryos (DUDLEY & CLIFF 1993). The reproductive cycle is at least 2 years, and includes biennial ovulation with a 1 year gestation (BRANSTETTER 1981, 1982, 1987; GARRICK 1982; CAPAPE et al. 2004). In Australian waters, the mean gonad indices are the highest during September and October. Males and females considered to be in mating condition were found predominantly in November and December. In contrast, in Bulls Bay, South Carolina, mating and ovulation occur from mid-May to early June. Implantation usually occurs during the 10th and 11th weeks of gestation when the embryos measure 17.8–19.4 cm TL. Parturition occurs in the following year from early May to early June in the shallow coastal waters of the Carolinas.

At birth, young have 40–60 cm TL. The young remain to fall in the shallow coastal water nurseries off Georgia and the Carolinas (SIMPFENDORFER & MILWARD 1993, CASTRO 1996, BORNATOWSKI 2008).

Food. Mainly the species feed on bony fishes, e.g. pilchard, herring, mackerels, flatfish, but also other sharks (*Mustelus* sp., *Rhizoprionodon* sp., juvenile dusky sharks, *C. obscurus*, none further specified skates and stingrays), and occasionally molluscs and cephalopods (TAVARES & PROVENZANO 2000).

Red List status: NT (IUCN 2010).

Carcharhinus longimanus (POEY 1861)

En: Oceanic whitetip shark, Whitetip whaler; Fr: Requin oceanique; Sp: Tiburon oceanico; Ge: Weißspitzenhochseehai, Longimanus

Squalus longimanus POEY (1861). Holotype: Institutional deposition of the holotype unknown. Holotype material is a male of 164 cm TL, from Cuba (mentioned by GARRICK, 1982). Paratype: MNHN 1138. Type material is a female embryo of 32.2 cm TL, from Coromandel, India.

Synonyms: *Squalus (Carcharias) maou* LESSON (1831), *Carcharias insularum* SNYDER (1904), *Pterolamia longimana* SPRINGER (1950), *Pterolamiops magnipinnis* SMITH (1958), *Pterolamiops budkeri* FOURMANOIR (1961).

Synonyms in recent use: *Carcharhinus maou* LESSON (1831), *Pterolamiops longimanus* POEY (1861).

Diagnostic features. Large, robust species, an interdorsal ridge is present but can be very low. Upper precaudal pit well developed, lower pit weak. Internarial width 0.9–1.3 times the preoral length. Snout short and broadly rounded, eyes circular and small (diameter 1.6 % TL). Eyes anterior margin above the front of mouth in small specimens and sub-adults but a little bit posterior to this level in bigger specimens. Nostrils strongly oblique, incurrent aperture large and "Boomerang"-shaped. Anterior nasal flaps low, straight, with short lobes. Upper labial furrows short and inconspicuous. Hyoman-

Figure 38. *Carcharhinus longimanus*: 281 cm TL female from Indo-Pacific; **a**, left side; **b**, ventral view; **c**, dermal denticles; **d**, enlarged left nostril; **e**, left upper and lower teeth (symphysis to the left), teeth beneath are the enlarged sixth upper and lower teeth; **f**, juvenile male of 71 cm TL from New Guinea.

dibular pores alongside corners of mouth inconspicuous. Gill slits long (2.5–3.4 % TL). First dorsal fin very large (height 12.3 % TL), its apex broadly rounded, the posterior margin curving anterior- and posterior-ventrally from fin apex. D1 origin above or halfway posterior to pectoral fin insertion. The inner margin of first dorsal fin short. Second dorsal fin large and high (height 3.4 % TL), its inner margin short. Occasionally, D2 origin over anal fin origin, but usually slightly anterior to anal fin origin. Pectorals very long and broad, (length of anterior margin 24.3 % TL), nearly straight, distally expanded, apices broadly rounded. Vertebrae: 126 precaudal centra and 109 caudal centra on average. Dermal denticles closed set, slightly to relative imbricate, ovate to slightly circular, wider than long, with three, five to seven longitudinal ridges and a corresponding number of posterior cusplets. Changes in the size of the main cusps in dependence of the body size are known.

Dental Formula:

$$\frac{14 - 2 - 14}{14 - 1 - 14} \quad \text{to} \quad \frac{13(14) - 1(2) - 13(14)}{13(15) - 1(3) - 13(15)}$$

Upper teeth are broad, crowns triangular, erect to slightly oblique. Distal median edges weak concave (AL to L rows) to distinct concave (LP to C rows). Mesial edges straight to slightly convex. Both edges are strongly serrated, serration slightly coarser basally, roots arched. One or two rows of symphysial teeth are present. Lower teeth are narrow, crowns erect to slightly oblique and a little bit more oblique in the LP rows. Both median edges concave. Crowns abruptly tapered towards their tips, fine serrated apically, smooth-edged basally. Towards angles of jaw the serration extends more onto bases of crowns and is complete in the penultimate and ultimate rows. Roots are arched, deeply notched. One to three rows of symphysial teeth are present. Files in upper and lower jaw are arranged as mixed alternate i. e. imbricate overlapping dentition.

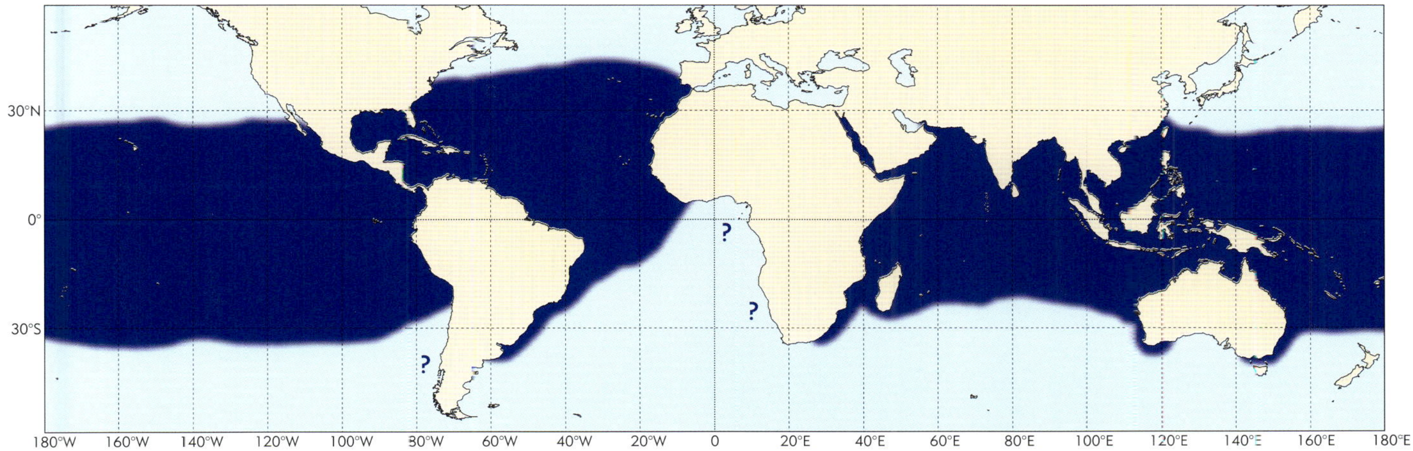

Colour. Grey to bronze-grey dorsally, white ventrally. D1 and D2 are with prominent mottling white tips or spots, which are mostly present on pectorals, pelvic fins and upper and lower caudal fin lobe too. Occasionally, young have black blotches or tips on most fins, especially on D2, pelvics and anal fin, and on ventral caudal lobe. Frequently, black saddles are distinct on the second dorsal fin insertion, on the upper caudal fin origin and occasionally between the dorsal fins. Often an inconspicuous white band on flanks is distinct.

Size. Maximum of 390–395 cm TL, and an average of 250–300 cm TL. The maximum body mass for this species was determined to be 170 kg (RANDALL et al. 1990, IGFA 2001).

Misidentifications. The species can occasionally be mistaken for the silvertip shark, *C. albimarginatus*, due to a slightly similar fin colouration (because white tips on D1 are present but the trailing margin is not white mottled), and comparable body size. However, oceanic whitetip sharks are easy identifiable and separable from all other species of the genus by its fin-shape (broadly rounded) and colouration. Moreover, *C. longimanus* differs from the silvertip shark in the average number of the precaudal vertebrae (126 in *C. longimanus*, *versus* 120 in *C. albimarginatus*).

Geographical Distribution. Western Atlantic: from Maine along the coastline to Argentina (including Gulf of Mexico and the Caribbean Sea, Cuba, Bahamas). – Eastern Atlantic: Madeira, southern coast of Portugal to the Gulf of Guinea. – Western Indian Ocean: Red Sea, Mozambique, Madagascar, Mauritius and Seychelles, South Africa, Pakistan, India, Sri Lanka. – Western Pacific: Indonesia, Thailand, Vietnam, China, Taiwan, Philippines, Australia, including Southern Australian coast), New Guinea, New Caledonia. – Central Pacific: Hawaiian Islands, south to Samoan Islands, Tahiti, Tuamoto Archipelago. – Eastern Pacific: Southern California to Peru (including Gulf of California and Clipperton Island), Galapagos Islands. – Not distributed in the Mediterranean. – (HUBBS 1951; BACKUS et al. 1956; BASS et al. 1973, 1986; FOURMANOIR 1976 a, b; GARRICK 1982; COMPAGNO 1984; CHIARAMONTE 1998; LESSA et al. 1999 a, b).

Habitat. It is a common, oceanic-pelagic living species but occasionally found along coasts of tropical waters also. Oceanic whitetip sharks occur from the surface to 150 m depth, and swim quite often directly at or near the surface (COMPAGNO 1984).

Biology. The species prefers water temperatures of > 20 °C but tolerates temperatures from 15 to 28 °C. It seems the species withdraw themselves from waters that are cooling below this minimum value, e. g. in the Gulf of Mexico in winter. When the water is warming up in regions where *C. longimanus* is normally not distributed, the species can migrate into these areas, and occur there in large numbers. Normally, *C. longimanus* is slow-swimming but the species can be quite active when hunting and specimens can suddenly dash out of water for a short distance. Together with the silky shark, *C. falciformis*, and the blue shark, *Prionace glauca*, *C. longimanus* is one of the most abundant oceanic-pelagic shark species, often found alongside these two species. *Carcharhinus longimanus* is a more leisurely species than *C. falciformis* but oceanic whitetip sharks are more aggressive and perennial in its behaviour when competing for food with silky sharks. *Carcharhinus falciformes* profits from its greater speed and mobility, especially when food is drifting, but flees when *C. longimanus* appears.

Sexual and body size segregation within populations of *C. longimanus* is known. The species can live together in larger numbers and can form huge aggregations around food sources but these aggregations are not truly schools. It is one of the most dangerous shark species in the world, and was made responsible for the most fatal attacks on humans in World War II, e. g. after aircraft or ship disasters in the Pacific Ocean.

[...] Oceanic whitetip sharks are often accompanied by suckerfish, Dolphin fishes, and pilot fishes. An unusual behaviour of the oceanic whitetip shark is its association with the shortfin pilot whale *(Globicephala macrorhynchus)* in Hawaiian waters (SOMMER et al. 1996).

Although the reason for such behaviour is unknown it is suspected to be food-related, because the pilot whales are efficient at locating squid upon which the oceanic whitetip sharks feed also (BESTER, source on FMNH web-site).

A theoretical longevity of 25 years was calculated for the species.

Reproduction. Males reach maturity at 175–195 cm TL, corresponding to an age of 6–7 years, and at least 240 cm TL. Females reach maturity at 180–200 cm, and at least 270 cm TL (LESSA et al. 1999 a, b). Reproductive mode is placental viviparous (DULVY & REYNOLDS 1997). At birth, young have 65–75 cm TL. The litter size can vary from 1–15. The number of young varies in dependence of the body size of the mother, because larger females can have a higher number of young per litter than smaller ones. The species has a gestation of 12 months. In the western Atlantic a seasonality in reproduction cycle and time for giving birth is known but not for the population in the Pacific Ocean. The young are born in shallower water and are slow-growing (LESSA et al. 1999 a, b).

Food. Mainly pelagic living bony fishes, e.g. lancet fish, oarfish, barracudas, tunas, scombrids, and marlins, also dolphins, sea turtles, stingrays (probably *Dasyatis violacea*), birds, crustaceans, cephalopods (squid) and marine gastropods, whale carcasses and offal/garbage (KATO 1964).

Red List status: VU A2ad + 3d + 4ad (IUCN 2010).

Carcharhinus macloti (MÜLLER & HENLE 1839)

En: Hardnose shark, Maclot's shark; Fr: Requin a nez rude; Sp: Tiburon trompudo; Ge: Hartnasenhai

Carcharias (Hypoprion) macloti MÜLLER & HENLE (1839). Holotype: RMNH D 2591. Type material is an adult male of 72 cm TL, caught at New Guinea.

Synonyms: *Hemigaleus heterodus* PHILIPPI (1887), *Hypoprion heterodus* PHILIPPI (1887, 1893), *Hypoprion macloti* NAKAYA (1984).

Diagnostic features. Small, slender species, without strongly distinct interdorsal ridge (mostly a very low interdorsal ridge is present only). Upper precaudal pit well developed, lower pit weak. Internarial width 1.6–2.1 times the preoral length. Snout long, narrowly rounded or slightly pointed with a hyper-calcified rostrum, which can easily be felt by pinching the snout. This hard rostrum is the name giving reason for the common / vernacular name(s) of the species. Eyes circular and large (diameter 2.3 % TL), eyes anterior margin anterior to the front of mouth. Nostrils strongly oblique, incurrent apertures broad and ovate-shaped, anterior nasal flaps with narrow, elongated, nipple-shape lobes. Upper labial furrows short and inconspicuous. Hyomandibular pores alongside corners of mouth inconspicuous. This opinion is in contrast to GARRICK (1985) where *C. macloti* has discrete longitudinal rows of enlarged hyomandibular pores lateral to each corner of mouth. Gill slits mid-sized (2.7–3.1 % TL). First dorsal fin falcate and low (height 8.0 % TL), but its rear tip very long. Its apex rounded or pointed and posterior margin curving ventrally from fin apex. D1 origin above pectoral free rear tips. Second dorsal fin small and low (height 1.9 % TL), with a very long inner margin. D2 origin behind anal fin origin. Pectorals short and falcate (length of anterior margin 14.9 % TL), with pointed apices. Vertebrae: 70 precaudal centra and 81 caudal centra on average. Dermal denticles imbricate, sub-circular, ovate, slightly wider than long, with three well developed longitudinal ridges and three posterior cusplets in smaller specimens. Dermal denticles of adult specimens have the same shape as the juveniles, but few denticles have five posterior cusplets.

Dental Formula:

$$\frac{14 - 2 - 14}{13 - 1 - 13} \quad \text{to} \quad \frac{14(15) - 1(3) - 14(15)}{13(15) - 1(2) - 13(15)}$$

Upper teeth are narrow, with high and erect crowns, increasingly oblique towards angles of jaw. Distal median edges straight on two third of crowns, deeply notched, convex on basal third. Mesial edges are straight or slightly convex apically, on two third of crowns. Deeply notched to straight or slightly concave basally in the rows towards angles of jaw. Distal edges smooth-edged, basally always strongly serrated, mesial edges often strongly serrated basally. Roots are arched, deeply notched. Differences in the strength of serration could reflect a sexual dimorphism, and is presumably related to size and age, but this opinion is not proved at the moment. One or three rows of symphysial teeth are present. Lower teeth are narrow, high crowns, erect to oblique. Both median edges deeply notched, concave, smooth-edged. Roots nearly straight, deeply notched, divide the roots almost into two root lobes. One to two rows of symphysial teeth are present. Files in upper jaw are arranged as mixed alternate i.e. imbricate overlapping dentition and in lower jaw as alternate overlapping dentition.

Colour. Grey to grey-brown dorsally, white below. Normally, fins lacking conspicuous markings but sometimes narrow, dark margins on D2 and on dorsal lobe of caudal fin can be present. Pectorals are normally not conspicuously marked but occasionally trailing margins, also on pelvics and anal fin are distinct. The lower caudal lobe is white or pale-bright. An inconspicuous white band on flanks can be present.

Size. Maximum of 110 cm TL, and an average of 70–85 cm TL for males, and of 70–90 cm TL for females (LAST & STEVENS 1994).

Misidentifications. The species could be confused with *C. hemiodon*, *C. signatus* and *C. borneensis* but *C. macloti* differs from Pondicherry sharks in snout length and shape, and is lacking prominent fin colouration patterns. Both species have a very similar body size, dental formulae, number of precaudal vertebrae and geographical distribution are almost equal but in contrast to *C. signatus* the first dorsal rear tip of hardnose sharks is much longer. *Carcharhinus borneensis* differs from the hardnose shark in snout length and in tooth-morphology. Furthermore, *C. macloti* differs from all mentioned species in the average number of precaudal vertebrae (70 in *C. macloti versus* 62 in *C. borneensis* and 101 in *C. signatus*).

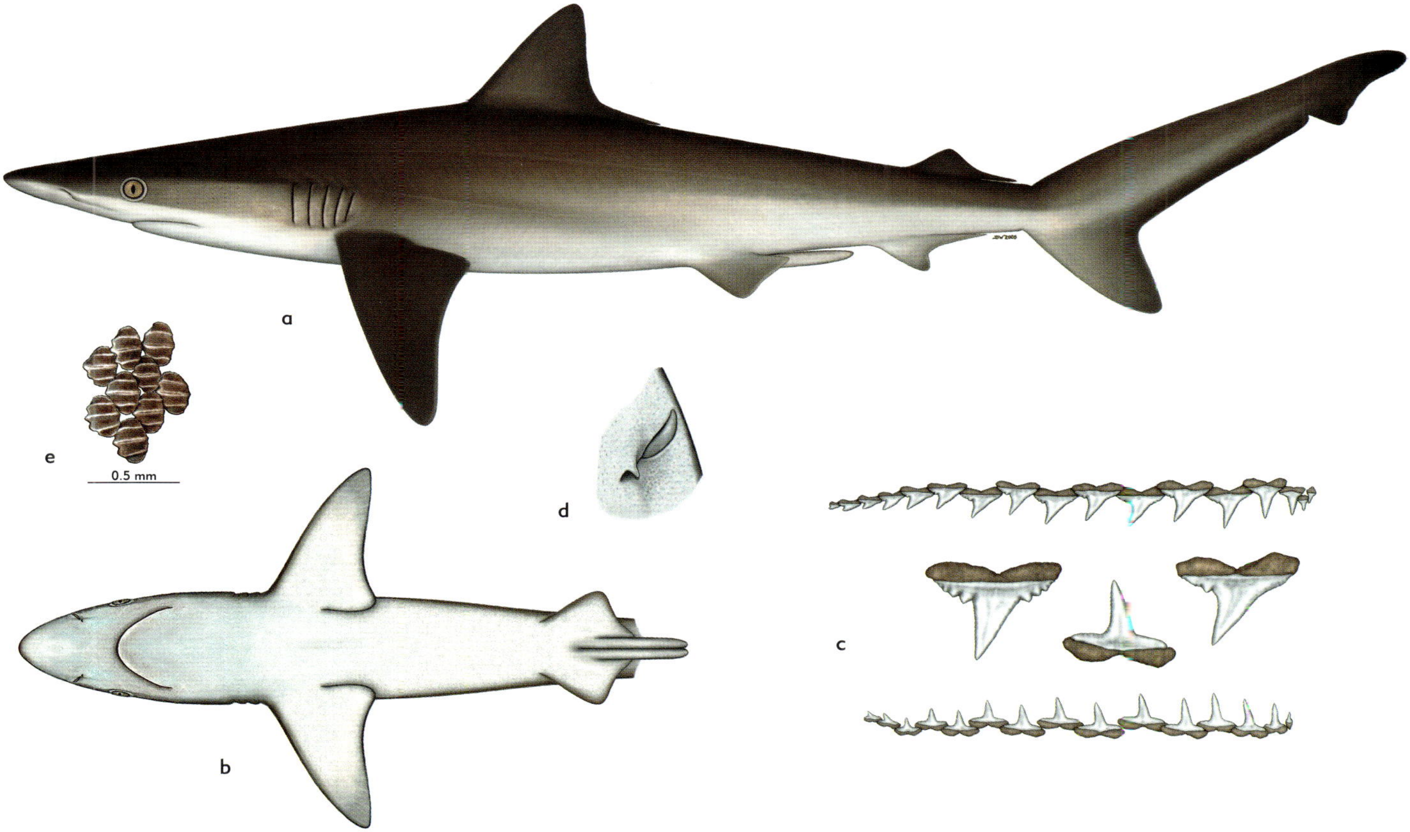

Figure 39. *Carcharhinus macloti*: 73 cm TL male from Taiwan; **a**, left side; **b**, ventral view; **c**, dermal denticles; **d**, enlarged left nostril; **e**, left upper and lower teeth (symphysis to the left), inset are the enlarged fifth upper (left) and the fifth lower tooth (middle) of the Taiwanese specimen, and the sixth upper tooth of a female of 72 cm TL from Burma (right).

Geographical Distribution. Indo-West Pacific: Kenya, Tanzania, Pakistan, India, Sri Lanka, Andaman Islands, Burma, Vietnam, China, Taiwan, Australia (Queensland and western Australia) and New Guinea. – (Compagno 1984, Garrick 1985, Ajazuddin & Ahmed 2002).

Habitat. It is a common and abundant species of tropical waters, inshore living but poorly known. *Carcharhinus macloti* is typically found over continental and insular shelves from surface to 170 m depth (Compagno 1984, Ajazuddin & Ahmed 2002).

Biology. Often the species was reported living in aggregations with *C. tilstoni* and *C. sorrah*. [...] In Bombay waters over 95 % of caught individuals are males indicating a strong sexual segregation within the population (Compagno 1984).

Reproduction. Males reach maturity at 70–90 cm TL. Females reach maturity at 75–90 cm TL. Reproductive mode is placental viviparous (Dulvy & Reynolds 1997). Litter size varies from 1–8 young. At birth young have 45–55 cm TL, and duration of gestation is 12 months (Compagno 1984, Garrick 1985).

Food. Mainly, bony fishes are taken as prey but hardnose sharks feed on crustaceans and cephalopods to a lower extend also.

Red List status: NT (IUCN 2010).

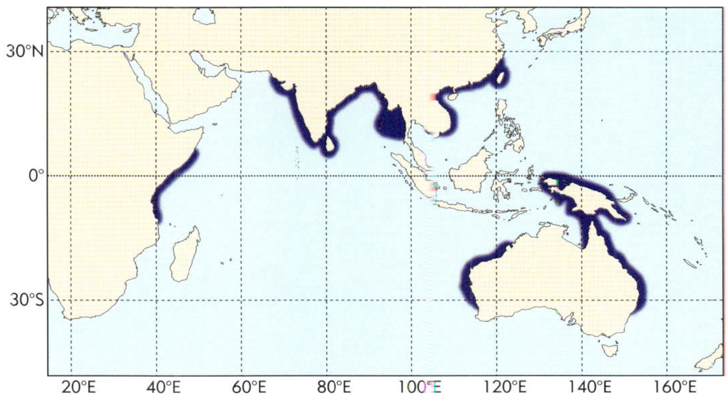

Carcharhinus melanopterus (Quoy & Gaimard 1824)

En: Blacktip reef shark, Guliman; Fr: Requin pointes noires; Sp: Tiburon de puntas negras; Ge: Schwarzspitzenriffhai

Carcharias melanopterus Quoy & Gaimard (1824). Holotype: MNHN 1129. Type material is an immature male of 59 cm TL, from Waigeo Islands.

Synonyms: *Squalus (Carcharhinus) commersonii* Blainville (1816, 1825), Boeseman (1956), *Hypoprion playfairy* Günther (1870), *Carcharias elegans* Ehrenberg (1871), *Carcharias marianensis* Engelhardt (1912).

Synonyms in recent use: *Mapolamia melanoptera* Whitley (1934), *Mapolamia spallanzani* LeSueur (1818, 1822), *Hypoprion playfairy* Günther (1870).

Diagnostic features. Mid-sized species, without interdorsal ridge. Upper precaudal pit well developed, lower pit weak, internarial width 0.8–1.3 times the preoral length. Snout bluntly rounded and short. Eyes horizontally oval-shaped and relatively large (diameter 2.5 % TL), eyes anterior margin above or slightly anterior to the front of mouth. Nostrils large, incurrent apertures hook-shaped, and falcate. Anterior nasal flaps elongated, with bluntly pointed lobes. Upper labial furrows short and inconspicuous. Hyomandibular pores alongside corners of mouth inconspicuous. Gill slits mid-sized (2.5–3.0 % TL). First dorsal fin large (height 9.1 % TL), falcate, its apex sharply rounded or pointed. The posterior margin curving ventrally from fin apex, the inner margin relatively long. D1 origin over pectoral free rear tips. Second dorsal fin very large and high (height 3.8 % TL), with a short inner margin. D2 origin over anal fin origin. Pectorals large, narrow and falcate (length of anterior margin 18.2 % TL), its apices narrowly rounded or pointed. Vertebrae: 116 precaudal centra and 89 caudal centra on average. Dermal denticles closed set, imbricate, sub-circular in smaller specimens, with three longitudinal

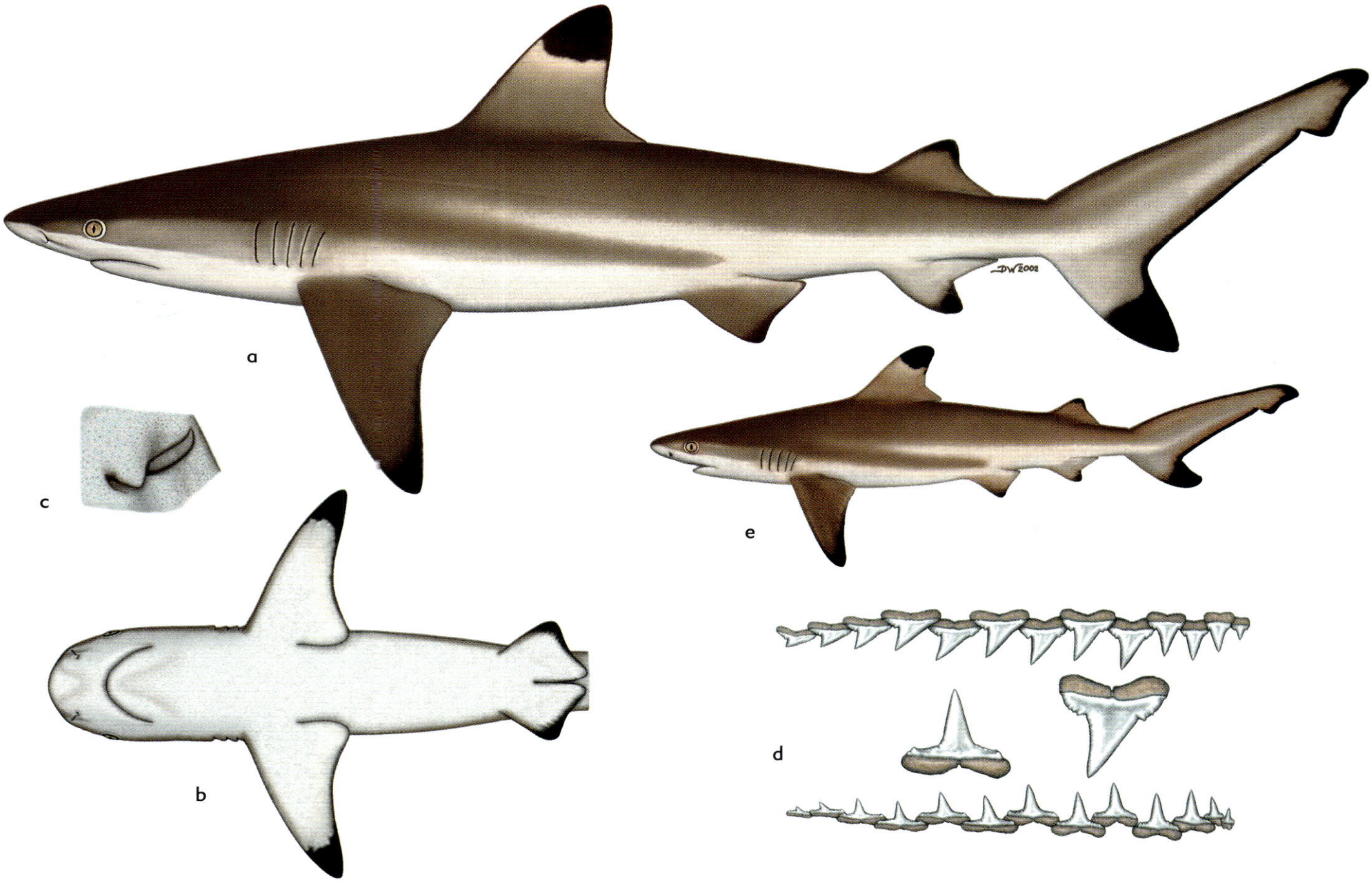

Figure 40. *Carcharhinus melanopterus*: 158 cm TL female from the Red Sea; **a**, left side; **b**, ventral view; **c**, enlarged left nostril; **d**, left upper and lower teeth (symphysis to the left), and inset are the enlarged fifth upper and lower teeth; **e**, juvenile female of 53 cm TL from Great Barrier Reef, Australia.

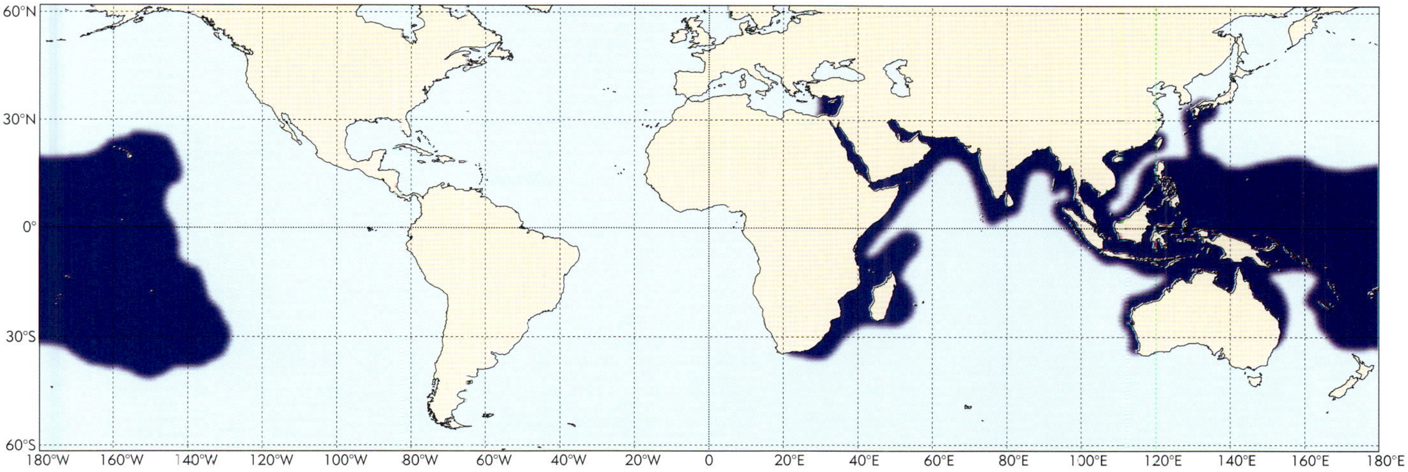

ridges and three posterior cusplets. Denticles of larger individuals are more rhomboid in outline, with five to seven ridges, and a corresponding number of posterior cusplets.

Dental Formula:

$$\frac{12 - 2 - 12}{11 - 3 - 11} \quad \text{to} \quad \frac{11(13) - (1)2 - 11(13)}{10(12) - (1)3 - 10(12)}$$

Sexual dimorphism in tooth-shape and dental formula is known and proven within the species. Variations in tooth-shape seem to occur as adaptation on local food sources in dependence of specific habitat. Upper teeth are narrow, high crowns, erect to oblique. Distal median edges are concave on two thirds of crowns, deeply notched, convex on basal third. The shape of mesial edges can vary from almost straight to weakly concave, slight to deeply notched. Both edges are finely serrated, a little bit coarser and sometimes irregularly serrated basally, roots are arched, slightly notched. Normally, two or three rows of symphysial teeth are present. Lower teeth are narrow, crowns erect to slightly oblique. Both median edges concave, slightly serrated and notched. Roots are arched (S to L rows) to almost straight (LP to C rows), deeply notched, dividing the roots almost into two root lobes. One to three rows of symphysial teeth are present. Files in upper and lower jaw are arranged as alternate overlapping dentition.

Colour. Usually, light brown to grey-brown or back, white ventrally. D1, anal fin and ventral caudal lobe have a conspicuous black apical blotch, brilliantly highlighted proximally with white. Other fins have less prominent black tips. On pelvic and caudal fin a black margin, and usually a conspicuous white band on flanks is present.

Size. Maximum of 200 cm TL for both sexes, and an average of 125–140 cm TL (COMPAGNO et al. 1989). The maximum reported body mass for the species is 13.6 kg (IGFA 2001).

Misidentifications. Only *C. cautus* could be confused with blacktip reef sharks, *C. melanopterus*, but nervous sharks have a much narrower dark edging on the leading margin of the first dorsal fin only, and dusky margins on some other fins. Furthermore, both species differs in the average number of precaudal vertebrae (89 in *C. cautus* versus 116 in *C. melanopterus*).

Geographical Distribution. Eastern Mediterranean (apparently as an invader via Suez Canal from the Red Sea). – Not distributed in the Atlantic Ocean. – Indian Ocean: Red Sea, along the coastline of East Africa to Mauritius, Seychelles and Madagascar, South Africa, Oman, Pakistan, India, Sri Lanka, Andaman and Maldives Islands. – Western and central Pacific: Gulf of Thailand to China, Taiwan, Japan, Philippines, Australia (Queensland, Northern Territory and Western Australia), New Caledonia. Wide-ranging distributed in the island-groups of the western central Pacific, from Hawaiian, Marshall Islands, south to the Tuamoto Archipelago. Apparently rare or absent in more easterly island-groups including the Marquesas, Pitcairn, Tubuai (Rapa) Islands, Austral Ridge, Johnston Atoll, Marcus and Easter Islands. – (HOBSON 1961; GARRICK 1962a, b, 1982; GOHAR & MAZHAR 1964a, b; BASS et al. 1973, 1986; FOURMANOIR 1961, 1976a, b; JOHNSON 1978; COMPAGNO 1984; STEVENS 1984; LYLE 1987; LYLE & TIMMS 1987; YANO et al. 1999; COMPAGNO et al. 2005; PORCHER 2005; PAPASTAMATIOU et al. 2009).

Habitat. It is a common species of estuarine and inshore waters of northern Australia (tropical and sub-tropical waters off Queensland), from the central Pacific to the Red Sea and the east African coast living (LYLE 1987). The species has an affinity to shallower waters, and lives from close inshore and near to coral reefs, to offshore, and was reported from 20 to 75 m depth (HOBSON 1964).

Biology. *Carcharhinus melanopterus* is known to be particularly abundant on reef drop-offs and can tolerate freshwater. In several species reports it is assumed the species is able to penetrate estuarine and river mouths but cannot live in freshwater as bull shark, *C. leucas* (COMPAGNO 1984). It is an active, fast-swimming species which can occur as single specimen, in small groups or in bigger aggregations, but not truly in schools. The species is known to show an agonistic or hunching display and is responsible for some attacks on humans (RANDALL & HELFMAN 1973). In northern Australia waters, body size of the specimens is smaller than in other parts of its geographical distribution what indicates the species does not grow as large as possible there.

Reproduction. Females and males usually reach maturity at 95–97 cm TL (LYLE 1987). Reproductive mode is placental viviparous (DULVY & REYNOLDS 1997). At birth, young have 48–60 cm TL. The litter size of this species varies from 1–8 but has a mean of 4 young per litter. Mating occurs between January and March, ovulation in February–March, and parturition in October–November. Duration of the gestation was calculated to be between 7–9 months. The duration of the reproduction cycle of the blacktip reef sharks is depending of its geographical distribution. At Aldabra, individual females breed every second year and have a slightly longer gestation of 10–11 months (STEVENS 1984) *versus* a gestation of 16 months in the Red Sea. Some other sources gave reports of two periods of parturition for this area during the year, one is in January and the other one is in June (GOHAR & MAZHAR 1964 a, b) but this is not clear yet. Biannual birth seasons may also apply from Madagascar (FOURMANOIR 1961, 1976 a, b) and French Polynesia. In the latter mentioned area, some underwater visual and photographic observations over a longer period (4 years) yielded the following results.

Mating begins there in November and continues until the end of March as each female follows her own temporal cycle. Correspondingly, parturition begins in September and continues until January. Each female mate again 1.5–2.5 months after parturition, thus completes an annual reproductive cycle (JOHNSON 1978, PAPASTAMATIOU et al. 2009). Gestation of 9–10 months is known with slight individual differences and resident specimens under observation have followed this pattern. Evidence of reproductive events presented by transient females conformed to the pattern of the residents (PORCHER 2005).

Food. Mainly bony fishes, e. g. mullets, groupers, jacks, wrasses, surgeonfish, cuttlefish, but also crustaceans (shrimp and mantis shrimp), molluscs (squid and octopus) and sea snakes. Often blacktip reef sharks are prey of bigger sharks (HOBSON 1961, LYLE & TIMMS 1987).

Red List status: NT (IUCN 2010).

Carcharhinus obscurus (LESUEUR 1818)

En: Dusky shark, Black whaler, Bronze whaler; Fr: Requin sombre; Sp: Tiburon arenero; Ge: Düsterer Hai, Schwarzhai

Squalus obscurus LESUEUR (1818). Holotype: MNHN 0000-3464. Type material is a female of 760 cm TL, from the Canary Islands.

Synonyms: *Prionodon obvelatus* VALENCIENNES (1844), *Galeolamna greyi* OWEN (1853), *Carcharias macrurus* RAMSAY & OGILBY (1887), *Galeolamnoides macrurus* WHITLEY (1934), *Galeolamna eblis* WHITLEY (1944 a, b), *Galeolamnoides eblis* WHITLEY (1944 a, b), *Carcharhinus iranzae* FOURMANOIR (1961), *Carcharhinus obscurella* DENG et al. (1981).

Synonyms in recent use: *Carcharhinus lamiella* JORDAN & GILBERT (1882), *Galeolamna macrurus* RAMSAY & OGILBY (1887).

Diagnostic features. Large to fairly slender species, a low interdorsal ridge is present. Upper precaudal pit well developed lower pit weak. Internarial width 0.9–1.5 times the preoral length. Snout short to moderately long, broadly rounded, eyes circular and moderately large (diameter 1.8 % TL), eyes anterior margin slightly forward to the front of mouth. Nostrils oblique, slit-shaped, incurrent apertures ovate. Anterior nasal flaps low, weakly pointed. Upper labial furrows short and inconspicuous. Hyomandibular pores alongside corners of mouth inconspicuous. Gill slits mid-sized (2.4–3.4 % TL). First dorsal fin small and relatively low (height 7.9 % TL), with a short inner margin, falcate to semi-falcate. Its apex pointed or narrowly rounded, the posterior margin curving ventrally from fin apex. D1 origin over or slightly anterior to pectoral free rear tips. Second dorsal fin equal to the first dorsal fin small and low (height 2.0 % TL) but its inner margin long. D2 origin over anal fin origin. Pectorals large and long (length of anterior margin 18.1 % TL), falcate, its apices narrowly rounded or pointed. Vertebrae: 91 precaudal centra and 96 caudal centra on average. Dermal denticles closed set, imbricate, sub-circular in smaller specimens, and with three longitudinal ridges and three fairly long and prominent posterior cusplets. In adults dermal denticles with five to seven longitudinal ridges and a corresponding number of posterior cusplets.

Dental Formula:

$$\frac{14 - 1 - 14}{14 - 1 - 14} \quad \text{to} \quad \frac{14(15) - 1(2) - 14(15)}{13(15) - 1(2) - 13(15)}$$

Upper teeth are broad, crowns oblique, except symphysial and anterior rows. Distal median edges concave, mesial edges slightly concave basally and convex apically. Both edges are strong and uniformly serrated, while serration is slightly coarser basally, roots straight. One or two rows of symphysial teeth are present. Lower teeth are narrow, slightly laterally expanded crowns in larger specimens, erect to slightly oblique. Both median edges are convex and deeply notched basally, straight to convex apically, fine and uniformly serrated. Roots are arched to straight, slightly notched. One, occasionally two rows of symphysial teeth are present. Files in upper and lower jaw are arranged as mixed alternate i. e. imbricate overlapping dentition.

Colour. Bronze-grey to dark grey above, white below. With an inconspicuous white band on flanks which begins in pectoral fin region. Dusky tips on most fins present but they are not black.

Size. Maximum of 420 cm TL, and an average of 240–330 cm TL. The average body mass was estimated to be 160–180 kg, and the maximum reported body mass is 346.5 kg (COMPAGNO et al. 1989, IGFA 1991, NATANSON et al. 1995, SIMPFENDORFER 2000).

Misidentifications. Confusion of this species with *C. galapagensis*, *C. plumbeus* and *C. altimus* is conceivable because these species

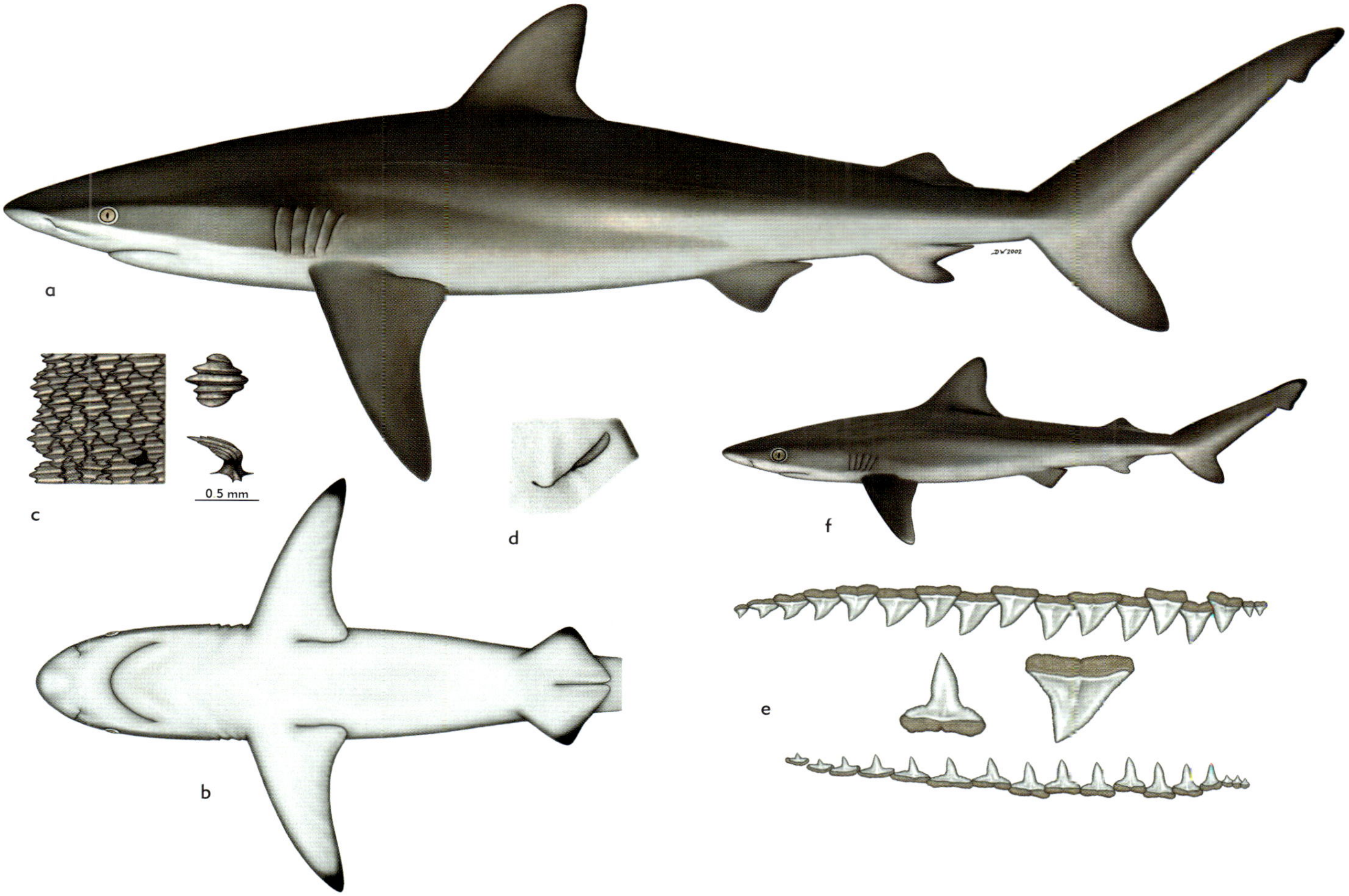

Figure 41. *Carcharhinus obscurus*: 387 cm TL female from Mauritius; **a**, left side; **b**, ventral view; **c**, dermal denticles; **d**, enlarged left nostril; **e**, left upper and lower teeth (symphysis to the left), and inset are the enlarged fifth upper and lower teeth; **f**, juvenile male of 98 cm TL from the Atlantic coast of the United States.

have a very similar combinations of diagnostic features: presence of an interdorsal ridge, without distinct conspicuous colouration on fins, similar tooth-shape and dental formulae. *Carcharhinus obscurus* differ from *C. plumbeus* and *C. altimus* in having D1 origin over or nearer the pectoral free rear tip. Furthermore, the species differs in tooth-morphology, and in height of the second dorsal fin which is higher in *C. altimus* than in *C. obscurus*. Galapagos sharks have a different relationship of D1:D2 height than dusky sharks, and the relationship of height and length of dorsal fins : rear tips of both species are fairly different. Moreover, *C. obscurus* and *C. galapagensis* have distinct differences in their spatial distribution and in habitat preference. It is of advantage separating one species from another due to comparing the average number of precaudal vertebrae (107 in *C. galapagensis versus* 91 in *C. obscurus*, 90 in *C. plumbeus* and 109 in *C. altimus*).

Geographical Distribution. Western Atlantic: Southern Massachusetts to Florida, Bahamas, Cuba, northern Gulf of Mexico, along the coastline of Middle and South America to Brazil. – Eastern Atlantic and Mediterranean: Spain, Portugal and northern Africa, western Mediterranean, Canary and Cape Verde Islands, Senegal to Sierra Leone. – Western Indian Ocean: Red Sea, Mozambique, Madagascar, Mauritius, South African. – Western Pacific: Vietnam, China, Japan, Australia (Western Australia, Queensland, New South Wales), New Caledonia. – Eastern Pacific: Southern California to the Gulf of California, Revillagigedo Islands. – (BASS et al. 1973, 1986; FOURMANOIR 1976a, b; CAPAPÉ et al. 1979; BRANSTETTER 1981; GARRICK 1982; COMPAGNO 1984, 2000; SMALE 1991; VILLAVICENCIO-GARAYZAR 1996; FERGUSSON & COMPAGNO 1999; SIMPFENDORFER et al. 2002; HEMIDA & LABIDI 2002; COMPAGNO et al. 2005; HUSSEY et al. 2009).

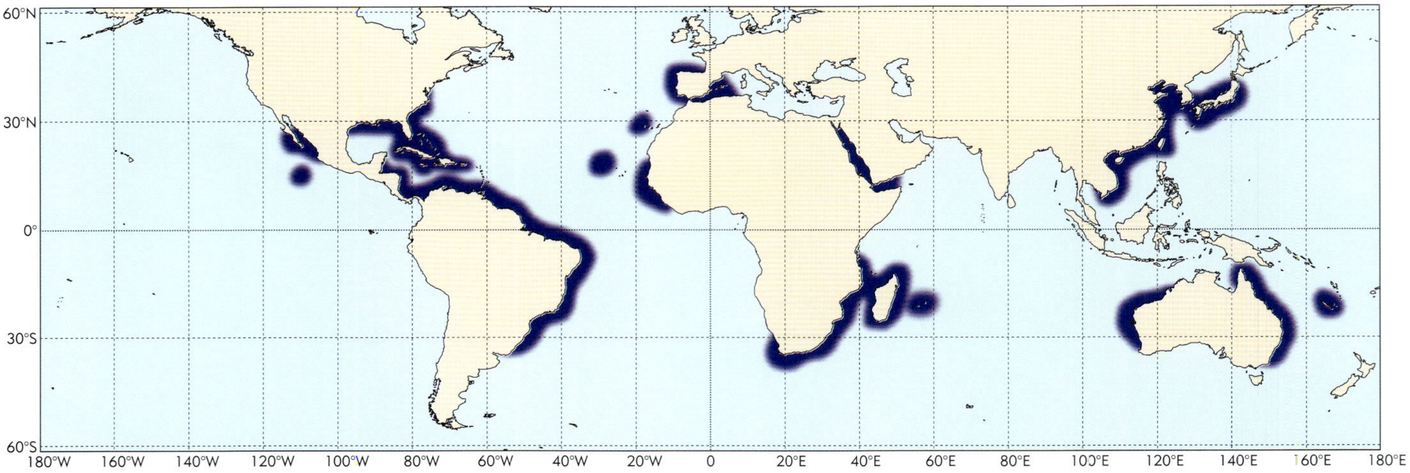

Habitat. It is a common and coastal-pelagic living species of tropical waters, typically found over continental and insular shelves and the adjacent oceanic waters. The species is reported to occur from surface to 400 m depth, from far offshore to close inshore (MUSICK et al. 2000).

Biology. The species show a strong migration pattern in temperate and subtropical areas in the eastern North Pacific and western North Atlantic, moving northwards during the warmer months of summer, and retreating south when the water cools. Adults tend toward regions with reduced salinity, and were often seen to follow ships (presumably to consume food which was thrown overboard ...?).

[...] Off the southern coast of Natal, South Africa, a nursery area occurs, where newborn sharks of 80 to 90 cm TL are resident. Larger, immature sharks over 90 cm TL moves out of this area, females tending to move northwards and males southwards, but there is some overlap in this partial sexual segregation. This pattern is complicated by seasonal, temperature-related migrations (COMPAGNO 1984). The juveniles are known to live together in bigger aggregations or schools. A new nursery area in the American North Pacific was found for dusky sharks. The young occur in shallower waters than adults, whereas adults are commonly found at depths of 200–400 m (BASS et al. 1986). The young are slow growing and although the species is pelagic, the young are born close to the coast in shallower water. After birth females depart nursery area.

The oldest caught specimen was a female of 274 cm TL and an age of 32, i. e. 34 years (NATANSON & KOHLER 1996, SIMPFENDORFER et al. 2002).

Reproduction. Males reach maturity at 230–243 cm TL, corresponding to an age of 20–23 years. Females reach maturity at 220–250 cm TL, corresponding to an age of 17–22 years (SIMPFENDORFER et al. 2002). In contrast, other sources give a different age at maturity of 6–7 years (SHARKINFO). Reproductive mode is placental viviparous (DULVY & REYNOLDS 1997). Normal are 3–14 young per litter, and at birth young have 40–75 cm TL. Seldom birth length is about 80 cm TL (BASS et al. 1973, 1986, NATANSON et al. 1995, SEKI et al. 1998). Duration of gestation was not clearly estimated yet because the caught pregnant females had two size-classes of young in their uteri (45–70 cm and 70–100 cm TL) what indicates a biannual staggered birth season. Parturition may occur over a long time span of several months and gestation could be 8–9 or 16 months. Usually, birth is taking place in late winter to summer. In South African waters birth may occur all year-round, with a peak in the number of birth in autumn. Neonates (80–123 cm TL) and adult females (313–347 cm TL) with embryos are present during February–August in coastal lagoons. Sex ratio of neonates is nearly 1 : 1 (VILLAVICENCIO-GARAYZAR 1996, HUSSEY et al. 2009).

Food. Mainly bony fishes, reef fishes, pelagic and demersal, e.g. sardines, herring, anchovies, eels, lizardfish, needlefish, mullets, barracudas and mackerels, goatfish, groupers, porgies, grunts, croakers, bluefish, tunas, soles, flatfishes (KATO 1964). Moreover, the species feeds on other shark species, skates and rays *(Squatina* sp., *Mustelus* sp., *Squalus* sp., *Centrophorus* sp., *Halaelurus* sp., *C. limbatus* and *C. brevipinna).* In addition, the species feeds on crabs, shrimp, cephalopods (octopi, squid and cuttlefish), starfish, barnacles, bryozoans, whale meat and carcasses, as well as offal. Newborn and juvenile dusky sharks are often eaten by bigger sharks (e.g. *Carcharodon carcharias*, *Galeocerdo cuvier*, *C. leucas* and *Eugomphodon taurus*).

Red List status: VU A2bd (IUCN 2010).

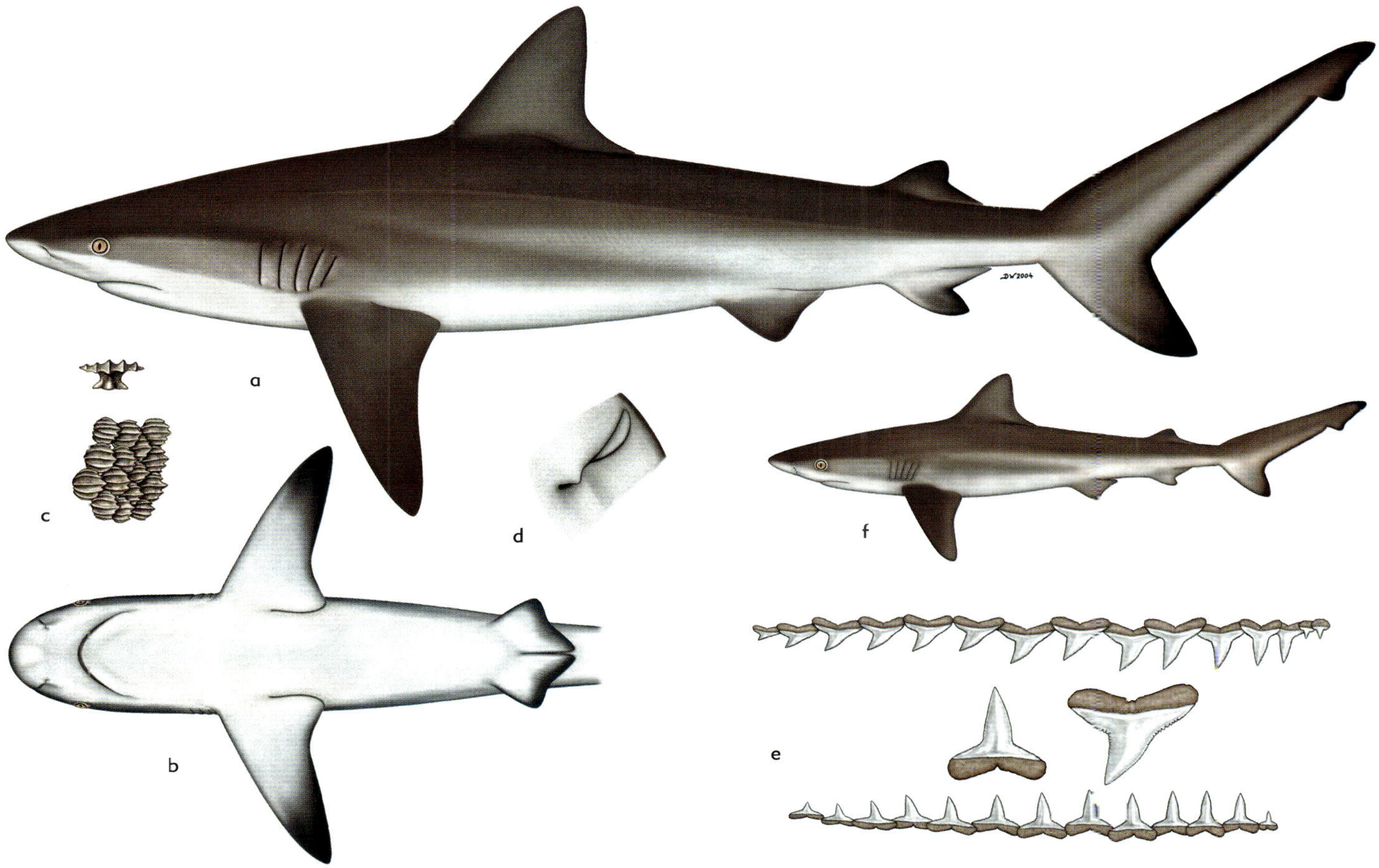

Figure 42. *Carcharhinus perezii*: 215 cm TL female from the Bahamas; **a**, left side; **b**, ventral view; **c**, dermal denticles; **d**, enlarged left nostril; **e**, left upper and lower teeth (symphysis to the left), and inset are the enlarged sixth upper and lower teeth; **f**, juvenile male of 8[illegible] cm TL from Mexico.

Carcharhinus perezii (Poey 1876)

En: Caribbean reef shark; Fr: Requin de recif; Sp: Tiburon coralino; Ge: Karibischer Riffhai

Platypodon perezii Poey (1876). Holotype: Institutional deposition of the holotype unknown. Type material of six original specimens is mentioned by Garrick (1982): three females from 86–130 cm TL, two males from 78–80 cm TL, plus a head of a female of 200 cm TL, from Cuba.

Synonyms: *Eulamia springeri* Bigelow & Schroeder (1944). Holotype: U.S. National Museum, Washington, USNM 37141. Type material is a female of 86 cm TL, from Cozumel Island, Yucatan, Mexico.

Diagnostic features. Large and robust species, an interdorsal ridge is present, which is normally very low. Upper precaudal pit well developed, lower pit weak. Internarial width 0.9–1.3 times the preoral length. Snout short, broadly rounded, eyes circular and relatively large (diameter 2.0 % TL). Eyes anterior margin above or slightly anterior to the front of mouth. Nostrils strongly oblique, strongly ellipsoid to slit-like. Anterior nasal flaps low, with weakly developed, pointed lobes. Upper labial furrows short and inconspicuous. Hyomandibular pores alongside corners of mouth inconspicuous. Gill slits large (2.4–3.2 % TL). First dorsal fin large (height 10.8 % TL), semi-falcate, with a narrowly pointed or rounded apex, its posterior margin curving ventrally from fin apex. D1 origin over or slightly anterior to pectoral free rear tips, inner margin of the first dorsal fin relatively long. Second dorsal fin large and high (height 3.1 % TL), with a short inner margin. D2 origin over or a little bit anterior to anal fin origin. Pectorals large and long (length of anterior margin 20.4 % TL), semi-falcate, its apices narrowly rounded or pointed. Vertebrae: 106 precaudal centra and 104 caudal centra on average. Dermal denticles closed set, imbricate, sub-circular, normally wider than long, with three to five longitudinal ridges and a corresponding number of posterior cusplets in smaller specimens. Larger specimens (adult and sub-adult) have seven longitudinal ridges and posterior cusplets.

Dental Formula:

$$\frac{13 - 1 - 13}{12 - 1 - 12} \text{ to } \frac{12(13) - 1(2) - 12(13)}{10(12) - 1 - 11(12)}$$

Upper teeth are narrow, crowns triangular, semi-erect to oblique. Distal median edges concave, strongly notched, mesial edges concave, weakly notched. Both edges are fine and uniformly serrated apically but with slightly coarser serration basally, roots are arched, deeply notched. One or two rows of symphysial teeth are present. Lower teeth are narrow, high crowns, erect to slightly oblique. Both median edges concave, notched, fine and uniformly serrated. Roots are arched (S to LP rows) to almost straight (P and C rows). One row of symphysial teeth is present. Files in upper jaw are arranged as mixed alternate i.e. imbricate overlapping dentition and in lower jaw as mixed alternate overlapping dentition.

Colour. Dark-grey or grey-brown dorsally, with a tinge of bronze in colouration laterally, ventrally white. Underside of paired fins, anal fin and upper caudal lobe are dusky (not black!) but often colouration is inconspicuous. Young have mostly dark coloured fins. Occasionally, an inconspicuous white band on flanks is present.

Size. Maximum of 300 cm TL for males and of 295 cm TL, and an average of 200–250 cm TL (LIESKE & MYERS 1994). Maximum reported body mass of 69.9 kg (IGFA 2001).

Misidentifications. The species could be confused with Galapagos sharks, *C. galapagensis*, with *C. obscurus*, *C. plumbeus* and *C. altimus*. Indeed, all mentioned species share similar body-features but *C. perezii* differs from the other species in the average number of precaudal vertebrae (106 in *C. perezii versus* 107 in *C. galapagensis*, 91 in *C. obscurus*, 90 in *C. plumbeus* and 109 in *C. altimus*). In addition, the tooth-shape of Caribbean reef sharks is different compared to all other mentioned species.

Geographical Distribution. Western Atlantic: alongside off southern Florida, Gulf of Mexico, Caribbean Sea, Yucatan, Cuba, Jamaica, Bahamas, Bermudas, Virgin Islands, Puerto Rico, Venezuela to southern Brazil. – Western Pacific: Colombia. – (GARRICK 1982, COMPAGNO 1984, ACERO & SANTOS-MARTINEZ 1992, SAZIMA & MOURA 2000, COMPAGNO et al. 2005, GARLA et al. 2006a, b).

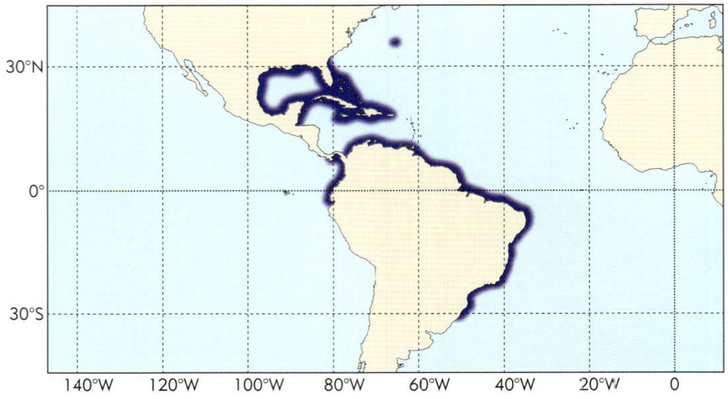

Habitat. *Carcharhinus perezii* occur inshore near the sea bed, and is typically found over continental and insular shelves, from surface to depths of at least 40–65 m depth (GARLA et al. 2006a, b).

Biology. This species is the most common and abundant shark species associated with coral reefs in the Caribbean Sea. Caribbean reef sharks inhabit mainly shallow waters, and are commonly found close to drop-offs on the outer edges of coral reefs. This species was reported to lie in caves and on the bottom, motionless, with its pharynx and gills evidently capable of pumping water for respiration. This behaviour has also been observed in caves off the coast of Mexico and off the Brazilian archipelago (PIKITCH et al. 2005, GARLA et al. 2006a, b).

Reproduction. Males reach maturity at 150–170 cm TL. Females reach maturity at 200–240 cm TL. Reproductive mode is placental viviparous (DULVY & REYNOLDS 1997). Normal are numbers of young per litter is 4–6. At birth, young are below 73 cm in TL (BONFIL 1989). Parturition seems to occur from February to April but few facts on the biology of the species are well known yet.

Food. Poorly known, but probably the species feed on bony fishes, e.g. Priacanthidae. Juveniles can be preyed by larger sharks, e.g. *Galeocerdo cuvier* and *C. leucas*.

Red List status: NT (IUCN 2010).

Carcharhinus plumbeus (NARDO 1827)

En: Sandbar shark, Thickskin shark; Fr: Requin gris; Sp: Tiburon trozo; Ge: Sandbankhai, Großflossenhai, Atlantischer Braunhai

Squalus plumbeus NARDO (1827). Holotype: No specimens mentioned (GARRICK 1982). Paratype: RMNH 7381. Type material is a female embryo of 47.7 cm TL, from Aru Islands, Indonesia.

Synonyms: *Carcharias (Prionodon) milberti* VALENCIENNES (in MÜLLER & HENLE 1839), *Carcharias ceruleus* DEKAY (1842), *Squalus caecchia* NARDO (1847 a, b), *Carcharias (Prionodon) japonicus* TEMMINCK & SCHLEGEL (1850), *Carcharias obstusirostris* MOREAU (1881) (a misspelling), *Carcharias stevensi* OGILBY (1910 a, b, 1911), *Carcharhinus latistomus* FANG & WANG (1932), *Galeolamna dorsalis* WHITLEY (1944 a, b).

Synonyms in recent use: *Carcharhinus milberti* VALENCIENNES (in MÜLLER & HENLE 1839), TANIUCHI (1971), WASS (1973), *Eulamia milberti* VALENCIENNES (in MÜLLER & HENLE 1839), *Galeolamna stevensi* OGILBY (1910 a, b, 1911), *Carcharhinus japonicus* SCHLEGAL (1850), *Carcharhinus bleekeri* DUMERIL (1865), *Carcharhinus platyodon* POEY (1861).

Diagnostic features. Mid-sized to robust species, a low distinct interdorsal ridge is present. Upper precaudal pit well developed, lower pit weak. Internarial width 0.9–1.5 times the preoral length. Snout short, broadly rounded, eyes circular and large (diameter 2.2 % TL), eyes anterior margin above or slightly anterior to the front of mouth. Nostrils strongly oblique, incurrent apertures broad and ovate or "hook"-shaped. Anterior nasal flaps very short, with strongly

Figure 43. *Carcharhinus plumbeus*: 183 cm TL male from Hawaiian Islands; **a**, left side; **b**, ventral view; **c**, dermal denticles; **d**, enlarged left nostril; **e**, left upper and lower teeth (symphysis to the left), and inset are the enlarged fifth upper and lower teeth; **f**, juvenile female of 81 cm TL, from the Atlantic coast of the United States.

developed lobes. Upper labial furrows short and inconspicuous. Hyomandibular pores alongside corners of mouth inconspicuous. Gill slits large (3.1–3.3 % TL). First dorsal fin large (height 10.9 % TL), very high in adults, erect to semi-falcate, with bluntly pointed or rounded apex. Its posterior margin curving ventrally from fin apex. D1 origin over or slightly anterior to pectoral insertion, inner margin of the first dorsal fin relatively long. Second dorsal fin large and high (height 2.9 % TL), with a short inner margin. D2 origin over or a little bit anterior to anal fin origin. Pectorals large and long (length of anterior margin 18.8 % TL), semi-falcate and apices narrowly rounded or pointed. Vertebrae: 90 precaudal centra and 85 caudal centra on average. Dermal denticles of juveniles and sub-adults wide-spaced, not imbricate, and only three longitudinal ridges and a corresponding number of short posterior cusplets are distinct. In contrast, denticles of adults are more closed set and partially overlapping, large, ovoid in shape and wider than long, with three to five ridges and feeble and scarcely projecting main cusps.

Dental Formula:

$$\frac{14-1-14}{13-1-13} \quad \text{to} \quad \frac{13(15)-1(2)-13(15)}{12(15)-1(2)-12(15)}$$

Upper teeth are broad, crowns triangular, semi-erect to oblique. Distal median edges concave, slightly notched on the outermost three or four rows. Mesial edges straight to weakly convex on the central part of crowns. Both edges are strongly serrated, slightly coarser serrated basally, roots arched, deeply notched. One or two rows of symphysial teeth are present. Lower teeth are narrow, crowns semi-erect to oblique. Both median edges concave, very finely and uniformly serrated. The apical halves of crowns in anterior to lateral rows are finely serrated, and completely serrated in LP to C rows. Roots are slightly arched, deeply notched, dividing the roots almost into two root lobes. One or very seldom two rows of symphysial teeth are present. Files in upper jaw are arranged as alternate overlapping dentition and in lower jaw as mixed alternate i.e. imbricate overlapping dentition.

Colour. Grey-brown to bright brownish above, ventrally white. In contrast, other sources describe the body colouration as pale blue to greenish-grey. Tips on the underside of the paired fins, on anal fin and ventral caudal lobe are present. Fins posterior margins are often dusky but actually without prominent markings, and an inconspicuous white band on flanks can be present.

Size. Maximum of 300 cm TL, and an average of 240 cm TL, with an average body mass of 45–90 kg. The maximum reported body mass for the species is 117.9 kg (IGFA 1991).

Misidentifications. Sometimes sandbar sharks are mistaken for *C. obscurus*, *C. altimus* and *C. leucas* but *C. plumbeus* differ from the other mentioned species in having D1 origin over or in front of the pectoral axis. In comparison to the very similar bignose shark, *C. altimus*, the sandbar shark differs in tooth-shape and in length of the upper teeth. Furthermore, *C. plumbeus* differs from the other species in having less attenuated nasal lobes, in snout length and in the average number of precaudal vertebrae (90 in *C. plumbeus* versus 109 in *C. altimus*, 116 in *C. leucas* and 91 in *C. obscurus*).

Geographical Distribution. Western Atlantic: from Massachusetts to Florida, Bahamas, Gulf of Mexico, Caribbean Sea, Cuba, Nicaragua, Costa Rica, Venezuela and southern Brazil. – Eastern Atlantic: in the Mediterranean, Spain, Portugal, Canary and Cape Verde Islands, Morocco, along the North African coast (Senegal, Democratic Republic of Congo) to South Africa. – Western Indian Ocean: Red Sea, Gulf of Oman, Mozambique, Tanzania, Madagascar, Mauritius, Seychelles, South Africa. – Western Pacific: Indonesia, Vietnam, China, Taiwan, Korea, Japan, Australia, New Caledonia. – Central Pacific: Hawaiian Islands. Eastern Pacific: Galapagos Islands and Revillagigedo Islands. – (NARDO 1747 a, b; BASS et al. 1973, 1986; WASS 1973; FOURMANOIR 1976 a, b; BARANES & BEN-TUVIA 1978 a, b, c; BRANSTETTER 1981; GARRICK 1982; COMPAGNO 1984; CHIARAMONTE 1998; CARLSON 1999; HEMIDA & LABIDI 2002; COSTANTINI & AFFRONTE 2003; JOUNG SHOOU-JENG et al. 2004; COMPAGNO et al. 2005; SAIDÏ et al. 2007; LIPEJ et al. 2008).

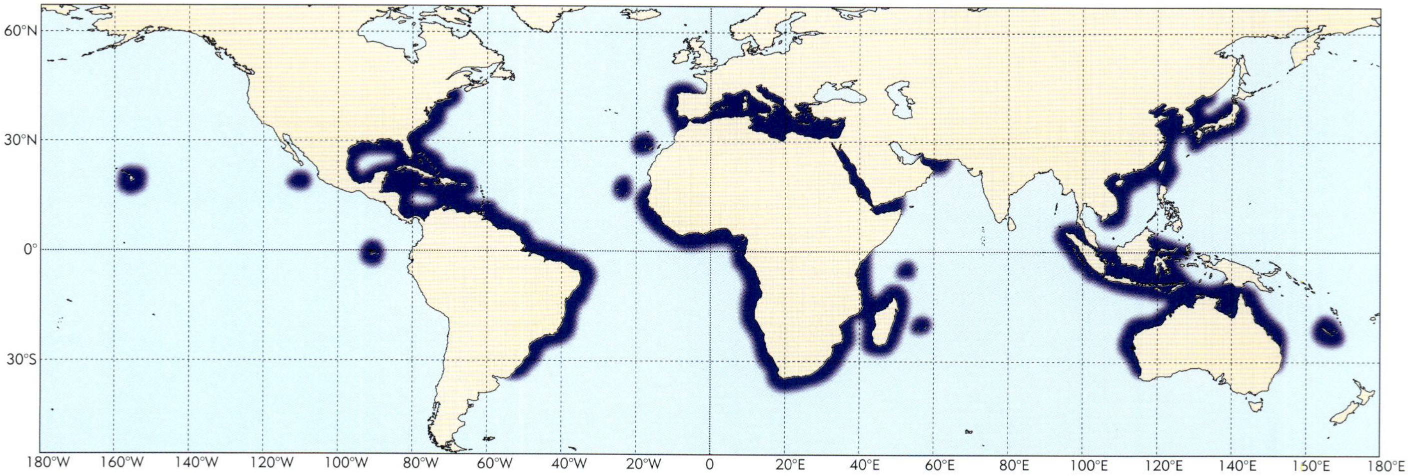

Habitat. *Carcharhinus plumbeus* is a common and abundant species in temperate and tropical waters, typically found over continental and insular shelves and in adjacent waters, coastal-pelagic from inshore to offshore living (MCAULEY et al. 2007).

Biology. Specimens of this species spend most of the time in water depths from 20–65 m and swim near the bottom up to 280 m depth. Sometimes individuals were observed near the surface, close to beaches or over sandy substrate. The species was reported from harbours, bays and river mouths, inside shallow muddy or grain-coarse substrate areas. Sandbar sharks invade river mouths and are able to tolerate decreasing salinity over a short time but are not really in brackish or freshwater living as e.g. bull sharks, *C. leucas*. The species feeds at day and night time but is more actively hunting at night.

[...] This species has an annual migration cycle along the western North Atlantic along the coastline off the US, heading south for the winter and north for the summer (RECHISKY & WETHERBEE 2003). Seasonal temperature changes apparently are a prime cause of these migrations but they are strongly influenced by the pattern of currents and locally by upwelling (COMPAGNO 1984). Off South Africa, similar migration movements along the coast are known.

Reproduction. Males reach maturity at 130–180 cm TL, females with a minimum of 140 cm TL. Reproductive mode is placental viviparous (DULVY & REYNOLDS 1997). Most embryos have attained a value of about 45–65 cm TL, after 10–12 months of development. At this size they could be readily separated from the placenta suggesting that they are full termed and ready to be born (JOUNG & CHEN 1995). Normal are 4–12 young per litter, with an average of 8 young, depending on the body size of the mother.

The young are born from February–April and the sex ratio of born individuals is 1 : 1. In the allopatric population of the western Atlantic off the US the ratio is 5 : 1 (females : males). It seems young are born only in every second year, and birth is close to the beach, whereby this is unusual because these places are completely different to the ordinary habitat of adult specimens. The young live separately to the adults in nursery grounds to avoid cannibalism by females before they depart from birth places (MERSON & PRATT 2001, MCAULEY et al. 2007).

Food. Mainly small, ground living bony fishes, but also bigger fishes are included in the food-spectrum, e.g. sea catfish, moray and snake eels, sardines, shad, anchovies, pipefish, barracudas, mullets, goatfish, bonitos, groupers, croakers, grunts, porgies, mackerels and flatfish, additionally molluscs, crustaceans (crabs), smaller shark species (*Rhizoprionodon* sp. and *Squalus* sp.), and rays.

Red List status: VU A2bd + 4bd (IUCN 2010).

Carcharhinus porosus (RANZANI 1839)

En: Smalltail shark; Fr: Requin tiqueue; Sp: Tiburon poroso; Ge: Atlantischer Zwerghai

Carcharias porosus RANZANI (1839). Holotype: MNHN 0000-1140. Type material is a male of 422 cm TL, from Cayenne, French-Guinea.

Synonyms: *Carcharias (Prionodon) henlei* VALENCIENNES (in MÜLLER & HENLE 1839), *Carcharhinus cerdale* GILBERT (1892, 1905), GILBERT & STARKS (1904).

Diagnostic features. Small, slender species, usually without interdorsal ridge. Upper precaudal pit well developed, lower pit weak. Internarial width 1.2–1.9 times the preoral length. Snout long and pointed. Eyes circular and large (diameter 2.1 % TL), eyes anterior margin above or slightly anterior to the front of mouth. Nostrils large and strongly oblique, incurrent apertures ovate and slit-shaped. Anterior nasal flaps expanded to slender pointed lobes. Upper labial furrows short and inconspicuous. Hyomandibular pores alongside corners of mouth inconspicuous. Gill slits mid-sized (2.3–3.2 % TL). First dorsal fin large, short (height 8.6 % TL) and falcate. Its apex bluntly pointed and the posterior margin curving ventral-laterally from fin apex, with a long inner margin. D1 origin over or slightly anterior to pectoral free rear tips. Second dorsal fin small and low (height 2.5 % TL), its inner margin relatively long. D2 origin over or slightly behind anal fin mid-base. Pectorals short and falcate (length of anterior margin 14.9 % TL), apices narrowly rounded or pointed. Anal fin deeply notched. Vertebrae: 59 precaudal centra and 65 caudal centra on average. [...] Dermal denticles scarcely or not overlapping, sub-circular in small specimens, but noticeably wider than long in large specimens, each with three longitudinal ridges and a corresponding number of posterior cusplets, the latter greatly reduced in denticles from adults (GARRICK 1982).

Dental Formula:

$$\frac{14-1-14}{13-0-13} \quad \text{to} \quad \frac{13(15)-1(2)-13(15)}{12(15)-0(2)-12(15)}$$

Upper teeth are narrow, crowns semi-erect to strongly oblique, except symphysial and anterior rows. Distal median edges straight on apical part of crowns, convex on basal part, strongly notched. Mesial edges are straight to slightly concave. Both edges finely and uniformly serrated, except the distal edges basally where serration is coarser. Roots are arched, deeply notched. One or two rows of symphysial teeth are present. Lower teeth are narrow, crowns semi-erect (S to LP rows) to oblique (P to C rows). Distal median edges concave, deeply notched on basal third of crowns, fine and uniformly serrated. Mesial edges are concave, finely and uniformly serrated also, but a little bit coarser serrated basally. Roots are arched, slightly or deeply notched. Normally, one or two, occasionally even none row(s) of symphysial teeth are present, but because the two central symphysial teeth are much smaller than those laterally adjacent and hence they could easily be overlooked (DIAZ 1979, GARRICK 1982). Files in upper and lower jaw are arranged as mixed alternate i.e. imbricate overlapping dentition.

Figure 44. *Carcharhinus porosus*: 114 cm TL female from Florida; **a**, left side; **b**, ventral view; **c**, dermal denticles; **d**, enlarged left nostril; **e**, left upper and lower teeth (symphysis to the left), and inset are the enlarged sixth upper and lower teeth; **f**, juvenile male of 38 cm TL from Brazil.

Colour. Grey above, white ventrally, without prominent markings on fins, but the apices of pectorals, pelvic fins, dorsal and caudal fin can be dark to black coloured. Often the anal fin is coloured with a tinge of red and sometimes an inconspicuous white band on flanks is present.

Size. Maximum of 150 cm TL, and an average of 90–120 cm TL (COMPAGNO 1984).

Misidentifications. It is easier to confuse *C. porosus* with *Rhizoprionodon* sp. than with other species of the genus *Carcharhinus*. The smalltail shark, *C. porosus*, differs from *Rhizoprionodon* sp. in lacking enlarged lines of hyomandibular pores directly behind mouth corners, in tooth-morphology and having in shorter labial furrows (0.7 % TL) which are not visible when mouth is closed, whereas in *Rhizoprionodon* sp. they are never less than 1.0 % TL and always visible. Moreover, only two requiem shark species can be confused with *C. porosus*, *C. acronotus* and *C. borneensis*. In comparison to *C. porosus*, blacknose sharks differ in the position of the D2 origin which is more reward and slightly posterior to anal fin origin. Borneo sharks differ in having noticeably enlarged lines of hyomandibular pores directly behind corners of the mouth, also in the height of the second dorsal fin, in length of the fins rear tip, and in dental formula. Moreover, the species has a different average number of precaudal vertebrae (59 in *C. porosus versus* 84 in *C. acronotus* and 62 in *C. borneensis*).

Geographical Distribution. Western Atlantic: Florida, Northern Gulf of Mexico, off the Maranhao coast, along the coastline to northern Brazil. – Eastern Pacific: California to Peru but not distributed at offshore islands. – (DIAZ 1979; GARRICK 1982; COMPAGNO 1984; BATISTA & SILVA 1995; LESSA et al. 1997, 1998, 1999c; COMPAGNO et al. 2005).

Habitat. Commonly it is an inshore, in tropical waters, near the American continental shelves living species. Normally, found near the bottom but was reported from surface (inshore) down to at least

40 m depths. The species has an affinity to muddy bottom, is particularly able to invade estuaries and river mouths and can tolerate low salinity over short time (LESSA et al. 1997, 1998, 1999c).

Biology. Although it is an abundant species only few facts are known about it. The largest specimen caught ever (its age calculated from vertebrae rings) was a female of 101 cm TL and an age of 12 years (BATISTA & SILVA 1995, LESSA et al. 1999c).

Reproduction. Both sexes reach maturity with 70–80 cm TL, corresponding to an age of 6 years. Reproductive mode is placental viviparous (DULVY & REYNOLDS 1997). The common number of young per litter is 2–7, with a maximum of 16. A significant relationship between the number of embryos and female size was proven (LESSA et al. 1997, 1998, 1999c). At birth, young have 30–40 cm TL. A minimum duration of 10 months for gestation is known. In Brazilian waters young are born in late spring or summer, close to the coast, and are slow growing (LESSA et al. 1999c).

Food. Mainly pelagic, but demersal bony fishes also, e. g. sardines, sea catfish, croakers, jacks, grunts, other small shark species (juvenile *Sphyrna* sp. and *Rhizoprionodon* sp.) and shrimps.

Red List status: DD (IUCN 2010).

Carcharhinus sealei (PIETSCHMANN 1916)

En: Blackspot shark; Fr: Requin a taches noires; Sp: Tiburon alinegro; Ge: Schwarzfleckhai

Carcharias sealei PIETSCHMANN (1916). This is the replacement name for *Carcharias borneensis* SEALE (1910). Holotype: Bureau of Science, Manila, the Philippines, No. 2720. Type material is a specimen (sex unknown) of 37.2 cm TL, from Sandakan, Borneo (but apparently lost, GARRICK 1982).

Synonyms: *Carcharias borneensis* SEALE (1910), *Platypodon coatesi* WHITLEY (1939).

Diagnostic features. Small, slender species, very seldom an interdorsal ridge is present which is usually not pronounced or very low only. Upper precaudal pit well developed, lower pit weak. Internarial width 1.1–1.9 times the preoral length. Snout short, pointed to rounded, eyes normally large (diameter 2.4 % TL), horizontally oval. Anterior margin of the eyes slightly posterior to the front of mouth. Nostrils strongly oblique, incurrent apertures broadly ovate. Anterior nasal flaps elongated to broad, triangular pointed lobes. Upper labial furrows short and inconspicuous. Hyomandibular pores alongside corners of mouth inconspicuous. Gill slits short (2.2–2.8 % TL). First dorsal fin high (height 9.7 % TL), falcate, its inner margin short, and its apex normally pointed or narrowly rounded. D1 origin over or slightly anterior pectoral free rear tips (occasionally slightly posterior to it). Second dorsal fin short and high (height 3.6 % TL), with a short inner margin. D2 origin slightly or distinct posterior to anal fin origin. Pectorals small, short and falcate (length of anterior margin 15.5 % TL), apices narrowly pointed. Vertebrae: 78 precaudal centra and 76 caudal centra on average. Dermal denticles closed set, arranged a little bit imbricate (more in adults than in juveniles), sub-circular, with three longitudinal ridges and three to five fairly strong posterior cusplets in smaller specimens. In contrast, adults have more rhomboid or ovoid-shaped denticles, and five ridges and posterior cusplets.

Dental Formula:

$$\frac{12-2-12}{12-1-12} \quad \text{to} \quad \frac{12(13)-(0)2-12(13)}{11(13)-0(2)-11(13)}$$

Sexual dimorphism in tooth-morphology is known within the species. Upper teeth are narrow in mature males, crowns oblique to relatively broad in females and immature males. Distal median edges convex on apical half of crowns, convex on basal part of crowns also, deeply notched, smooth-edged in smaller specimens, strongly serrated in larger individuals. Mesial edges are convex basally or straight to concave apically, finely serrated, a little bit coarser basally. Roots are slightly arched, slightly notched. One or two rows of symphysial teeth are present, seldom without. Lower teeth are narrow, crowns oblique, except S and A rows. Distal median edges straight apically, convex basally, deeply notched, serrated. Mesial edges concave, fine and uniformly serrated or smooth-edged. Some of the LP rows in adult females are large and irregularly serrated basally on the distal median edges, foreshadowed in smaller specimens. Roots are slightly arched, slightly notched. One, two or even none symphysial row(s) present. Files in upper and lower jaw are arranged as mixed alternate i. e. imbricate overlapping dentition.

Colour. Grey to brownish dorsally, bright-pale ventrally. A prominent and very conspicuous black blotch on the apex of D2 is present, with a narrow, bright anterior margin. Posterior edges of the other fins are pale but not black. A white band over flanks is commonly present but sometimes it is inconspicuous.

Size. Maximum of 100 cm TL, with an average of 60–90 cm TL (TORRES 1991).

Figure 45. *Carcharhinus sealei*: 89 cm TL female from the Philippines; **a**, left side; **b**, ventral view; **c**, enlarged left nostril; **d**, left upper and lower teeth (symphysis to the left), and inset are the enlarged sixth upper and lower teeth; **e**, juvenile female of 36 cm TL from Natal, South Africa.

Misidentifications. *Carcharhinus sealei* could easily be confused with *C. dussumieri* due to the similar D2 colouration but *C. sealei* differs in shape and height of D1 which is more falcate. Further, both species differ in dental formula from another. Usually, enlarged lateral cusplets on the upper teeth are distinct in the dentition of *C. sealei*, not serrated in themselves. The mouth of blackspot sharks are narrower than in *C. dussumieri*, and the species has a different ratio of pectoral fin height : pectoral fin width. Confusion in regard of tooth-shape is also possible with *C. hemiodon*, *C. macloti*, and *C. wheeleri* but all mentioned species have no prominent smooth-edged lateral cusplets. Moreover, confusion among *C. sealei* and *C. wheeleri* is conceivable because both species have a comparable colouration of fins. However, only the apical spot of D2 of *C. sealei* is dark whereas D1 of *C. wheeleri* has a white spot and apices of D2 and anal fin are dark. All mentioned species differ in the average number of precaudal vertebrae (78 in *C. sealei versus* 62 in *C. dussumieri*, 70 in *C. hemiodon*, also 70 in *C. macloti* and 114 in *C. wheeleri*).

Geographical Distribution. Indo-West Pacific: Mozambique, Tanzania, Kenya, Seychelles, Mauritius, Madagascar, South Africa, Pakistan, Sri Lanka, India, Andaman Islands, Thailand, Malaysia, Singapore, Vietnam, China, Sumatra, Java, Sulawesi, Borneo, Philippines, New Guinea, Australia (Western and Northern Australia, Queensland). – (Bass et al. 1973, 1986; Garrick 1982; Compagno 1984; Compagno et al. 2005).

Habitat. It is an abundant, commonly coastal living species which occur over continental and insular shelves. Mostly found close to the

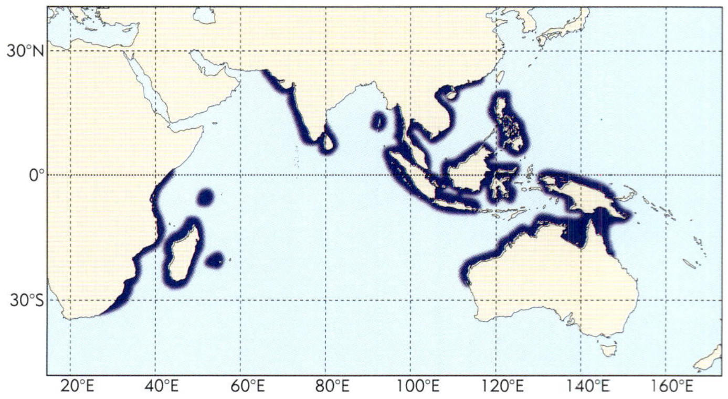

shore, from surface and intertidal zones to 40 m depths. *Carcharhinus sealei* seems to have an affinity to shallower waters (COMPAGNO 1984, COMPAGNO & NIEM 1998).

Biology. Although this is a common and abundant species only a few facts are known about it. It seems blackspot sharks tend to move out of an area when salinity is decreasing. Off South Africa specimens of this species are resident the whole year with increasing numbers during the summer (DARRACOTT 1977).

Reproduction. Males reach maturity at 70–80 cm TL. Females reach maturity at 70–75 cm TL, corresponding to an age of 1–2 years. Reproductive mode is placenta viviparous (DULVY & REYNOLDS 1997). The average number of young per litter is 1–2. At birth, young have 30–36 cm TL (GARRICK 1982). In contrast to this opinion, BASS et al. (1973) gave values for the length at birth of up to 45 cm TL. Duration of nine months for gestation was calculated. It is a rapid growing species, and a maximum age of at least five years was calculated.

Food. Mainly small bony fishes (including sea horses) and invertebrates (squid and prawns).

Red List status: NT (IUCN2010).

Carcharhinus signatus (POEY 1868)

En: Night shark; Fr: Requin de nuit; Sp: Tiburon nocturno; Ge: Nachthai

Hypoprion signatus POEY (1868). Holotype: MNHN 55-4915. Type material is a female of 159 cm TL, from French Guinea and Portuguese Guinea.

Synonyms: *Hypoprion longirostris* POEY (1876), *Hypoprion bigelowi* CADENAT (1956).

Diagnostic features. Large species, an interdorsal ridge is present. Upper precaudal pit well developed, lower pit weak, internarial width 1.0–2.2 the preoral length. Snout long, very pointed (considering ventrally snout is V-shaped), eyes circular and large (diameter 2.3 % TL), eyes anterior margin slightly anterior to the front of mouth. Nostrils strongly oblique, incurrent apertures long and slit-shape. Anterior nasal flaps low, with narrow pointed lobes. Upper labial furrows short and inconspicuous. Hyomandibular pores alongside corners of mouth inconspicuous. Gill slits short (1.9–3.1 % TL). First dorsal fin small and low (height 7.6 % TL), triangular in shape, with a short inner margin. Apex narrowly rounded or pointed, its posterior margin curving posterior-ventrally from apex. D1 origin varies from above middle of inner pectoral margin to being behind pectoral free rear tips. Second dorsal fin low but long (height 1.8 % TL), its inner margin long. D2 origin over or slightly posterior to anal fin origin. Pectorals long and falcate (length of anterior margin 17.9 % TL), with rounded apices. Vertebrae: 101 precaudal centra and 86 caudal centra on average. Dermal denticles imbricate, sub-circular to ovoid in shape, a little bit wider than long, with three strong longitudinal ridges and with a corresponding number of posterior cusplets in smaller specimens. Adults have five ridges and corresponding number of posterior cusplets.

Dental Formula:

$$\frac{15-2-15}{15-1-15} \quad \text{to} \quad \frac{15(16)-2-15(16)}{15-1-15}$$

[...] The upper jaw tooth-morphology of the night shark, *C. signatus*, undergoes marked ontogenetic change, with symphysial serrations increasing in both size and number relative to the commissural edge of each tooth. Similarly, placoid scale keel number increases with increasing total length (RASCHI et al. 1982). Upper teeth are narrow, crowns erect (S and A rows) to more oblique (AL to C rows). Distal median edges straight or oblique apically, convex basally, deeply notched. Mesial edges are straight to slightly convex basally, convex apically (P to C rows). Both edges are strongly or irregularly serrated basally, apically more finely serrated or smooth-edged. Median edges of newborn specimens smooth-edged. Roots are arched, deeply notched, dividing the roots into two root lobes. Two rows of symphysial teeth are present. Lower teeth are narrow, with high crowns, erect to slightly oblique. Both median edges concave, deeply notched in the basal third of crowns, smooth-edged. Roots are arched, deeply notched, dividing the roots almost into two root lobes. One row of symphysial teeth is present. Files in upper and lower jaw are arranged as alternate overlapping dentition.

Colour: Grey above, white ventrally. Sometimes small black spots are scattered on the whole body. Dusky apices on the undersides of pectorals are present, increasingly darker from inner parts to outer tips. Usually, juveniles lacking conspicuous markings but often with dusky tips on apical margins of D1 and D2, on anal fin, on pectorals and on upper caudal lobe are present.

Size. Maximum of 280 cm TL, and an average of 200–250 cm TL (COMPAGNO 1984). Maximum reported body mass of 76.7 kg (IGFA 2001).

Misidentifications. Night sharks are frequently confused with *C. hemiodon* and *C. macloti* because the three species share similar body size features. *Carcharhinus signatus* differ from the mentioned species in its green eye colour (the eyes illuminate in the dark), and in tooth-shape. Furthermore, the average number of precaudal vertebrae of the mentioned species is different (101 in *C. signatus* versus 70 in *C. hemiodon* and 70 in *C. macloti*)

Geographical Distribution. Western Atlantic: Delaware to Florida, Bahamas and Cuba, southern Brazil and Argentina. – Eastern Atlantic: Senegal, Ivory Coast, Ghana, Democratic Republic of Congo, Angola. – Maybe the species is distributed in the eastern Pacific Ocean too, then they have travelled to the area via Panama Canal but this issue to distribution pattern has not been confirmed yet. – (BRANSTETTER 1981; RASCHI et al. 1982; COMPAGNO 1984; GARRICK 1985; HAZIN et al. 2000; COMPAGNO et al. 2005; CARLSON et al. 2008).

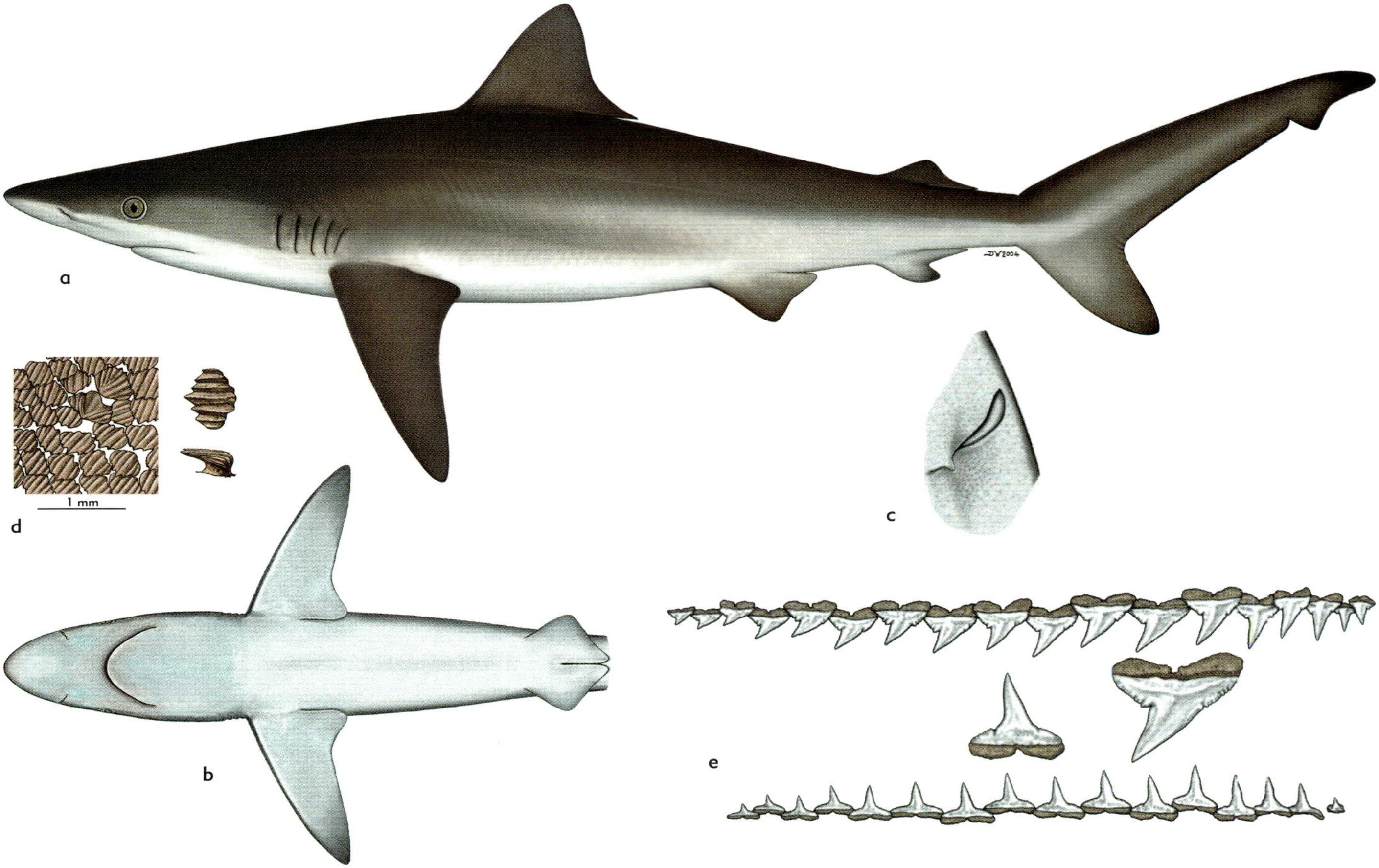

Figure 46. *Carcharhinus signatus*: 179 cm TL female from Senegal, West Africa; **a**, left side; **b**, ventral view; **c**, enlarged left nostril; **d**, dermal denticles; **e**, left upper and lower teeth (symphysis to the left), and inset are the enlarged sixth upper and lower teeth.

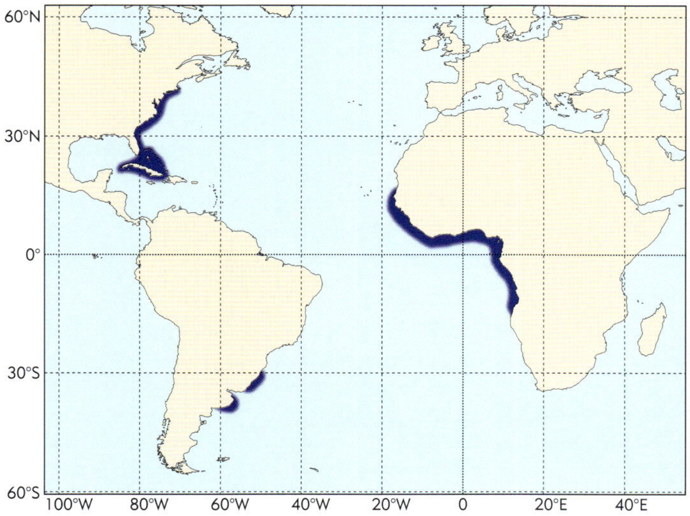

Habitat. It is a common coastal and semi-oceanic living species, which occur typically over deep waters along the outer continental and insular shelves of the tropical and warm-temperate waters of the Atlantic Ocean (maybe the species is in eastern Pacific distributed also). Night sharks show distinct vertical migration patterns. They occur in 275–365 m depth at day, and 185 m by night, but some single specimens were caught in depths of more than 600 m (Hazin et al. 2000, Santana & Lessa 2004).

Biology. Only few facts about this species are known. Night sharks live together in larger schools, and the species is known to be active and fast swimming (Compagno 1984).

Reproduction. Males reach maturity at 185–190 cm TL (corresponding to an age of 8 years), females reach maturity at 200–205 cm TL (age of 10 years) (Santana & Lessa 2004). Reproductive mode is placental viviparous (Dulvy & Reynolds 1997). The number of young per litter is 4–18. At birth, young have 10–40 cm TL, whereas other sources give the length at birth of 60–70 cm TL (Hazin et al. 2000, SANTANA & LESSA 2004). Copulation occurs in summer (HAZIN et al. 2000). The duration of gestation is unknown. The young are born in deep water and they are rapid growing. Other facts of reproduction cycle are poorly known.

Food. Mainly small and fast-swimming bony fishes, e.g. flying fishes, perch, scombrids, butterfishes, and sea basses, also squid and shrimps (FROESE & PAULY 2010).

Red List status. VU A2abd + 3bd + 4cd (IUCN 2010).

Carcharhinus sorrah (VALENCIENNES 1839)

En: Spottail shark; Fr: Requin à queue tachetee; Sp: Tiburon rabo manchado; Ge: Sorrahhai

Carcharias (Prionodon) sorrah VALENCIENNES (in MÜLLER & HENLE 1839). Lectotype: RMNH 4294. Type material is a female of 57 cm TL, from India (designated by GARRICK 1982).

Synonyms: *Squalus spallanzani* PERON & LESUEUR (1822), *Carcharias (Prionodon) bleekeri* DUMERIL (1865), *Carcharias taeniatus* HEMPRICH & EHRENBERG (1899), *Galeolamnoides isobel* WHITLEY (1947), *Carcharhinus spallanzani* DOR (1984).

Diagnostic features. Mid-sized to fairly large species, robust to spindle-shaped. Usually, a low, narrow, interdorsal ridge is present. Upper precaudal pit strongly developed, lower pit weak. Internarial width 1.2–1.7 times the preoral length. Snout long, rounded or pointed. Eyes fairly large (diameter 2.1 % TL), eyes anterior margin slightly anterior to the front of mouth. Nostrils strongly oblique, slit-shaped, incurrent apertures large, ovate-shaped. Anterior nasal flaps narrow, elongated and pointed to nipple-shaped lobes. Upper labial furrows short and inconspicuous. Hyomandibular pores alongside corners of mouth inconspicuous. Gill slits short (length 2.0–2.6 % TL). First dorsal fin falcate, low (height 8.5 % TL), pointed or narrowly rounded apex, its posterior margin curving ventrally from fin apex, and with a short rear tip. D1 origin can vary from slightly anterior (seldom) to posterior of free pectoral rear tips, its inner margins of first dorsal moderately long. Second dorsal fin small and low (height 1.8 % TL), with a long rear tip. D2 origin slightly behind origin of anal fin to about one-third back along anal base. Pectorals small (length of anterior margin 15.6 % TL), falcate, narrowly rounded or pointed apices. Vertebrae: 70 precaudal centra and 88 caudal centra on average. The number of vertebrae of individuals from different locations can vary strongly. Dermal denticles closed set, imbricate, sub-circular in smaller specimens. Nearly rhomboid in larger individuals, each with three to five longitudinal ridges and a corresponding number of low posterior cusplets (in smaller specimens) and seven posterior cusplets in larger individuals.

Dental Formula:

$$\frac{12 - 1 - 12}{12 - 1 - 12} \quad \text{to} \quad \frac{12(13) - 1(3) - 12(13)}{(11)12 - 1(2) - 11(12)}$$

Upper teeth are broad, with high crowns, oblique (except S and A rows). Distal median edges straight apically, convex basally, deeply notched halfway longitudinally. Mesial edges are straight to convex. Both edges are fine and uniformly serrated apically but more coarsely serrated on central part of crowns and on distal edges basally. Serration is more distinct in juveniles and sub-adults than in adults. Roots are arched, slightly notched. One to three rows of symphysial teeth are present. Lower teeth are narrow, crowns erect (S to L rows) to more oblique (LP to C I rows). Both median edges concave, slightly notched on mesial edges, more deeply notched on distal edges. Roots are arched, slightly notched. One or two rows of symphysial teeth are present. Files in upper and lower jaw are arranged as alternate overlapping dentition.

Colour. Dark grey to brownish-grey above, pale to white ventrally, with a large, prominent and conspicuous black tip on ventral caudal lobe, on pectoral fins, and on the apex of D2. Only the apex of the first dorsal fin has a very narrow, black margin.

Size. Maximum of 130–160 cm TL, and an average of 110–130 cm TL for males, i.e. of 130–160 cm TL for females (RANDALL et al. 1990). The average total length of the species varies strongly in dependence of specific habitat where the measured and examined individuals are from (STEVENS & WILEY 1986). The maximum reported body mass was estimated for the species to be 28.0 kg (KAILOLA et al. 1993).

Misidentifications. The species could be confused with *C. falciformes* because of having a very low and attenuate second dorsal fin, but it is easily to distinguish by its colour pattern, tooth-shape and dental formula. Furthermore, both species have a different average number of precaudal vertebrae (70 in *C. sorrah versus* 102 in *C. falciformes*).

Geographical Distribution. Indo-West Pacific: Red Sea, Mozambique, South Africa, Madagascar, Mauritius, Seychelles, Pakistan, India, Sri Lanka, Indonesia, Singapore, Malaysia, Thailand, Vietnam, China, Philippines, Australia (western and northern Australian coast), Solomon Islands. – (BASS et al. 1973, 1986; FOURMANOIR 1976 a, b; GARRICK 1982; COMPAGNO 1984; LYLE 1984, 1987; STEVENS & WILEY 1986; COMPAGNO et al. 2005).

Habitat. It is a common coastal species, living near to shores in shallow, tropical waters over continental and insular shelves, from the surface to 140 m depth. The species is seldom found in offshore waters (SOMMER et al. 1996).

Biology. The species has a stronger affinity to habitats around reefs with a water depth range from intertidal to 70 m. Young individuals of *C. sorrah* occur in quiet, shallow water, separated from adults, but a distinct sexual segregation is not known within the species (SIMPFENDORFER & MILWARD 1993). [...] Live near the seabed during the day and near the surface at night. Individual movements are often short (50 km) but may be more than 1000 km (KAILOLA et al. 1993).

Figure 47. *Carcharhinus sorrah*: 141 cm TL female from the Gulf of Thailand; **a**, left side; **b**, ventral view; **c**, enlarged left nostril; **d**, left upper and lower teeth (symphysis to the left), and inset are the enlarged sixth upper and lower teeth; **e**, juvenile female of 70 cm TL, from Queensland, Australia.

Reproduction. Males reach maturity at 90–100 cm TL. Females reach maturity at 110–120 cm TL. Sexual differences in weight-length regression curves between both sexes are known, depending of specific habitat (Kasim et al. 1996). Reproductive mode is placental viviparous (Dulvy & Reynolds 1997). Number of young is 1–8 per litter, with a mean of four. At birth, young have 50–60 cm TL (Stevens & Wiley 1986, Davenport & Stevens 1988). Mating is in February–March, ovulation in March–April, main parturition period in January (depends on specific area). Testes weight is low between May–December, increasing rapidly to a peak in February and declined fast to May (Lyle 1984). Off Bombay birth season is in spring (March–May). Duration of gestation of 10–11 months was calculated.

Food. Mainly bony fishes, includes fast-swimming bonitos, also cephalopods, molluscs and crustaceans. *Carcharhinus sorrah* feed near reefs, and a change in food-spectrum associated to the increasing body size is known.

Red List status: NT (IUCN 2010).

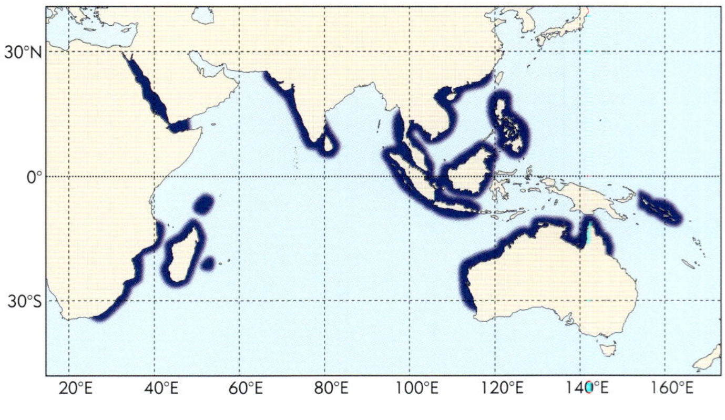

Carcharhinus tilstoni (WHITLEY 1950)

En: Australian blacktip shark; Fr: Requin borde de Australia; Sp: Tiburon macuira de Australia; Ge: Australischer Schwarzspitzenhai

Galeolamna pleurotaenia tilstoni WHITLEY (1950). Holotype: AMS, IB 2421. Type material is a gravid female of 154.5 cm TL, weight 55 lbs, caught near the Van Cloon Reef, Joseph Bonaparte Gulf, West Australia, by S. Fowler.

Synonyms: *Carcharias (Prionodon) pleurotaenia* BLEEKER (1852), JORDAN & SEALE (1906), *Eulamia (Platyodon) pleurotaenia* FOWLER (1940).

Authors note. Some morphometric data of the species cannot be given (Tab. 4) due to the unusual kind of measurements of the type material by WHITLEY (1950). The attempt was made to reconstruct the measurements of WHITLEY (1950) and to convey the data according to the method of measurement of BASS et al. (1973), GARRICK (1982, 1985), and COMPAGNO (1984, 1988), but in some cases this attempt failed. Hence, values for D1, D2 and anal fin posterior margins length, as well as for pelvic anterior margin length are deviant to the measurement method as shown in Figure 4 and 13, and in Table 4 those different values are marked with an asterisk.

Diagnostic features. Large, stout-tough species, without interdorsal ridge. Upper precaudal pit strongly developed, lower pit weak. Internarial width 1.1–1.6 times the preoral length. Snout long and acutely pointed to gothic-arched from above (WHITLEY 1950). Eyes circular and large (diameter 1.7 % TL), pupils a vertical slit, eyes anterior margin a little bit posterior to the front of mouth. Nostrils oblique and slit-shaped. Anterior nasal flaps low, triangular, with pointed lobes. Upper labial furrows short and inconspicuous. Hyomandibular pores alongside corners of mouth inconspicuous. Gill slits long (3.2–3.5 % TL). First dorsal fin large and falcate (height 10.4 % TL), with a pointed or narrowly rounded apex, its posterior margin curving ventrally from fin apex. D1 origin usually over or slightly posterior halfway of pectoral inner margin, but near pectoral free rear tip, its inner margin short. Second dorsal fin moderately high (height 2.9 % TL), its inner margin short. D2 origin over or slightly anterior to anal fin origin. Pectoral fins large narrow and falcate, narrow apices, rounded or pointed. Vertebrae: 88 precaudal centra and 88 caudal centra on average. Dermal denticles closed set, a little bit imbricate-overlapping and sub-circular in smaller specimens with three longitudinal ridges and three corresponding posterior cusplets. In larger specimen dermal denticles are more rhomboid, with five

Figure 48. *Carcharhinus tilstoni*: 168 cm TL female from Northern Territory, Australia; **a**, left side; **b**, ventral view; **c**, enlarged left nostril; **d**, fifth left upper and lower teeth.

to seven ridges and a corresponding number of weakly developed posterior cusplets, arranged more regularly in diagonal rows.

Dental Formula:

$$\frac{15 - 3 - 15}{15 - 1 - 15} \quad \text{to} \quad \frac{15 - (2)3 - 15}{(13)14 - 1(2) - 13(14)}$$

For the paratype specimen F (WHITLEY 1950), a different dental formula is given with

$$\frac{14 - 1 - 14}{12 - 1 - 12}$$

This specimen is an immature male of 76.5 cm TL, some teeth could not be placed in this stadium of age, and could be the reason for the mismatch.

Upper teeth are narrow, crowns erect to slightly oblique. Distal median edges concave and notched halfway longitudinally, mesial edges are weakly concave. Serration on both edges is fine and uniformly. On the central part of crowns, serration is coarser and often somewhat irregular basally. Roots are arched, slightly notched. Three, i. e. two rows of symphysial teeth are present. Lower teeth are narrow, with high crowns. Both median edges concave, notched, very finely and uniformly serrated. In newborn and juveniles the margins are smooth-edged. Roots are arched, deeply notched. One or two rows of symphysial teeth are present. Files in upper jaw are arranged as alternate overlapping dentition and in lower jaw as mixed alternate i. e. imbricate overlapping dentition.

Colour. Bright-grey or dusky grey-brown dorsally, with a tinge of bronze-brown-greyish, white ventrally. Almost all fins have more or less prominent black tips or edges, except pelvic fins and anal fin, which are normally plain. A black blotch on the apex of D2 and on the lower caudal lobe is, i. e. can be present. Laterally, a broad lateral bronze band is distinct. The dark upper tone is reaching the middle of the eye anteriorly and the top of eye posteriorly (WHITLEY 1950).

Size. Maximum of 200 cm TL, and an average of 150–180 cm TL (COMPAGNO & NIEM 1998). Maximum reported body mass for the species was 52 kg (KAILOLA et al. 1993).

Misidentifications. This species, the second form of the blacktip shark, was high frequently mistaken for *C. limbatus* because of the co-occurrence of both species in the same area and the very similar body-colouration. *Carcharhinus tilstoni* (the Australian blacktip shark) could also be confused with *C. amblyrhynchoides* or *C. brevipinna* but those species are different in their body-contour and in morphometric features, in tooth-morphology and in dental formula. *Carcharhinus tilstoni* differ from *C. limbatus* in colouration, in dental formula, and in the average number of the precaudal vertebrae (96 in *C. limbatus*, versus 82 in *C. brevipinna*, 91 in *C. amblyrhynchoides* and 88 in *C. tilstoni*). Both forms of blacktip shark co-occur in the subtropical and tropical areas of Australia and were divided after genetic relationship investigations (OVENDEN et al. 2009, 2010).

Carcharhinus limbatus was believed to be relatively rare in the northern part of Australia (occurrence ratio of *C. limbatus* to *C. tilstoni* of 1:300) but recent studies have revealed *C. limbatus* is not rare around Australia, and a species abundance ratio of approximately 50:50 was estimated (OVENDEN et al. 2010). Furthermore, studies of OVENDEN et al. (2010) and others provide genetic evidence *C. limbatus* and *C. amblyrhynchoides* are more closely related to each other than to *C. tilstoni*. This species seems to be more endemic than known before i. e. as suspected by COMPAGNO et al. (2005) and LAST & STEVENS (2009).

The Australian blacktip shark, *C. tilstoni*, differs from all mentioned species in the average number of precaudal vertebrae (88 in *C. tilstoni versus* 96 in *C. limbatus*, 91 in *C. amblyrhynchoides* and 82 in *C. brevipinna*).

Geographical Distribution. Western Pacific: Bigger and lesser Sunda Islands, Flores Sea, north-western shelf and northern coast of Australia, Gulf of Carpentaria, Great Barrier Reef (Northeast and eastern coast, Western Australia, Northern Territory, Queensland, New South Wales. – (STEVENS & WILEY 1986; DAVENPORT & STEVENS 1988; LAVERY & SHAKLEE 1989, 1991; SALINI et al. 1991; STOBUTZKI et al. 2002; COMPAGNO et al. 2005).

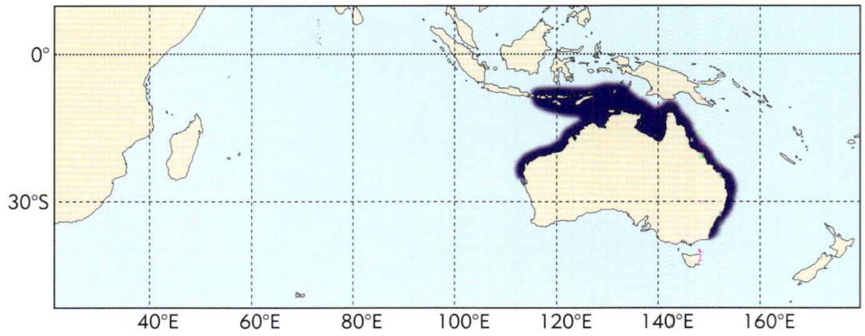

Habitat. It is an endemic species in tropical waters off Australia, pelagic and inshore living, and has an affinity to intertidal waters and to pelagic areas. The species is seldom found at depths of 150 m living, and occur over continental and insular shelf close to the bottom (LAVERY & SHAKLEE 1989, 1991).

Biology. *Carcharhinus tilstoni* can occur in large groups, sometimes composed predominantly by one sex or body size range. These groups have mostly a very limited spatial distribution. However, tagging studies have shown the maximal travelled distance was estimated to be 1345 km, whereas normally 60 % of tagged individuals were recaptured within 50 km of their tagging site (LAVERY & SHAKLEE 1989, SIMPFENDORFER & MILWARD 1993, STOBUTZKI et al. 2002).

It's a rapidly growing species, reaching sexual maturity in 3–4 years. Juveniles are not spatial restricted to nursery grounds as newborn and are widely distributed both in inshore and offshore areas. Significant differences between body size regression curves of males and females were not detectable (LAVERY & SHAKLEE 1989, 1991).

The maximum longevity of the species was estimated to be 20 years.

Reproduction. Males reach maturity at 110 cm TL. Females reach maturity at about 115 cm TL (STEVENS & WILEY 1986). Both sexes are fully matured with 3–4 years in age. Smallest matured virgin caught ever was a specimen of 120 cm TL. The reproductive mode is placental viviparous (DULVY & REYNOLDS 1997).

The duration of gestation was calculated to be 10 months. Average of young per litter is 3, with a range of 1–6. At birth, young have 60–75 cm TL (STEVENS & WILEY 1986).

Mating occurs in February–March, ovulation is in March–April and the main parturition period in January, but is depending on individuals age and the specific area. Testes weight is low between May–December, increasing rapidly to a peak in February and declining fast in May. [...] The length of claspers (in Paratype A, described by WHITLEY 1950) was measured with 11 cm, but this male was not fully mature. In Paratype D which is an immature specimen also, a clasper length of 18 cm was measured (DAVENPORT & STEVENS 1988).

Food. Mainly fish, but food-spectrum depends on body size of the shark specimen and density of the prey organisms. Changes in food-spectrum associated with an increasing body size are known. The species feeds actively on surface and mid-water living fishes, including fast-swimming pelagic species, e. g. *Thynnus tonggol*, also demersal prey, cephalopods (but here predation strongly depends of prey density), and other sharks, e. g. *Hemipristis* sp., *Rhizoprionodon* sp. and *C. sorrah* (KATO 1964, SALINI et al. 1991).

Red List status: LC (IUCN 2010).

Carcharhinus wheeleri GARRICK (1982)

En: Blacktail shark, Short-nosed Blacktail shark; Fr: Requin a queue noire; Sp: Tiburon rabinegro; Ge: Schwarzschwanz Riffhai, Schwarzschwanzriffhai

Carcharhinus wheeleri GARRICK (1982). Holotype: USNM 197418. Type material is a male of 132.2 cm TL, from the Red Sea (collected by E. CLARK, 1962).

Synonyms: *Carcharhinus amblyrhynchos* Bleeker (1856), *Carcharhinus spallanzani* Peron & LeSuer (1822).

Diagnostic features. Mid-sized, robust species, normally without an interdorsal ridge. Upper precaudal pit well developed, lower pit weak. Internarial width 0.9–1.4 times the preoral length. Snout long and broad bluntly rounded, eyes circular and fairly large (diameter 2.5 % TL), eyes anterior margin above or slightly anterior to the front of mouth. Nostrils strongly oblique, incurrent apertures broad and ellipsoid-shaped. Anterior nasal flaps expanded as very low, triangular lobes. Upper labial furrows short and inconspicuous. Hyomandibular pores alongside corners of mouth inconspicuous. Gill slits large (2.4–3.0 % TL). First dorsal fin large (height 9.7 % TL), semi-falcate, its inner margin long. Its apex sharply rounded or acute, the posterior margin curving ventrally from fin apex. D1 origin over or slightly anterior halfway of pectoral inner margin. Second dorsal fin large and high (its height 3.3 % TL), with a very long inner margin. D2 origin over or slightly anterior to anal fin origin. Pectorals large and falcate (length of anterior margin 18.6 % TL), apices narrowly rounded or pointed. Vertebrae: 114 precaudal centra and 105 caudal centra on average. Dermal denticles closed set, imbricate, sub-circular in small specimens, and almost rhomboid in larger individuals. With three to five longitudinal ridges and a corresponding number of short and sharp-pointed posterior cusplets. Adults have seven to nine ridges and cusplets.

Dental Formula:

$$\frac{13-1-13}{12-1-12} \quad \text{to} \quad \frac{13(14)-1(2)-13(14)}{(12)13-1-12(13)}$$

Upper teeth are narrow, high crowns, semi-erect to oblique. Distal median edges slightly concave apically, convex basally, deeply notched in basal third of crowns, and strongly serrated. Mesial edges are weakly concave to sinuous-shaped, coarsely serrated on central part of crowns, roots are arched and slightly notched. One or two rows of symphysial teeth are present. Lower teeth are narrow, crowns erect. Both median edges concave, slightly notched on mesial edges, weakly serrated on the apical parts and upper half of crowns (A to LP rows), extending to the bases in the rows towards angles of jaw. Roots are arched (S to LP rows) to almost straight (P to C rows), deeply notched. Usually one, but occasionally two rows of symphysial teeth are present. Files in upper and lower jaw are arranged as alternate overlapping dentition.

Colour: Grey dorsally, with a tinge of bronze, white ventrally. Laterally, colour fades into white, with a broad strip of bronze passing obliquely down from pectoral fin region and fades towards to pelvic region. On caudal fin a broad black margin extending from terminal margin to ventral tip is present. D1 is grey-bronze like the back with a white apex and a narrow whitish trailing margin which is curving from apex posteriorly. The apices of the pectorals, pelvics, anal fin and D2 dusky or black tips are distinct.

Size. Maximum length of 170 cm total, and an average length of 130–150 cm TL (COMPAGNO 1984, SHARKINFO).

Misidentifications. Frequently the species is confused with blacktip reef sharks, *C. melanopterus*, but *C. wheeleri* differs in having a white tip on the first dorsal fin, whereas *C. melanopterus* have a prominent black tip on the fin apex. In addition, the blacktail shark can be confused with *C. amblyrhynchos* but differs in colouration of D1 apex, in its shorter prenarial and preoral length and in dental formula – but (!) none of these features are exclusive, they are overlapping in range. The blacktail shark differs only slightly in the average number of precaudal vertebrae from other mentioned species (114 in *C. wheeleri* *versus* 116 in *C. melanopterus* and 115 in *C. amblyrhynchos*), but in the number of caudal vertebrae (105 in *C. wheeleri versus* 89 in *C. melanopterus* and 110 in *C. amblyrhynchos*).

Geographical Distribution. Western Indian Ocean: Red Sea, Gulf of Aden, Kenya, Mozambique, Tanzania, Madagascar, Seychelles, Comoros, Aldabra Islands, St. Brandon Reef, South Africa, Maldives. – (BASS et al. 1973, 1986; GARRICK 1982; COMPAGNO 1984; COMPAGNO et al. 2005).

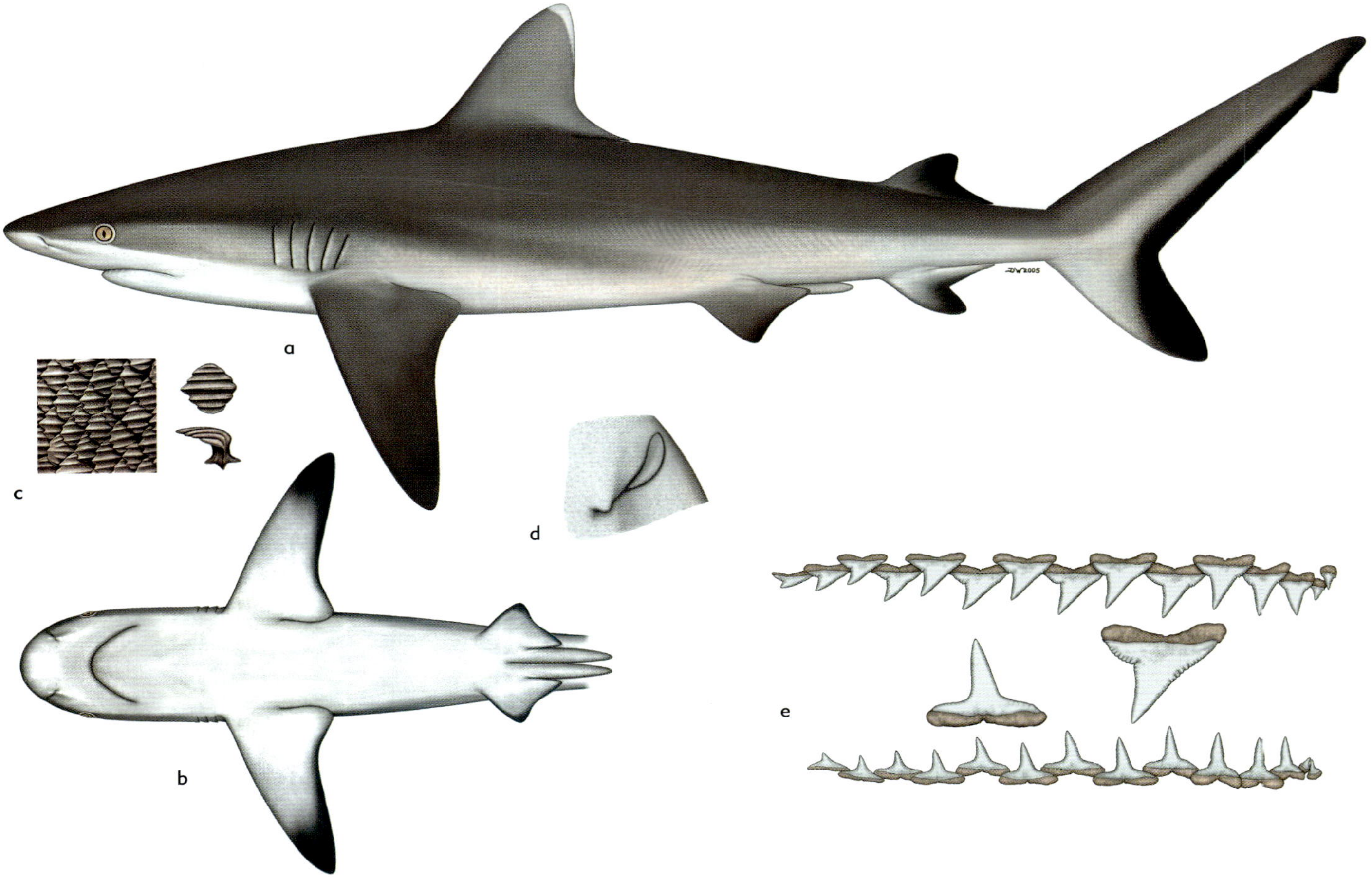

Figure 49. *Carcharhinus wheeleri*: 143 cm TL male from the Maldive Islands; **a**, left side; **b**, ventral view; **c**, dermal denticles; **d**, enlarged left nostril; **e**, left upper and lower teeth (symphysis to the left), and inset are enlarged sixth upper and lower teeth.

Habitat. The species is common coastal, abundant, inshore to offshore living, and occur from surface to at least 140 m depth. Normally, blacktail sharks are found on deeper banks, far away from major reefs, along fringing reefs, but can live close to shores and in shallower waters too (COMPAGNO 1984).

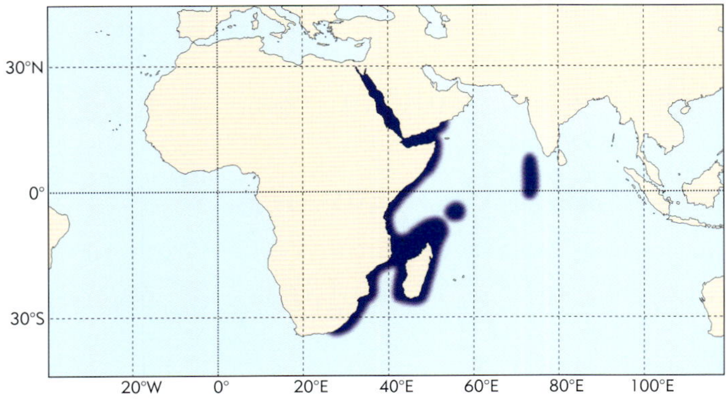

Biology. Only few facts about the biology of the species are known. Usually, juveniles live in shallower waters, separated from adults, but distinct sexual segregation is not known within the species. The species is less active at day but seems to be more active nocturnally (COMPAGNO 1984).

Reproduction. Males reach maturity at 110–130 cm TL. Females reach maturity at 120–140 cm TL. Reproductive mode is placental viviparous (DULVY & REYNOLDS 1997). Normally, the number of young per litter has a range of 1–4.

Duration of 12 months for gestation was estimated, and at birth, young have 65–75 cm TL. Mating and time for giving birth is in winter and spring (GARRICK 1982, COMPAGNO 1984).

Food. Mainly small bony fishes, cephalopods (squid) and shrimps (COMPAGNO et al. 1989).

Red List status: NE (IUCN 2010).

Species in question

Carcharhinus sp. A. COMPAGNO et al. (2005)

En: False smalltail shark; Fr: –; Sp: –; Ge: –

Carcharhinus sp. A COMPAGNO et al. (2005). Holotype: No holotype material specified.

Authors note. GARRICK (1982) stated this species as a local population of *C. porosus* represented due to ANSP-77121, a female of 36.5 cm TL, and other specimens from Thailand, Vietnam and Borneo. In contrast to this opinion, COMPAGNO (1988) separated the population by its cranial and morphometric differences from *C. porosus* and proposed it as a new species, *Carcharhinus* sp. A (COMPAGNO et al. 2005). In our opinion it is not comprehensible if this is truly a new species covered by the genus *Carcharhinus* or not.

Due to the fact the mentioned sample material of COMPAGNO (1988) and COMPAGNO et al. (2005) was not available we could not compare it with type material of *C. porosus* or other species, and hence the descriptions of GARRICK (1982) and COMPAGNO (1988) are reflected as diagnostic features only. The species was not considered in key to species, in creating tables of morphometric data and graphs.

Diagnostic features. [...] Small slender shark, without interdorsal ridge. Moderately long bluntly, pointed snout, large eyes, short labial furrows. Hyomandibular pores alongside corners of mouth inconspicuous. Large first dorsal, D1 origin over rear third of pectoral inner margin. Second dorsal fin small, D2 origin over first third of anal base. Moderately large pectoral fins broadly notched anal fin (COMPAGNO et al. 2005).

[...] A small, low second dorsal fin, and a transverse upper precaudal pit, is clearly a species of *Carcharhinus* and remarkable similar to *C. porosus*. Two other Asian specimens (ANSP 76859, from Saigon, and NMV61-463, from Bangkok) have the same facies, and although these and the paratype of *C. tephrodes* differ slightly from each other in some features, the differences are not greater than

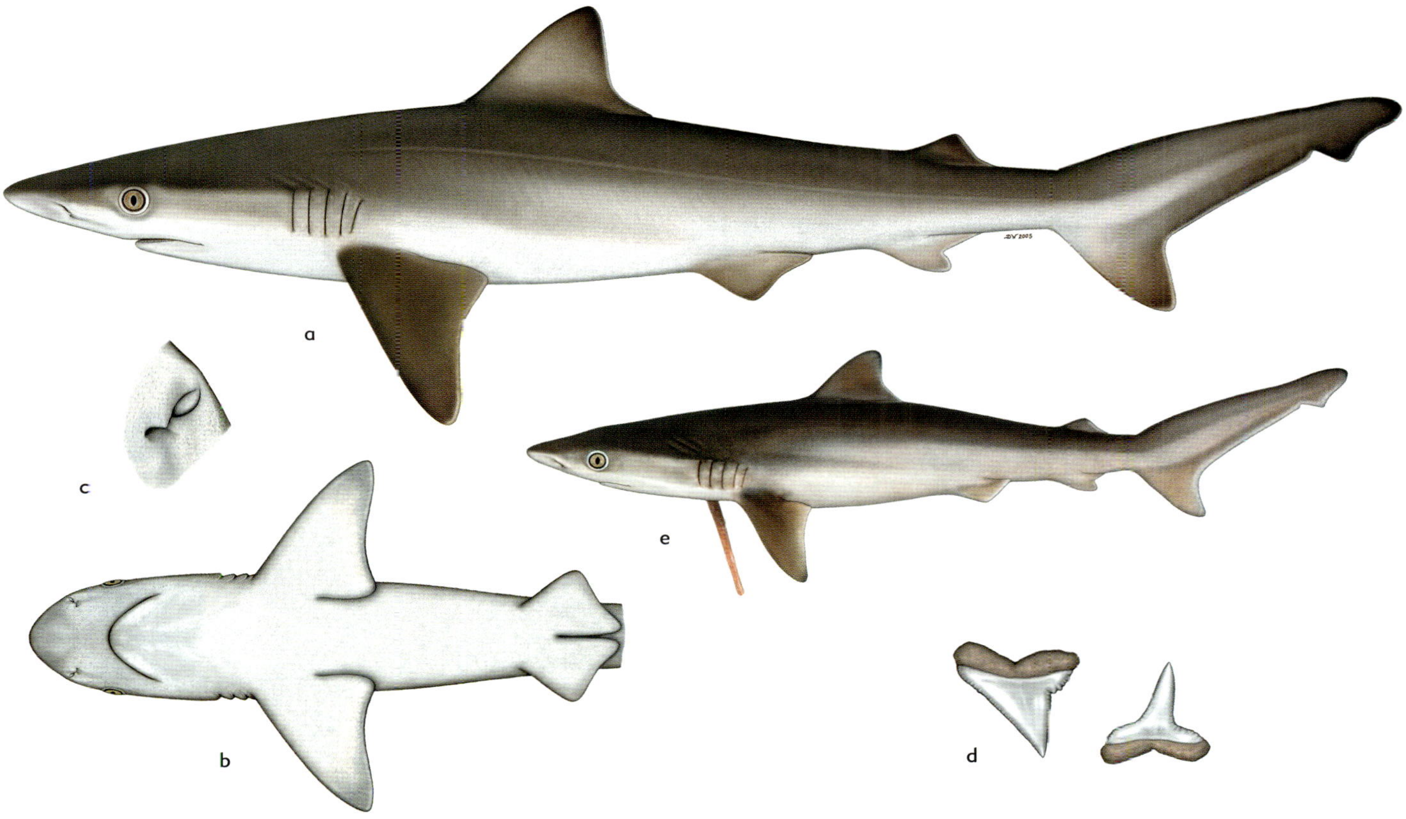

Figure 50. *Carcharhinus* sp. A: 43.3 cm TL female from Bangkok; **a**, left side; **b**, ventral view; **c**, enlarged left nostril; **d**, fifth left upper and lower teeth; **e**, 27 cm TL late embryo from Vietnam.

those between specimens of *C. porosus* from different localities. Comparison of the Asian "*C. porosus*" with American specimens of equivalent size indicates minor differences in proportions. In particular the Asian specimens are, on average, relatively longer tailed and broader headed and have longer pectorals and higher first dorsal fins. Because of their longer tails, their proportions measured from snout tip to various regions of the snout, head, and body are smaller relative to total length. In addition, the first dorsal fin origin is slightly farther forward relative to the pectoral base (from above the pectoral axis to halfway along the pectoral inner margin in the Asian specimens, but from halfway along the pectoral inner margin to above the pectoral inner corner in the American specimens). In considering the value which should be placed on the above minor differences for determining the status of the Asian specimens, it must be borne in mind that with only one exception (head width 13.1–13.4 % vs. 11.6–12.6 % TL) there is overlap with the American specimens. Furthermore, most (but not all) of the listed proportions vary with ongoing growth, and hence, the ranges will not generally apply when specimens of larger size were considered. No adults of the Asian form were found for study, and hence no firm criteria, other than geographic locality, can be put forward to distinguish the Asian from the American forms.

In view of this, and the fact that vertebral counts of the Asian specimens lie within the middle of the total range of counts for the American specimens, we favour recognizing the Asian and American forms as conspecific, despite the seemingly improbable distribution for a small, essentially tropical, littoral shark which results from the decision (GARRICK 1982).

Dental Formula:

$$\frac{13 - 2 - 13}{14 - 1 - 14}$$

Upper teeth are narrow, crowns semi-erect to strongly oblique, except symphysial rows. Distal median edges straight on apical part of crowns, convex on basal part, strongly notched. Mesial edges are slightly concave. Both edges finely and uniformly serrated, except the basal parts of crowns which are slightly more coarsely serrated on mesial edges, strongly serrated on distal edges. Roots are arched, slightly notched. One or two rows of symphysial teeth are present. Lower teeth are narrow, crowns semi-erect (S to LP rows) to oblique (P to C rows). Distal median edges deeply notched on basal third of crowns, straight basally, slightly convex apically, finely serrated, a little bit coarser basally. Mesial edges are concave, fine and uniformly serrated, a little bit coarser serrated basally. Roots are arched, deeply notched. One or two, but occasionally even none row(s) of symphysials present. Files in upper and lower jaw are arranged as mixed alternate i. e. imbricate overlapping dentition.

Colour. Light ventrally, and with an inconspicuous white band over flanks. Pectoral fins yellowish-brown, dorsal and caudal fins are often dusky but not conspicuously marked.

Size. Maximum length unknown, holotype specimen with 43.3 cm TL.

Misidentifications. Formerly placed in the American species *Carcharhinus porosus* but this species has a shorter and blunter shaped snout, position of first dorsal fin differs, pectoral fins are larger, and the structure of the skull is different in comparison to *C. porosus* (COMPAGNO et al. 2005). Moreover, this species is easily to confuse with *C. borneensis* but differs in shape of the nasal lobe (which is more pointed in *C. borneensis*), in body colouration, in tooth shape and formula, and the average number of caudal vertebrae (62 in C. sp. A vs. 58 in *C. borneensis*). In addition, Borneo sharks differ from *Carcharhinus* sp. A in having a row of 5–12 enlarged hyomandibular pores alongside mouth which are absent in *Carcharhinus* sp. A.

Geographical Distribution. South China Sea: Northern Borneo (Sarawak), Vietnam (Ho Chi Minh City) and Thailand. – (GARRICK 1985, COMPAGNO 1988, COMPAGNO et al. 2005).

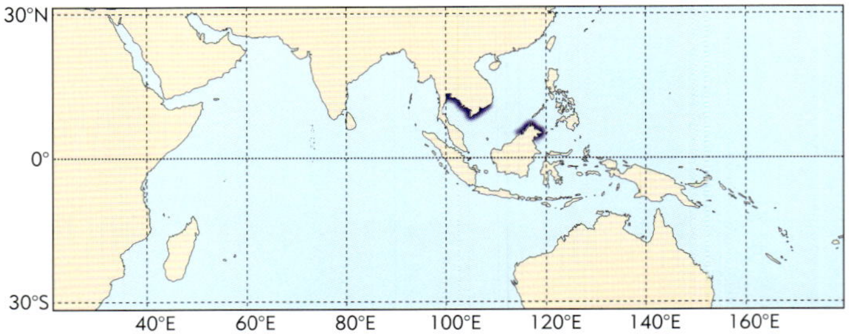

Habitat. Essentially unknown, presumably inshore living.

Biology. No relevant data available.

Reproduction. Length at reaching maturity unknown. At birth young are <34 cm TL. The reproductive mode is (probably) placental viviparous, but relevant data are not available.

Food. No data available but probably the species feed (in regard of the body size) on smaller bony fishes.

Red List status: NE (IUCN 2010).

Table 4. Values of proportional dimensions in percentage of TL. For every species are given the Mean, Min and Max values. Number of used data sets for calculation; see Table 2.

		C. acronotus		*C. albimarginatus*		*C. altimus*	
		Mean	Min–Max	Mean	Min–Max	Mean	Min–Max
Snout tip to	outer nostrils	3.2	2.9–3.5	3.4	2.7–4.0	3.9	3.5–4.4
	eye	6.9	6.5–7.3	7.6	6.5–8.8	8.1	7.6–8.7
	mouth	7.3	6.5–8.1	8.2	7.2–9.1	8.2	7.7–8.7
	1st gill opening	17.3	16.2–18.4	19.3	17.4–21.0	19.4	18.7–20.9
	3rd gill opening	19.6	18.3–21.0	21.2	18.6–23.0	22.3	21.1–24.0
	5th gill opening	21.4	19.3–23.1	23.4	21.7–25.0	24.6	23.1–26.8
	pectoral origin	20.7	19.3–22.6	22.3	20.2–23.9	23.4	21.7–25.6
	pelvic origin	47.9	46.7–49.2	49.8	49.1–51.0	52.1	51.4–52.7
	1st dorsal origin	30.7	29.2–32.2	31.3	29.3–32.7	31.2	29.6–31.9
	2nd dorsal origin	62.5	61.0–64.2	62.9	60.5–65.5	64.1	63.4–65.4
	anal fin origin	61.9	59.8–64.2	62.3	59.9–64.6	64.8	63.8–66.0
	upper caudal origin	73.8	71.6–75.6	73.2	71.3–76.2	75.2	74.1–77.4
	lower caudal origin	73.0	70.8–75.1	72.4	70.5–75.3	74.5	73.5–75.8
Nostril	distance between inner corners	4.7	4.4–5.1	6.7	6.6–6.9	5.8	5.3–6.2
Mouth	wide	6.8	6.3–7.4	8.8	7.9–9.9	8.3	7.9–8.6
	length	3.9	3.5–4.5	4.8	4.4–5.3	4.6	4.2–5.1
Labial furrow lengths	upper	0.5	0.4–0.5	0.5	0.3–0.8	0.5	0.3–0.8
	lower	0.5	0.4–0.5	0.5	0.3–0.8	0.6	0.5–0.6
Gill opening lengths	1st	2.3	2.2–2.5	2.6	2.0–3.0	3.1	2.7–3.4
	3rd	2.9	2.4–3.3	3.0	2.5–3.5	3.5	3.1–3.9
	5th	2.3	2.0–2.5	2.2	1.7–2.4	2.4	2.3–2.6
Eye	horizontal diameter	1.9	1.6–3.0	2.5	1.8–3.0	1.7	1.4–2.3
1st dorsal fin	length of base	9.2	8.6–10.2	10.6	10.0–11.2	10.9	9.8–11.9
	length posterior margin	3.3	3.0–3.9	3.3	2.8–4.0	4.8	4.5–5.0
	heigth	9.1	7.8–9.8	8.8	7.1–10.1	10.2	8.3–11.9
2nd dorsal fin	length of base	3.8	3.6–4.1	3.3	2.8–4.4	4.2	3.8–4.6
	length posterior margin	3.6	3.1–3.8	3.4	2.8–4.2	3.9	3.6–4.0
	heigth	2.7	2.6–2.9	2.0	1.5–2.3	3.1	2.8–3.4
Anal fin	length of base	4.6	4.2–5.0	4.0	3.2–4.6	4.0	3.8–4.3
	length posterior margin	3.4	3.1–3.7	3.4	3.1–4.0	3.7	3.5–4.0
	heigth	3.5	3.2–3.7	2.7	2.2–3.1	3.6	3.2–4.1
Pectoral fin	length of base	5.5	5.2–6.2	5.7	5.3–6.6	6.7	6.2–7.1
	length anterior margin	15.0	14.6–15.3	18.6	16.2–22.0	21.2	19.6–22.4
	length distal margin	11.6	10.5–12.6	14.5	10.1–19.2	16.9	11.8–18.9
	greatest width	8.4	7.8–9.6	8.8	7.5–10.3	11.0	10.1–11.3
Pelvic fin	length of base	5.1	4.4–5.6	4.8	4.1–5.9	5.4	4.8–5.8
	length anterior margin	5.9	5.7–6.1	5.0	4.6–5.7	6.6	6.2–7.0
	length distal margin	4.9	4.7–5.3	5.0	4.3–5.7	6.0	5.0–6.4
	length of claspers	6.9	4.7–9.1	8.6	–	6.0	3.3–8.7
Caudal	length of upper lobe	26.7	25.2–28.3	27.1	24.8–28.4	25.9	23.3–26.9
	length of lower lobe	11.6	11.0–12.3	12.5	10.8–13.7	12.8	12.1–13.7
Trunk at pectoral origin	width	10.9	10.0–11.2	12.9	11.4–13.8	12.4	11.9–13.1
	heigth	11.0	9.7–11.7	12.0	10.5–13.7	11.7	11.0–12.6
Vertebrae	precaudal	84	81–88	120	115–122	109	101–110
	caudal	88	84–93	103	101–107	95	64–96
	total	172	165–181	223	216–227	204	194–206

Table 4. (continued).

		C. amblyrhynchoides		*C. amblyrhynchos*		*C. amboinensis*		*C. borneensis*	
		Mean	Min–Max	Mean	Min–Max	Mean	Min–Max	Mean	Min–Max
Snout tip to	outer nostrils	2.9	2.5–3.3	3.6	3.2–4.2	2.7	2.5–2.9	5.3	3.7–6.0
	eye	6.1	5.6–6.6	7.6	6.7–8.1	6.0	5.8–6.3	8.1	7.1–9.3
	mouth	6.2	5.3–6.9	8.2	7.2–8.7	6.5	5.7–7.0	8.5	7.4–9.6
	1st gill opening	18.3	17.5–19.2	20.2	18.3–21.2	15.6	15.0–16.1	19.4	18.4–22.5
	3rd gill opening	20.6	19.7–21.3	22.3	20.4–24.1	19.1	18.9–19.3	21.9	21.0–25.0
	5th gill opening	23.0	21.2–24.7	24.1	22.1–26.0	21.1	19.8–22.4	23.7	22.3–26.8
	pectoral origin	21.7	20.3–22.6	22.9	21.3–24.9	19.3	17.6–20.8	22.6	21.1–26.1
	pelvic origin	48.1	46.8–49.8	48.5	46.5–50.9	49.2	48.4–49.7	47.0	44.6–50.3
	1st dorsal origin	29.0	27.9–30.4	31.7	30.4–33.0	28.2	26.3–31.3	30.5	27.5–34.0
	2nd dorsal origin	60.5	58.5–63.0	60.5	58.8–62.3	59.6	59.3–59.7	62.6	60.9–66.7
	anal fin origin	60.7	59.1–63.2	60.6	59.2–62.5	60.9	60.5–61.3	59.8	56.9–64.2
	upper caudal origin	72.4	70.7–74.4	72.0	70.8–73.4	72.9	71.0–76.1	74.2	71.0–77.8
	lower caudal origin	71.6	70.2–73.6	71.3	70.0–72.9	71.1	–	72.9	69.4–76.3
Nostril	distance between inner corners	5.8	5.4–6.2	6.3	5.7–6.8	7.0	6.6–7.3	6.1	5.0–6.4
Mouth	wide	9.2	8.1–10.3	8.6	7.7–9.5	10.2	9.6–11.2	8.0	7.2–8.6
	length	5.4	4.8–6.3	5.5	4.9–6.0	4.5	4.2–4.8	4.6	4.0–5.2
Labial furrow lengths	upper	0.6	0.5–0.8	0.5	0.3–0.7	0.3	–	0.3	0.1–0.6
	lower	0.5	0.3–0.8	0.5	0.3–0.7	0.5	–	0.5	0.2–0.7
Gill opening lengths	1st	4.1	3.5–4.6	2.8	2.4–3.4	3.1	2.9–3.2	2.2	1.8–2.9
	3rd	4.8	4.0–5.6	3.3	2.8–4.2	3.2	3.2–3.6	2.7	2.2–3.4
	5th	3.2	2.8–3.9	2.5	2.2–3.3	2.5	2.4–2.6	2.7	2.2–3.2
Eye	horizontal diameter	1.7	1.2–2.0	2.5	2.0–2.8	1.4	1.3–1.6	2.1	1.6–2.5
1st dorsal fin	length of base	11.8	10.5–12.9	9.5	8.5–10.2	12.1	11.3–13.1	10.6	9.0–12.9
	length posterior margin	3.9	3.3–4.6	4.0	3.5–4.7	4.5	3.9–5.4	5.0	4.1–5.7
	heigth	10.8	8.6–12.6	10.3	9.1–10.9	11.6	11.1–12.0	8.0	6.9–9.9
2nd dorsal fin	length of base	4.8	4.4–5.3	4.3	4.0–4.5	4.5	4.4–4.6	4.3	3.1–5.6
	length posterior margin	4.0	3.7–4.2	4.2	3.7–4.8	3.6	3.4–3.8	4.7	3.5–5.4
	heigth	3.4	3.1–3.7	3.1	2.7–3.4	3.1	2.9–3.3	2.0	1.6–2.8
Anal fin	length of base	5.1	4.8–5.4	4.6	4.1–5.5	4.5	4.2–4.9	5.0	3.8–6.8
	length posterior margin	3.4	3.1–3.7	3.8	3.4–4.4	3.2	2.8–3.4	4.0	3.4–4.8
	heigth	3.6	3.1–3.9	3.3	2.9–3.8	3.2	2.7–3.8	2.7	2.2–3.4
Pectoral fin	length of base	6.5	5.7–7.2	5.5	4.7–6.1	7.2	6.9–7.5	5.9	5.2–6.5
	length anterior margin	18.3	16.3–19.7	19.3	18.6–20.5	20.7	20.0–22.2	13.7	10.9–16.4
	length distal margin	14.3	11.3–17.5	15.9	14.4–18.6	17.2	–	9.7	7.9–14.0
	greatest width	9.9	9.3–10.4	9.3	8.5–10.2	11.5	–	9.4	8.0–11.3
Pelvic fin	length of base	5.5	4.5–6.0	4.9	4.3–5.2	6.0	5.6–6.4	4.5	3.8–5.0
	length anterior margin	6.4	6.0–7.1	5.4	5.2–5.7	6.5	6.3–6.7	5.2	4.4–5.8
	length distal margin	6.1	5.4–6.9	5.1	4.7–5.8	6.7	–	4.1	3.5–5.8
	length of claspers	2.3	2.0–2.4	1.9	1.8–2.1	1.9	–	4.3	1.8–6.5
Caudal	length of upper lobe	28.3	27.1–29.3	29.0	26.5–29.8	29.3	28.9–29.9	26.0	23.0–27.7
	length of lower lobe	13.6	11.8–14.9	14.1	13.3–14.7	13.6	13.3–13.8	10.8	9.1–11.8
Trunk at pectoral origin	width	12.9	12.1–14.2	12.2	10.5–13.6	13.8	–	10.8	8.7–12.7
	heigth	13.9	12.6–16.1	10.6	8.5–12.8	12.9	–	10.2	8.2–12.6
Vertebrae	precaudal	91	78–96	115	111–118	92	90–95	61	57–62
	caudal	95	90–101	100	96–104	96	92–100	58	56–60
	total	186	168–193	215	207–222	188	185–195	119	113–122

C. brachyurus		C. brevipinna		C. cautus		C. dussumieri		C. falciformes		C. fitzroyensis	
Mean	Min–Max	Mean	Min–Max	Mean	Min–Max	Mean	Min–Max	Mean	Min–Max	Mean	Min–Max
3.8	3.3–4.2	4.6	4.3–4.7	2.8	2.5–3.2	3.4	2.9–3.7	4.2	3.7–4.8	4.2	–
7.4	5.9–8.2	8.2	7.7–8.4	6.4	6.0–8.0	6.8	5.8–7.4	7.8	6.9–8.6	8.5	7.7–9.3
7.5	5.7–8.4	8.2	7.7–9.0	6.6	6.3–7.0	7.0	5.8–7.9	7.8	7.0–9.1	8.5	8.3–8.6
18.7	17.1–19.8	20.0	18.9–21.5	18.1	17.7–18.4	17.9	15.4–19.0	19.3	17.6–21.1	19.7	19.3–20.1
21.1	19.5–22.6	22.6	21.6–24.4	20.5	20.1–20.9	20.5	18.4–21.5	21.4	20.1–23.7	21.6	–
23.2	21.5–24.4	24.4	23.4–25.6	22.3	21.3–23.2	22.2	19.8–23.9	23.6	22.1–25.6	24.7	23.8–25.5
22.7	20.2–25.2	23.7	22.7–24.8	21.4	20.3–22.8	21.4	19.7–23.0	22.0	20.3–24.1	22.7	22.5–22.8
51.3	49.0–55.2	50.6	49.3–51.8	47.0	45.5–47.9	47.0	45.3–50.1	48.8	46.8–51.8	48.6	47.3–50.1
32.5	31.2–34.0	32.9	31.9–33.9	30.6	28.7–31.9	30.0	28.4–31.2	33.0	30.7–35.0	32.4	31.1–33.7
64.0	62.1–67.3	63.8	62.9–65.1	60.6	60.0–61.4	61.9	60.3–64.9	64.0	62.7–65.7	61.0	–
63.3	59.9–67.3	61.1	59.1–64.2	59.9	58.4–61.6	60.9	59.3–63.7	63.1	61.4–65.0	61.0	–
74.1	72.4–76.2	74.0	73.1–75.3	73.8	73.1–74.6	74.3	72.3–76.4	73.3	72.0–74.6	73.3	73.1–73.5
73.2	71.4–76.4	73.3	72.3–74.4	72.8	71.7–73.5	73.3	71.5–75.5	72.5	71.2–74.2	72.1	–
5.8	5.3–6.5	5.0	4.5–5.2	6.0	5.6–6.4	4.9	4.2–5.5	5.6	5.3–6.1	4.9	4.8–5.0
8.0	6.7–8.8	7.2	6.4–7.9	8.1	7.6–8.5	7.4	6.9–7.7	7.8	7.5–8.8	8.5	8.2–8.8
4.7	4.0–5.2	4.6	4.0–4.9	4.5	4.3–4.7	4.6	4.4–5.2	4.9	4.2–5.7	4.2	–
0.6	0.4–0.8	0.8	0.7–0.9	0.4	0.3–0.5	0.4	0.3–0.5	0.4	0.3–0.5	0.6	0.5–0.7
0.4	0.3–0.7	0.7	0.3–0.9	0.4	0.3–0.5	0.4	0.3–0.5	0.4	0.4–0.5	0.4	0.3–0.4
2.8	2.4–3.5	3.7	3.2–4.3	2.6	1.9–2.9	2.5	2.1–3.0	2.9	2.4–3.3	2.9	2.8–3.0
3.3	2.5–4.1	4.2	3.6–5.5	2.8	2.1–3.4	3.0	2.4–3.5	3.3	2.9–3.6	3.0	–
2.4	2.0–2.7	3.2	2.7–3.8	2.0	1.5–2.4	2.3	1.6–2.8	2.2	2.0–2.5	2.4	2.2–2.6
1.8	1.1–2.2	1.5	1.1–2.0	2.4	1.9–3.3	2.4	2.0–2.7	1.8	1.2–2.7	1.4	–
9.8	8.9–10.8	9.1	8.5–10.5	9.8	9.1–10.5	10.0	9.0–10.9	8.5	8.0–9.3	10.7	10.5–10.9
2.8	2.4–3.2	2.8	2.3–3.0	4.1	3.8–4.7	4.3	3.8–4.8	4.2	3.8–4.8	4.5	3.9–5.1
8.4	6.8–9.7	7.3	6.0–8.8	7.5	6.3–9.2	9.6	8.0–10.7	7.0	5.2–8.1	9.4	–
3.8	3.1–4.3	3.5	3.3–3.9	4.9	4.0–5.5	4.3	3.7–4.7	2.5	2.0–3.0	4.4	4.3–4.5
3.1	2.9–3.6	3.4	3.1–3.8	3.5	3.2–4.1	3.9	3.2–4.5	4.3	3.8–4.9	3.8	3.6–4.1
2.3	2.0–2.5	2.2	1.8–2.5	3.5	3.1–4.1	3.2	2.9–3.7	1.7	1.4–2.1	3.2	–
4.3	3.6–5.4	4.4	3.4–4.9	5.5	4.9–5.9	4.9	4.4–5.6	3.2	2.8–4.0	5.3	5.2–5.4
2.9	2.6–3.1	3.3	3.0–3.7	3.5	3.3–3.9	3.7	3.2–4.1	4.4	3.9–4.9	3.6	3.4–3.8
3.3	2.9–3.7	2.7	2.5–3.2	3.5	3.2–4.0	3.7	3.4–4.2	2.4	2.1–2.8	4.2	–
6.1	5.8–6.8	5.4	4.9–6.0	5.9	5.5–6.7	5.9	5.3–6.5	5.8	5.1–6.5	6.5	6.4–6.6
18.1	16.0–21.3	14.9	13.4–16.0	16.0	15.0–16.9	15.7	14.5–17.2	16.8	14.2–21.7	18.1	17.1–19.1
14.1	12.0–18.0	10.5	7.7–12.4	10.8	9.5–13.0	11.4	10.3–13.9	12.2	7.9–17.7	13.9	–
8.6	8.4–10.7	7.9	7.0–8.4	8.3	8.1–9.4	9.6	8.9–10.9	8.3	6.5–9.7	11.8	–
5.1	4.2–6.2	4.7	4.1–5.0	5.5	4.9–6.8	5.2	4.5–6.6	4.3	3.7–5.0	5.7	–
5.3	5.0–6.0	4.9	4.1–5.4	6.7	5.9–7.1	6.3	5.6–6.9	4.4	3.8–5.[illegible]	7.0	6.5–7.6
5.5	4.7–6.0	4.4	3.8–4.8	5.2	4.8–5.8	5.4	4.8–6.4	4.6	3.7–5.4	5.9	–
2.0	1.5–2.2	1.7	1.3–1.9	3.7	2.2–4.2	2.9	2.0–4.5	2.3	1.8–2.8	2.7	–
26.9	25.3–27.9	26.1	25.5–27.1	26.4	25.6–26.7	25.7	24.0–27.7	26.9	26.1–27.8	26.6	–
11.7	10.4–13.4	10.3	10.2–12.2	12.3	11.7–12.9	10.8	9.8–12.0	11.8	10.4–13.2	11.3	11.0–11.6
11.9	10.3–13.2	10.0	9.2–11.0	11.9	11.4–12.6	11.3	9.8–13.1	10.8	9.3–11.8	12.9	–
11.2	9.5–13.6	10.3	9.8–11.7	11.3	10.3–11.8	10.5	9.1–11.7	10.4	8.6–11.8	11.4	–
103	96–110	82	77–91	89	86–90	62	56–74	102	99–106	58	–
92	89–93	89	84–94	78	70–81	66	61–76	104	102–108	67	64–70
195	179–201	171	164–185	167	160–171	128	110–150	206	201–212	125	–

Table 4. (continued).

		C. galapagensis		*C. hemiodon*		*C. isodon*		*C. leucas*	
		Mean	Min-Max	Mean	Min-Max	Mean	Min-Max	Mean	Min-Max
Snout tip to	outer nostrils	3.3	2.5-3.9	3.4	3.3-3.9	3.6	3.1-4.1	2.1	1.7-2.6
	eye	7.1	5.7-8.0	7.1	6.5-7.3	6.9	6.8-7.6	5.4	4.5-6.2
	mouth	7.4	6.0-8.2	7.0	6.5-7.4	7.1	6.5-7.6	5.6	4.6-6.5
	1st gill opening	19.7	18.2-21.8	17.9	17.2-18.6	20.6	19.4-21.8	17.4	15.9-18.3
	3rd gill opening	22.0	20.5-23.9	20.5	19.7-21.2	23.4	22.5-24.0	19.7	18.5-21.0
	5th gill opening	24.0	22.0-25.6	22.7	21.7-23.6	25.1	24.7-25.6	21.5	20.4-23.1
	pectoral origin	23.0	21.4-24.5	21.9	21.0-22.9	24.3	22.7-25.4	20.2	18.8-21.3
	pelvic origin	49.3	48.4-50.4	46.8	46.1-47.8	48.8	46.7-53.2	49.9	46.5-53.0
	1st dorsal origin	31.1	29.7-32.3	29.6	29.0-30.0	32.2	31.5-33.1	29.2	27.5-31.4
	2nd dorsal origin	62.8	61.3-63.8	61.9	61.2-62.7	61.9	60.5-64.7	61.4	59.5-64.7
	anal fin origin	62.7	60.4-63.4	60.2	59.8-60.8	61.4	60.2-63.3	62.8	60.6-66.3
	upper caudal origin	73.3	71.8-74.9	72.2	71.9-72.3	72.9	72.0-74.2	73.8	72.0-76.5
	lower caudal origin	72.6	71.3-74.2	71.0	70.6-71.5	72.3	71.4-73.6	73.1	71.1-76.2
Nostril	distance between inner corners	6.3	6.0-6.7	5.7	5.6-5.8	5.4	5.0-5.8	6.5	5.9-6.9
Mouth	wide	8.6	7.8-9.2	8.2	7.7-9.2	8.9	8.2-9.7	9.8	8.6-11.5
	length	4.8	4.0-5.3	5.1	4.5-5.4	6.1	5.8-6.3	4.7	4.2-5.1
Labial furrow lengths	upper	0.5	0.4-0.6	0.5	0.5-0.6	0.9	0.8-0.9	0.5	0.4-0.7
	lower	0.6	0.4-0.7	0.5	0.4-0.6	0.8	0.8-0.9	0.6	0.4-0.9
Gill opening lengths	1st	2.8	2.5-3.2	3.0	2.8-3.2	4.9	4.2-5.8	3.1	2.9-4.4
	3rd	3.2	2.8-3.5	3.5	3.2-3.8	5.5	4.8-6.5	3.7	3.3-4.5
	5th	2.4	2.0-2.8	2.9	2.7-3.2	4.4	3.5-5.4	2.3	2.1-3.3
Eye	horizontal diameter	2.0	1.3-2.4	2.3	2.1-2.4	1.8	1.3-2.2	1.2	0.8-1.8
1st dorsal fin	length of base	9.5	8.9-10.0	10.3	9.9-10.5	9.6	8.8-10.1	11.8	11.2-12.2
	length posterior margin	3.8	3.5-4.3	6.1	5.9-6.5	3.9	3.5-4.3	3.3	2.9-3.6
	heigth	10.5	9.5-11.2	10.5	9.4-11.6	9.0	7.5-10.6	9.2	7.0-10.8
2nd dorsal fin	length of base	3.6	3.1-4.1	3.1	2.7-3.5	4.3	4.1-4.7	5.2	4.7-5.7
	length posterior margin	3.9	3.7-4.3	4.3	3.9-4.5	4.0	3.7-4.3	3.1	2.7-3.3
	heigth	2.7	2.5-2.8	2.3	1.9-2.6	2.7	2.5-2.9	3.8	3.2-4.5
Anal fin	length of base	4.1	3.8-4.8	4.4	4.2-4.7	4.9	4.6-5.3	5.0	4.7-5.5
	length posterior margin	3.6	3.2-3.9	4.0	3.8-4.2	3.6	3.3-4.0	3.0	2.6-3.1
	heigth	3.3	2.9-3.8	3.2	2.8-3.8	3.1	2.9-3.4	3.9	3.1-4.6
Pectoral fin	length of base	6.1	5.3-6.6	6.2	5.8-6.5	5.5	5.1-5.8	7.2	6.3-7.9
	length anterior margin	20.0	17.6-22.9	17.6	16.3-19.2	15.5	14.7-16.0	19.4	17.6-21.2
	length distal margin	16.4	12.5-19.1	12.8	10.8-15.2	10.4	9.9-11.4	15.1	11.4-18.6
	greatest width	9.8	8.6-11.0	9.3	8.7-10.2	8.3	8.1-8.5	10.7	9.3-11.9
Pelvic fin	length of base	5.3	4.4-6.8	5.0	4.6-5.4	5.1	–	5.8	5.1-6.2
	length anterior margin	6.0	5.2-6.4	5.5	4.8-5.9	5.3	4.7-5.8	6.6	6.1-7.5
	length distal margin	5.6	4.7-6.1	5.2	4.5-6.0	5.4	4.8-6.3	6.4	5.5-7.1
	length of claspers	5.0	1.8-9.3	2.5	2.4-2.7	2.4	–	3.1	1.6-8.8
Caudal	length of upper lobe	27.4	26.1-28.3	28.0	27.2-28.8	28.2	27.6-29.2	26.5	24.7-27.9
	length of lower lobe	13.4	12.5-14.8	12.8	12.6-13.2	11.8	10.8-12.6	12.1	10.8-12.9
Trunk at pectoral origin	width	11.9	11.1-13.2	11.3	10.8-11.7	10.5	10.4-10.6	14.0	12.8-15.3
	heigth	11.1	10.1-12.7	10.6	9.5-11.5	11.0	10.6-11.3	13.1	11.8-15.2
Vertebrae	precaudal	107	104-109	70	69-71	79	78-80	116	110-123
	caudal	101	94-106	82	78-84	86	85-87	101	99-104
	total	208	200-215	152	147-155	165	163-167	217	209-227

C. limbatus		*C. longimanus*		*C. macloti*		*C. melanopterus*		*C. obscurus*		*C. perezii*	
Mean	Min–Max	Mean	Min–Max	Mean	Min–Max	Mean	Min–Max	Mean	Min–Max	Mean	Min–Max
3.8	3.2–4.1	2.7	2.4–3.1	5.1	4.8–5.6	2.6	2.3–3.1	3.6	2.6–4.1	2.9	2.6–3.2
7.2	6.3–7.6	6.3	5.7–7.1	8.5	8.0–9.1	5.9	5.4–6.8	7.3	5.9–8.1	6.7	6.1–7.2
7.8	7.0–8.9	6.2	5.4–6.8	9.5	8.7–10.3	5.6	4.8–7.3	7.5	5.9–8.4	6.8	6.1–7.7
19.6	18.5–21.1	17.7	15.7–19.0	20.1	19.1–21.3	17.1	15.3–19.5	18.7	17.9–20.3	19.2	18.1–20.2
22.2	20.4–23.6	20.5	19.3–21.5	22.2	21.0–23.5	19.5	18.2–21.3	21.4	20.5–22.3	21.0	19.7–22.2
24.2	23.2–25.8	21.9	19.2–23.3	24.0	22.8–25.2	21.3	19.4–23.5	23.2	22.3–24.0	23.0	21.1–23.8
23.4	22.4–25.0	20.6	17.8–22.3	23.2	22.0–24.3	20.4	18.5–22.3	22.0	20.9–22.7	21.8	20.3–22.3
49.4	48.1–50.8	50.4	49.2–51.8	48.3	46.7–49.4	47.0	45.1–51.4	50.7	48.9–53.5	49.7	48.2–53.6
30.1	28.5–31.7	31.0	29.1–32.0	31.7	30.8–32.5	31.2	29.6–33.8	31.5	29.3–32.9	31.9	30.5–33.6
62.1	60.4–63.2	61.5	60.6–62.2	63.0	61.9–65.0	61.2	57.8–65.6	63.7	61.6–65.3	62.0	60.0–66.5
62.3	61.3–63.4	62.3	60.9–63.3	61.1	59.7–62.4	61.2	59.3–66.1	63.3	61.1–65.3	61.5	59.7–66.1
72.7	70.4–74.3	71.6	70.4–72.8	72.1	70.3–73.6	74.2	73.1–78.2	73.8	72.9–75.3	72.4	71.1–75.8
72.0	70.3–73.7	70.7	69.3–72.0	71.0	69.3–71.3	73.5	71.9–77.4	73.2	71.8–75.0	71.7	70.2–75.5
5.5	5.2–6.2	5.7	5.2–6.2	5.2	4.9–5.6	6.0	5.7–6.3	6.1	5.8–6.3	6.7	5.9–7.2
8.6	7.3–9.5	8.7	7.6–9.8	7.2	6.6–7.9	8.2	7.6–9.3	8.1	7.1–10.1	9.2	8.5–9.9
5.1	4.5–5.5	4.9	4.5–5.5	4.7	4.4–5.1	4.6	4.0–5.2	4.4	4.0–4.6	5.1	4.7–5.7
0.6	0.4–0.8	0.5	0.3–0.6	0.5	0.4–0.6	0.5	0.4–0.8	0.5	0.3–0.9	0.4	0.3–0.5
0.5	0.4–0.5	0.6	0.4–0.7	0.5	0.3–0.6	0.5	0.4–0.8	0.6	0.4–0.9	0.5	0.4–0.7
3.8	3.4–4.4	3.1	2.6–4.0	2.7	2.2–3.0	3.0	2.4–3.5	2.8	2.1–3.6	2.7	2.3–3.2
4.4	3.8–4.9	3.4	3.0–4.1	3.1	2.7–3.5	2.9	2.6–4.2	3.4	2.7–4.0	3.2	2.8–4.0
3.2	2.8–3.8	2.5	2.0–3.1	2.7	2.2–2.9	2.5	1.9–3.0	2.4	2.1–2.7	2.4	2.0–2.9
1.7	1.2–2.2	1.6	0.9–2.5	2.3	1.9–3.1	2.5	2.0–3.0	1.8	1.0–2.1	2.0	1.2–2.5
11.6	10.3–12.8	11.1	9.6–12.4	8.9	8.3–9.3	9.3	8.2–10.3	9.8	8.9–11.7	9.8	8.7–11.1
3.9	3.5–4.2	4.3	3.3–5.5	6.0	5.3–7.1	4.4	3.8–5.0	3.3	2.9–3.6	3.3	3.0–3.4
10.4	8.2–12.4	12.3	9.2–15.2	8.0	7.1–8.9	9.1	8.0–10.1	7.9	6.0–9.7	10.8	10.3–11.3
4.3	3.6–4.9	3.9	3.5–4.4	2.8	2.7–2.9	4.8	4.4–5.2	3.3	2.5–4.2	4.0	3.6–4.4
4.0	3.6–4.2	3.8	2.9–4.6	4.5	4.2–5.2	3.5	3.3–4.4	3.7	3.3–4.0	3.8	3.5–4.2
2.9	2.5–3.6	3.4	2.7–4.2	1.9	1.6–2.1	3.8	3.4–4.1	2.0	1.8–2.3	3.1	2.9–3.3
4.5	3.8–5.2	4.2	3.2–5.2	3.8	3.2–4.2	4.8	4.4–5.9	4.0	3.5–4.6	4.4	4.1–4.9
3.6	3.4–3.9	3.6	3.1–4.1	4.4	4.0–5.0	3.6	3.4–4.0	3.3	2.8–3.8	3.5	3.1–3.7
3.3	3.0–3.7	3.9	2.9–5.1	2.5	2.2–2.8	4.0	3.7–4.4	2.8	2.4–3.8	3.6	3.4–3.7
6.2	5.5–7.0	6.7	6.4–7.2	5.3	5.2–5.8	5.6	5.1–6.0	5.8	5.4–7.0	6.1	5.5–7.1
18.1	17.2–19.7	24.3	20.2–27.1	14.9	13.5–16.1	18.2	17.1–19.3	18.1	16.8–21.6	20.4	19.5–21.7
13.4	11.2–16.7	20.0	14.9–24.3	10.9	9.9–13.0	14.0	12.7–15.1	13.3	10.3–18.8	17.4	15.4–20.2
9.5	8.4–10.4	11.7	10.8–12.4	7.9	7.6–8.4	9.7	8.8–10.4	8.7	8.0–9.8	9.1	8.8–9.6
5.3	4.0–6.2	5.4	4.4–5.9	4.2	3.9–4.6	5.2	4.6–5.9	4.8	4.2–5.8	5.1	4.7–5.5
6.1	5.5–6.6	7.2	5.3–8.1	4.4	3.9–5.2	7.1	6.5–8.0	5.1	4.6–5.5	6.0	5.5–6.2
5.6	5.2–6.3	5.8	4.6–7.0	5.0	4.0–5.2	5.7	5.2–6.6	5.0	4.3–6.3	5.9	5.2–6.9
3.6	2.2–8.2	5.4	2.2–8.9	4.0	2.1–9.1	2.4	2.1–2.6	2.5	2.2–2.8	–	–
27.6	27.0–28.3	28.6	26.0–31.3	28.1	27.1–29.4	26.2	22.1–28.2	26.6	25.8–27.5	28.9	27.0–30.6
13.1	12.2–13.8	14.3	12.0–15.7	11.0	10.1–11.8	13.0	12.6–13.5	11.5	10.8–12.4	14.3	13.5–16.9
11.9	10.6–12.9	12.1	11.3–12.8	9.8	9.2–10.5	12.1	10.4–13.4	12.1	10.7–14.4	12.6	11.7–13.5
12.7	11.7–14.0	11.5	11.0–12.3	9.5	8.7–10.5	10.8	9.6–11.7	11.3	9.6–14.7	11.5	10.3–12.3
96	94–102	126	125–131	70	68–71	116	111–120	91	87–97	106	104–108
98	90–101	109	106–110	81	79–84	89	87–92	96	90–101	104	102–107
194	174–201	235	228–238	151	149–154	205	197–209	187	175–194	210	208–213

Table 4. (continued).

		C. plumbeus		*C. porosus*		*C. sealei*		*C. signatus*	
		Mean	Min–Max	Mean	Min–Max	Mean	Min–Max	Mean	Min–Max
Snout tip to	outer nostrils	3.3	2.4–3.8	4.8	3.7–5.4	3.0	2.2–3.7	5.2	4.9–5.5
	eye	7.2	5.6–7.9	8.6	7.1–9.3	6.7	5.9–7.6	8.8	8.4–9.2
	mouth	7.4	5.6–8.1	8.8	7.4–9.6	6.5	5.3–7.3	6.8	5.5–10.3
	1st gill opening	19.0	18.4–19.7	19.9	18.4–22.5	16.0	14.4–19.0	20.5	19.2–21.8
	3rd gill opening	21.5	21.1–22.4	22.4	21.0–25.0	18.6	17.6–21.0	23.1	22.2–23.9
	5th gill opening	23.7	23.1–24.5	24.7	22.5–26.8	20.4	19.5–21.8	25.0	24.1–25.5
	pectoral origin	23.6	21.3–23.8	23.9	21.5–26.1	19.9	19.0–20.9	24.0	23.0–24.7
	pelvic origin	49.9	47.6–53.3	48.8	46.4–50.3	45.2	43.0–47.2	50.1	47.6–52.5
	1st dorsal origin	28.7	28.0–29.4	32.6	27.9–34.0	29.2	26.3–31.3	32.7	31.3–33.7
	2nd dorsal origin	62.2	60.7–66.2	64.8	60.9–66.7	60.8	59.2–64.2	62.6	61.1–65.3
	anal fin origin	62.6	60.7–66.2	62.4	58.4–64.2	59.6	57.3–62.3	61.7	59.2–63.8
	upper caudal origin	74.7	72.9–77.9	76.2	71.0–77.8	74.2	72.3–76.8	73.1	71.2–74.3
	lower caudal origin	73.8	71.7–77.3	74.6	69.4–76.3	73.2	71.4–75.9	72.1	70.2–73.3
Nostril	distance between inner corners	6.1	5.8–6.4	5.5	5.0–6.0	4.4	3.8–4.9	5.3	4.8–5.8
Mouth	wide	8.5	7.8–9.1	7.7	7.2–8.3	6.3	5.6–7.9	7.2	7.0–7.7
	length	4.7	4.2–5.1	4.9	4.3–5.2	4.1	3.1–4.8	4.7	4.5–5.1
Labial furrow lengths	upper	0.6	0.4–0.8	0.5	0.3–0.6	0.5	0.2–0.5	0.5	0.4–0.6
	lower	0.5	0.4–0.7	0.5	0.4–0.7	0.4	0.2–0.5	0.5	0.4–0.6
Gill opening lengths	1st	3.1	2.3–3.7	2.6	2.1–2.9	2.3	1.9–2.6	2.6	2.2–3.0
	3rd	3.2	2.4–3.7	3.1	2.5–3.4	2.8	2.3–3.2	3.1	2.7–3.3
	5th	2.3	1.7–2.9	2.3	2.2–2.8	2.2	1.9–2.4	1.9	1.7–2.1
Eye	horizontal diameter	2.2	1.7–2.9	2.1	1.6–2.5	2.4	2.2–2.9	2.3	1.8–3.1
1st dorsal fin	length of base	12.0	11.1–13.8	10.2	9.1–11.2	9.0	8.3–9.7	8.9	8.6–9.3
	length posterior margin	4.3	3.6–5.3	4.7	4.4–5.4	4.3	3.9–4.9	4.1	3.6–4.6
	heigth	10.9	8.4–15.0	8.6	6.9–9.9	9.7	8.2–11.0	7.6	6.4–8.4
2nd dorsal fin	length of base	4.7	4.2–5.4	3.8	3.1–4.5	4.9	4.2–6.1	2.7	2.4–3.3
	length posterior margin	3.4	3.0–3.8	3.7	3.5–4.1	3.6	3.1–4.0	3.8	3.6–4.0
	heigth	2.9	2.6–3.3	2.5	2.0–2.8	3.6	2.9–4.3	1.8	1.6–1.9
Anal fin	length of base	4.7	4.2–5.8	4.4	3.8–5.0	5.4	4.3–6.2	3.5	3.0–4.3
	length posterior margin	3.3	2.8–3.6	3.7	3.4–4.0	3.5	2.6–4.0	3.6	3.4–3.9
	heigth	3.6	3.2–4.3	3.1	2.8–3.4	3.6	3.0–4.1	2.1	2.0–2.4
Pectoral fin	length of base	6.9	5.9–8.4	5.8	5.2–6.1	5.6	5.1–6.0	5.7	5.3–6.1
	length anterior margin	18.8	16.9–22.4	14.9	13.2–16.4	15.5	13.3–17.9	17.9	15.8–19.7
	length distal margin	14.0	8.5–19.1	11.3	8.0–14.0	11.4	9.9–12.7	13.8	9.6–15.8
	greatest width	10.3	9.2–11.8	9.2	8.0–10.5	8.8	8.0–9.3	9.1	8.2–9.7
Pelvic fin	length of base	5.5	4.9–6.0	4.5	3.8–5.0	5.5	4.6–5.7	4.7	4.5–5.1
	length anterior margin	5.9	5.1–6.9	5.3	4.8–5.8	6.5	5.7–7.3	5.2	4.7–5.3
	length distal margin	5.9	5.2–7.2	5.0	4.0–5.8	5.0	4.4–5.5	5.1	4.1–5.5
	length of claspers	3.7	2.0–10.0	2.1	1.8–2.2	5.3	2.0–11.5	1.9	–
Caudal	length of upper lobe	26.4	23.8–28.2	24.3	23.0–26.2	26.3	23.9–27.7	27.1	26.1–29.0
	length of lower lobe	11.0	10.1–12.0	10.9	9.1–11.4	11.7	9.4–12.5	11.9	11.2–12.7
Trunk at pectoral origin	width	12.6	11.1–13.8	11.9	11.1–12.7	10.7	9.7–11.6	11.4	10.7–11.8
	heigth	11.2	10.1–13.1	11.6	9.8–11.8	10.3	9.2–10.7	11.1	10.3–11.5
Vertebrae	precaudal	90	82–97	59	41–67	78	74–85	101	98–104
	caudal	85	70–92	65	55–73	76	71–82	86	84–87
	total	175	152–189	124	96–135	154	148–167	187	182–191

C. sorrah		C. tilstoni		C. wheeleri		C. acarenatus		C. leiodon	Carcharhinus sp. A	
Mean	Min–Max	Mean	Min–Max	Mean	Min–Max	Mean	Min–Max		Mean	Min–Max
3.4	3.2–3.8	3.6	–	2.8	2.4–3.2	4.4	4.2–4.5	3.2	4.3	3.7–4.6
7.1	6.5–7.8	8.9	7.9–9.7	6.8	6.2–7.5	6.9	6.2–7.4	6.7	7.9	7.1–8.5
7.7	7.1–8.4	7.7	7.1–8.3	7.2	6.4–7.9	7.6	7.4–7.8	6.8	8.5	7.4–9.2
17.3	15.9–18.4	20.5	19.9–21.0	19.3	18.6–20.3	20.3	19.5–21.2	19.8	20.1	18.8–21.3
19.5	18.1–21.1	–	–	21.6	20.8–22.3	–	–	22.5	22.4	21.0–24.1
21.3	19.5–23.6	24.9	24.1–26.2	23.4	22.1–24.3	–	–	25.3	24.2	22.5–25.8
19.8	18.2–22.0	24.5	23.4–25.8	23.4	21.8–23.6	23.8	22.9–24.6	24.9	23.2	21.5–24.4
45.9	44.4–50.8	50.4	48.7–52.6	48.7	46.2–52.1	51.9	51.0–53.6	50.2	47.8	46.4–49.0
29.0	27.7–31.5	33.2	30.8–36.9	30.5	29.3–31.6	32.1	31.2–33.0	32.6	29.5	27.9–31.8
60.2	58.4–65.8	68.7	59.4–71.0	61.3	57.4–64.6	65.4	64.6–66.6	64.4	62.3	60.9–63.3
58.8	57.1–63.6	63.7	60.7–72.3	61.8	58.8–64.8	64.4	63.6–65.0	64.0	60.0	58.4–60.8
70.0	68.4–74.2	71.7	69.5–74.3	72.9	70.1–76.5	–	–	76.1	72.9	71.0–75.0
69.4	68.0–73.8	–	–	71.8	69.1–75.9	–	–	75.7	71.6	69.4–73.6
5.5	4.9–6.1	5.8	5.1–6.5	6.3	5.6–7.1	5.8	5.5–5.9	6.3	6.3	6.0–6.6
7.3	6.3–8.3	8.9	8.4–9.9	8.8	7.5–10.8	8.2	7.5–8.6	9.6	8.2	7.8–8.5
4.3	3.9–6.0	–	–	5.2	5.0–5.5	–	–	5.5	4.7	4.3–5.1
0.3	0.2–0.6	0.4	0.2–0.6	0.5	0.4–0.5	0.7	0.6–0.8	0.4	0.4	0.3–0.6
0.4	0.2–0.5	0.3	0.2–0.4	0.5	0.4–0.8	–	–	0.6	0.5	0.4–0.6
2.1	1.8–2.6	3.5	2.6–4.2	2.6	2.3–3.0	–	–	4.0	2.3	2.1–2.7
2.6	2.1–3.3	–	–	3.0	2.5–3.3	–	–	4.5	2.8	2.5–3.3
2.0	1.6–2.4	3.2	2.6–3.3	2.4	2.0–2.6	–	–	3.2	2.3	2.2–2.4
2.1	1.7–2.4	1.7	1.3–2.1	2.5	1.9–2.9	1.1	1.0–1.4	1.9	2.3	2.2–2.4
9.4	8.6–10.4	10.2	8.6–11.4	9.6	8.3–10.2	10.1	9.6–10.8	11.2	10.9	9.5–11.8
3.8	3.3–4.1	13.6*	11.4*–15.8*	4.2	3.3–5.1	–	–	4.4	5.3	4.6–6.0
8.5	7.7–9.7	10.4	9.9–10.9	9.7	8.6–10.7	9.2	8.8–9.6	10.5	8.5	7.6–9.8
3.0	2.6–3.3	3.9	3.8–4.0	4.1	3.6–4.8	3.7	3.3–4.1	5.1	3.6	3.2–4.1
4.0	3.6–4.5	4.0*	3.8*–4.3*	4.0	3.3–4.5	–	–	3.5	3.7	3.5–4.1
1.8	1.5–2.2	2.9	2.7–3.4	3.3	2.8–4.1	2.6	2.3–2.8	4.1	2.2	2.0–2.3
4.2	3.7–4.7	5.0	3.9–6.1	4.3	4.0–4.7	3.9	3.8–4.0	5.6	4.6	4.5–4.7
4.0	3.7–4.6	4.9*	4.5*–5.7*	3.6	3.3–4.1	–	–	3.3	3.5	3.3–3.8
2.7	2.2–3.0	–	–	3.4	3.0–4.3	3.8	3.6–4.0	4.4	2.7	2.7–2.8
5.7	5.1–6.9	5.6	5.2–5.9	5.3	4.8–6.3	5.9	5.8–6.2	6.5	6.1	5.8–6.6
15.6	14.5–17.1	–	–	18.6	17.9–21.4	20.0	17.8–21.5	17.4	15.5	14.4–16.7
11.5	10.2–14.0	–	–	14.8	11.9–18.6	–	–	14.5	11.1	9.7–12.9
7.7	6.8–8.4	–	–	9.2	7.9–10.8	10.5	9.8–11.1	10.8	8.9	8.5–9.3
4.9	4.3–5.8	5.5	4.9–6.6	5.0	4.6–5.6	–	–	6.3	4.6	4.5–4.7
4.7	4.4–5.6	5.6*	5.1*–6.5*	5.3	4.8–6.0	–	–	6.7	5.1	4.8–5.3
4.4	3.9–4.9	–	–	5.3	4.4–6.8	–	–	7.1	4.9	4.4–5.2
2.3	2.0–2.9	4.6	2.1–7.1	4.0	1.7–7.5	–	–	2.8	–	–
30.1	27.0–31.7	28.0	25.2–29.9	27.8	23.8–29.9	27.7	26.9–28.1	24.8	27.3	25.2–28.8
11.8	10.7–12.5	13.5	12.7–14.3	14.2	13.0–15.5	12.6	12.0–13.2	13.1	11.2	10.5–11.5
10.3	8.8–11.1	11.2	–	11.9	10.5–14.9	–	–	14.2	12.1	11.8–12.6
10.6	8.7–13.4	–	–	10.7	9.3–12.1	–	–	12.8	11.0	10.4–12.1
70	66–79	88	86–91	114	111–115	–	–	115	56	54–57
88	86–93	88	85–90	105	102–112	–	–	83	62	59–66
158	153–175	176	173–178	219	213–227	–	–	198	118	116–120

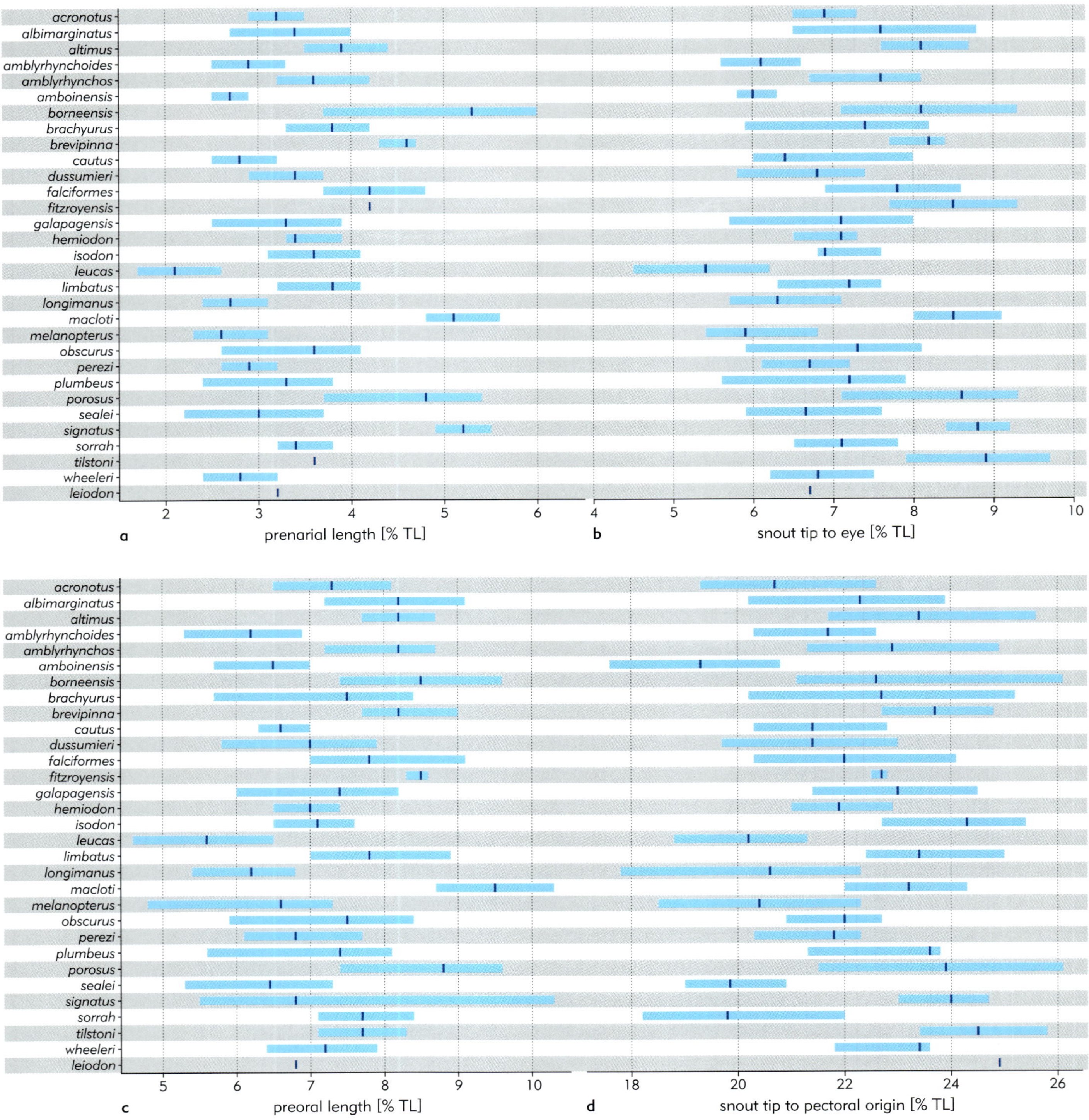

Figure 51. Proportional dimensions of 31 species of the genus *Carcharhinus*. **a**, prenarial length as percentage of total length (TL); **b**, snout tip to eye as percentage of TL; **c**, preoral length as percentage of TL; **d**, snout tip to pectoral origin as percentage of TL (horizontal lines represent ranges; vertical lines are means).

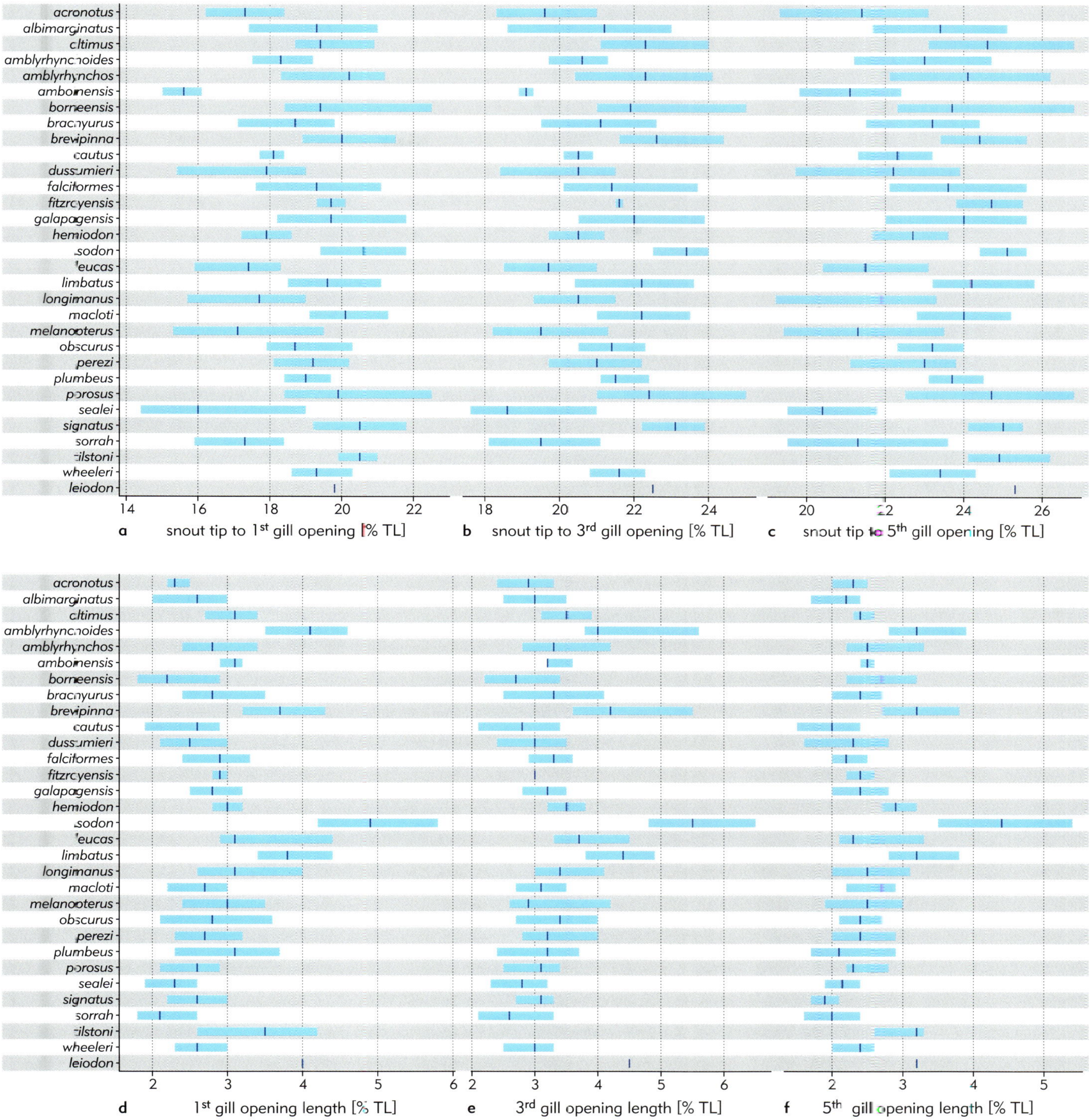

Figure 52. Proportional dimensions of 31 species of the genus *Carcharhinus*. **a**, snout tip to first gill opening as percentage of total length (TL); **b**, snout tip to third gill opening as percentage of TL; **c**, snout tip to fifth gill opening as percentage of TL; **d**, first gill opening length as percentage of TL; **e**, third gill opening length as percentage of TL; **f**, fifth gill opening length as percentage of TL (horizontal lines represent ranges; vertical lines are means).

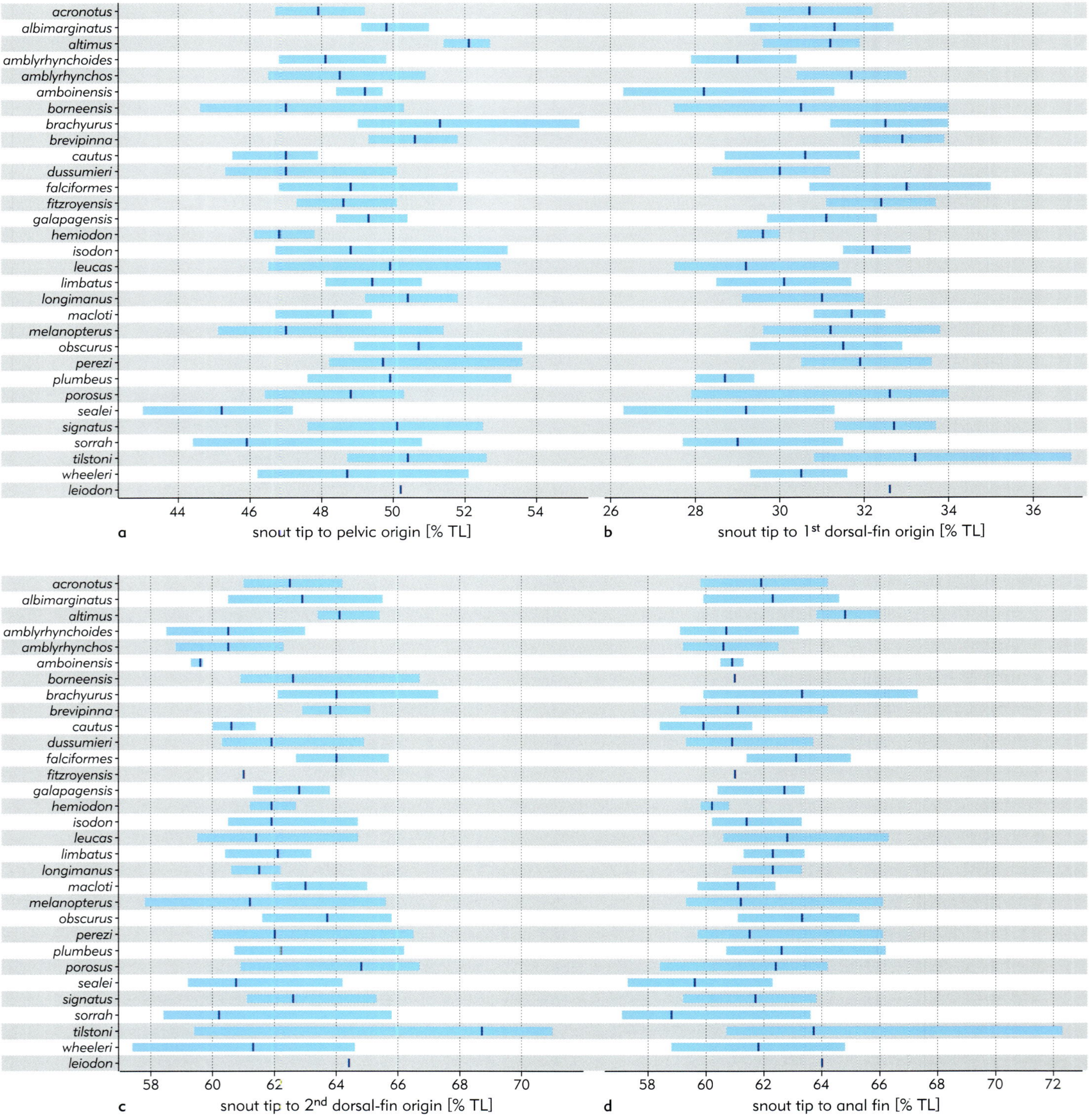

Figure 53. Proportional dimensions of 31 species of the genus *Carcharhinus*. **a**, snout tip to pelvic origin as percentage of total length (TL); **b**, snout tip to first dorsal fin origin as percentage of TL; **c**, snout tip to second dorsal fin origin as percentage of (TL); **d**, snout tip to anal fin as percentage of TL (horizontal lines represent ranges; vertical lines are means).

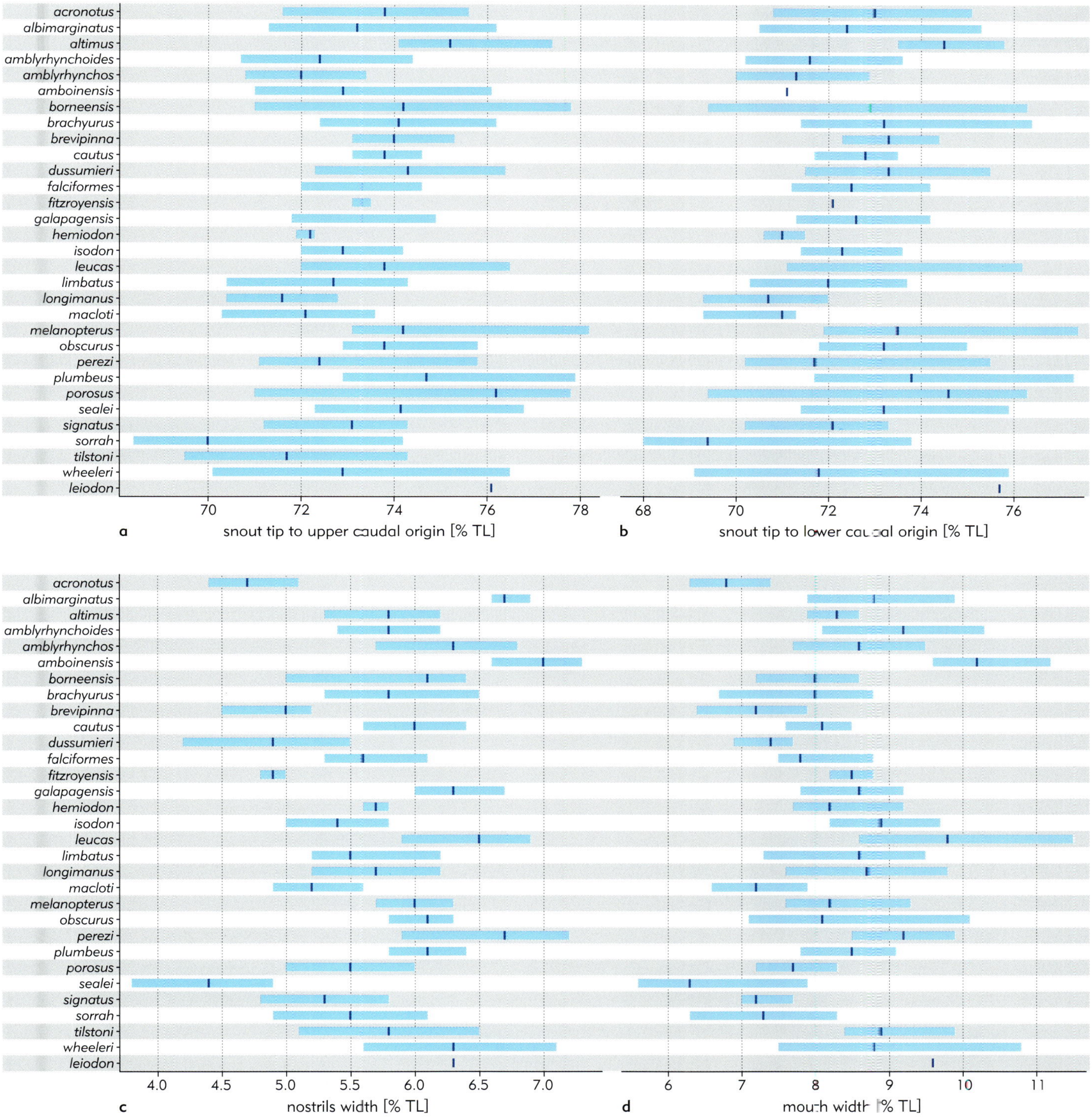

Figure 54. Proportional dimensions of 31 species of the genus *Carcharhinus*. **a**, snout tip to upper caudal origin as percentage of total length (TL); **b**, snout tip to lower caudal fin origin as percentage of TL; **c**, nostrils width as percentage of TL; **d**, mouth width as percentage of TL (horizontal lines represent ranges; vertical lines are means).

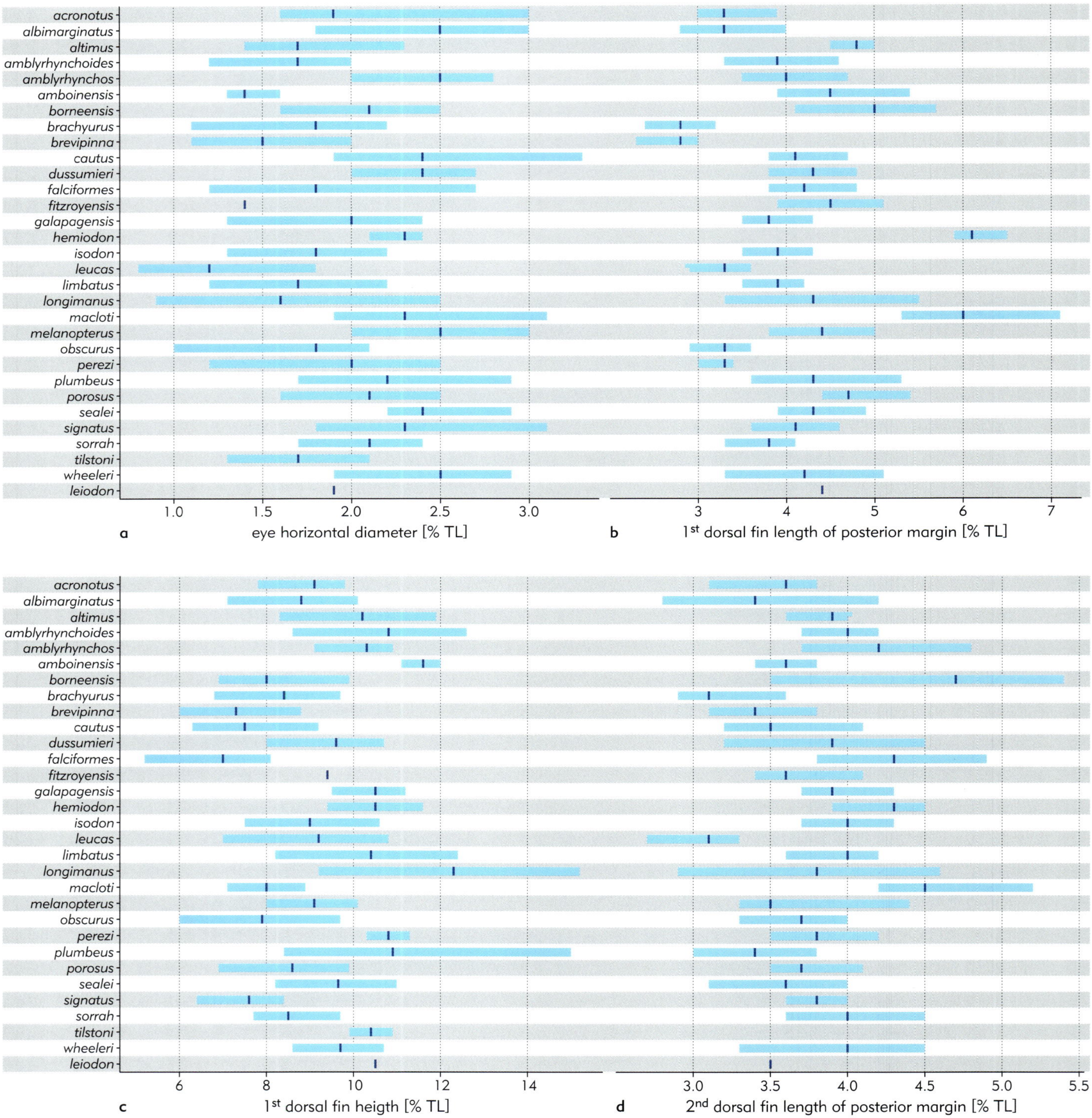

Figure 55. Proportional dimensions of 31 species of the genus *Carcharhinus*. **a**, eye horizontal diameter as percentage of total length (TL); **b**, first dorsal fin length of posterior margin as percentage of TL (**c**) first dorsal fin height as percentage of TL; **d**, second dorsal fin length of posterior margin as percentage of TL (horizontal lines represent ranges; vertical lines are means).

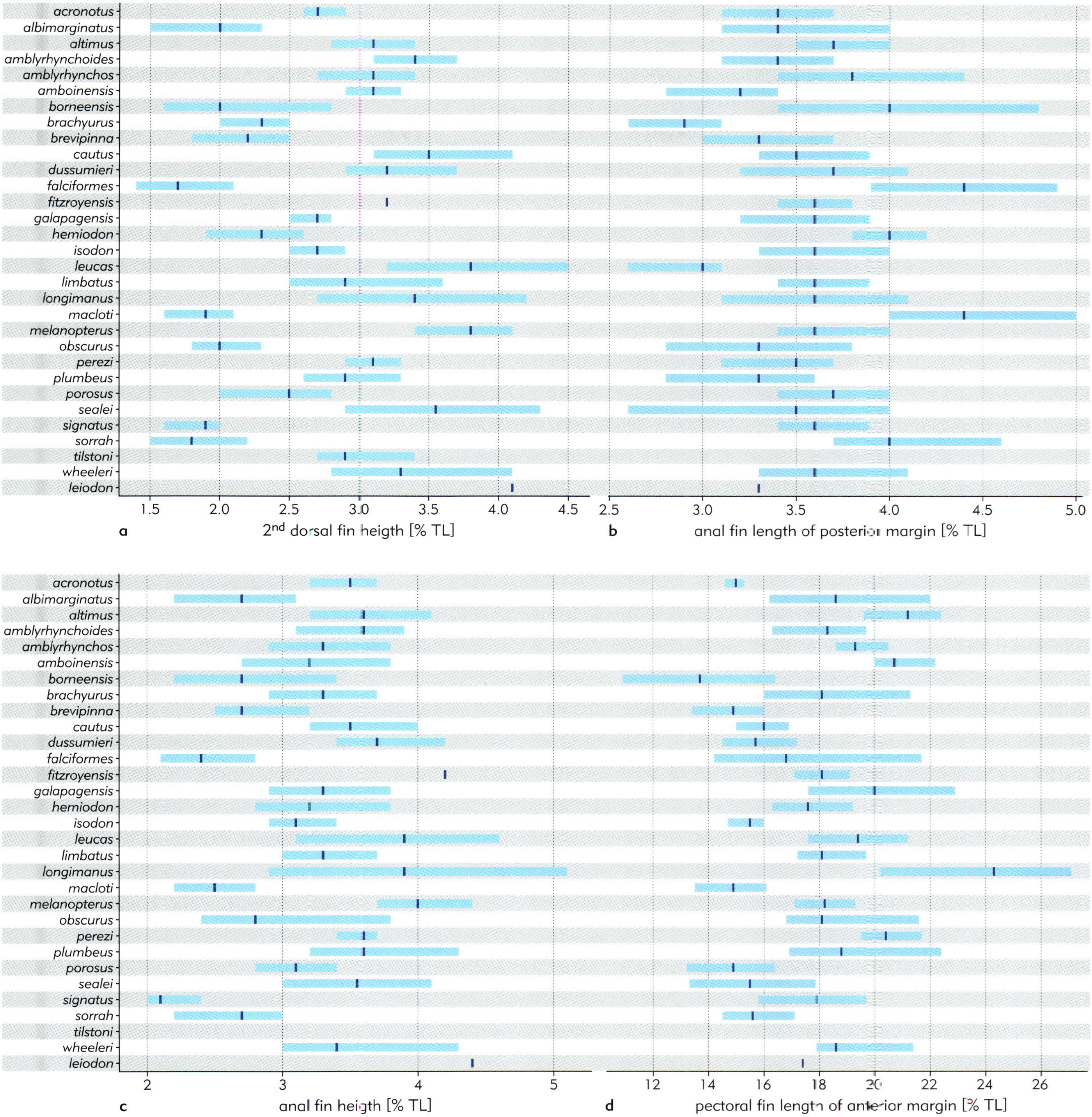

Figure 56. Proportional dimensions of 31 species of the genus *Carcharhinus*. **a**, second dorsal fin height as percentage of total length (TL); **b**, anal fin length of posterior margin as percentage of TL; **c**, anal fin height as percentage of TL; **d**, pectoral fin length of anterior margin as percentage of TL (horizontal lines represent ranges; vertical lines are means).

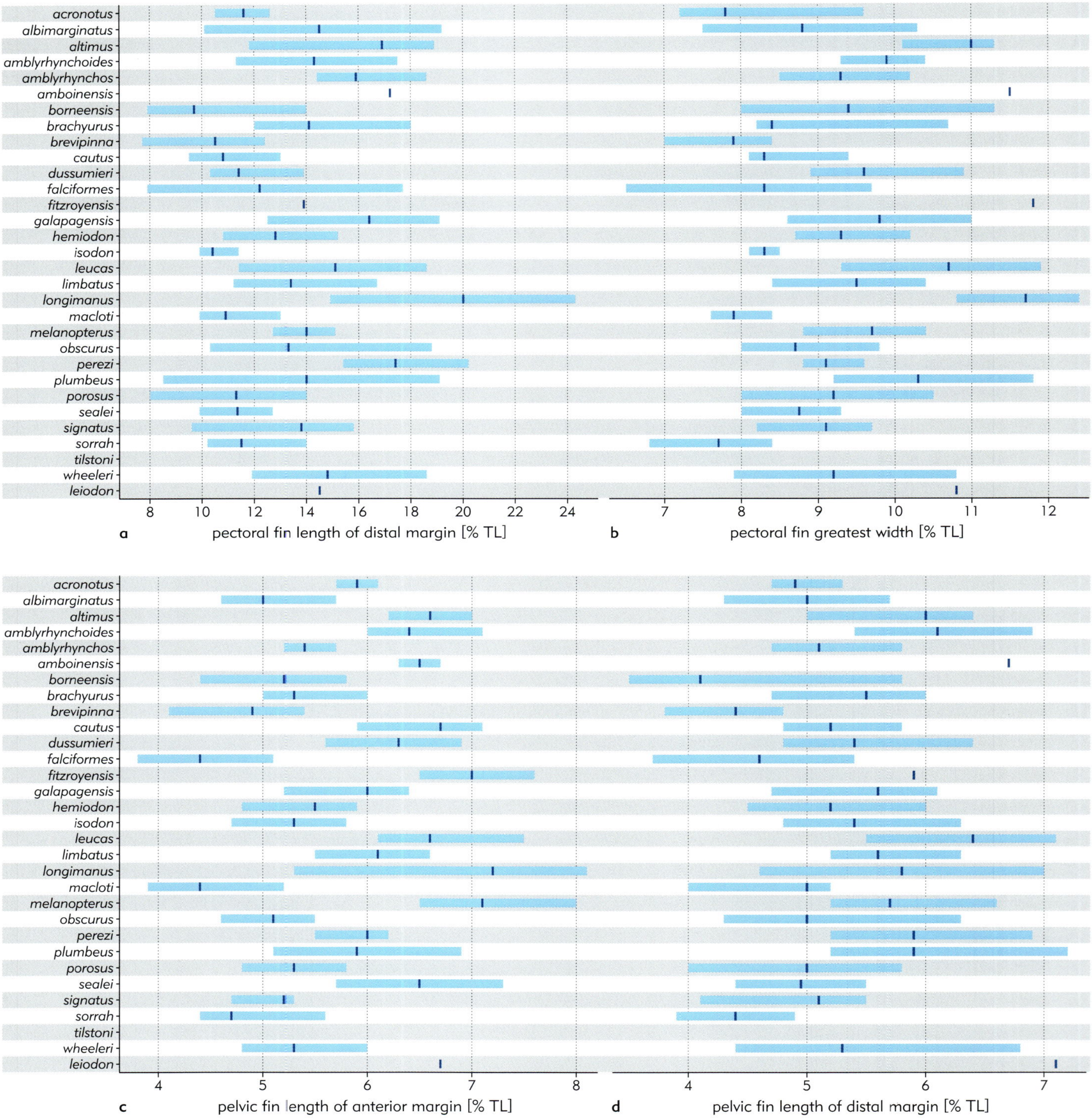

Figure 57. Proportional dimensions of 31 species of the genus *Carcharhinus*. **a**, pectoral fin length of distal margin as percentage of total length (TL); **b**, pectoral fin greatest width as percentage of TL; **c**, pelvic fin length of anterior margin as percentage of TL; **d**, pelvic fin length of distal margin as percentage of TL (horizontal lines represent ranges; vertical lines are means).

Figure 58. Proportional dimensions of 31 species of the genus *Carcharhinus*. **a**, caudal fin length of upper lobe as percentage of total length (TL); **b**, caudal fin length of lower lobe as percentage of TL; **c**, trunk at pectoral origin width as percentage of TL; **d**, trunk at pectoral origin height as percentage of TL (horizontal lines represent ranges; vertical lines are means).

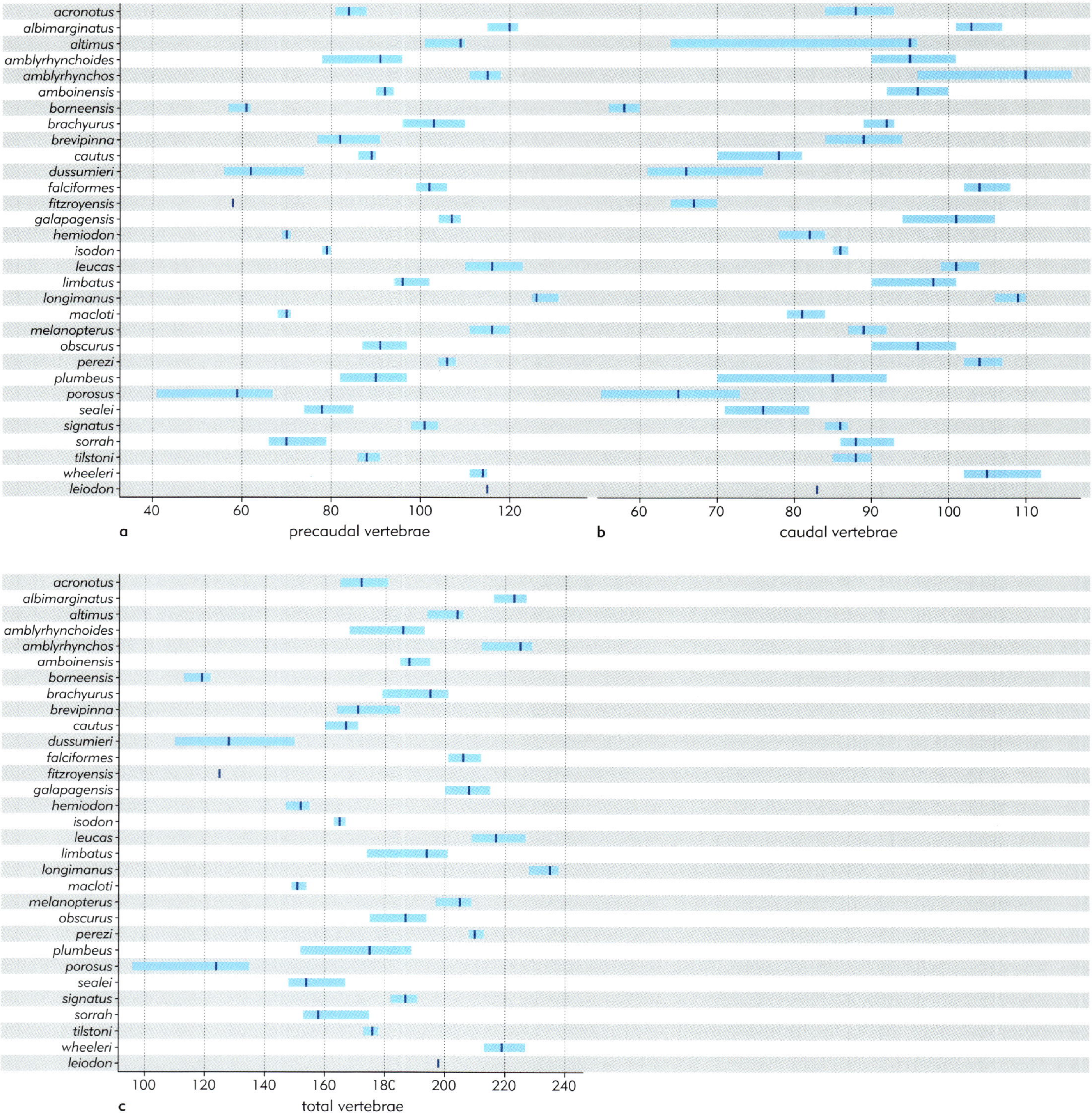

Figure 59. Proportional dimensions of 31 species of the genus *Carcharhinus*. **a**, number of precaudal vertebrae; **b**, number of caudal vertebrae; **c**, number of total vertebrae (horizontal lines represent ranges; vertical lines are means).

Plates 1-31

Plate 1	*Carcharhinus acronotus*
Plate 2	*Carcharhinus albimarginatus*
Plate 3	*Carcharhinus altimus*
Plate 4	*Carcharhinus amblyrhynchoides*
Plate 5	*Carcharhinus amblyrhynchos*
Plate 6	*Carcharhinus amboinensis*
Plate 7	*Carcharhinus borneensis*
Plate 8	*Carcharhinus brachyurus*
Plate 9	*Carcharhinus brevipinna*
Plate 10	*Carcharhinus cautus*
Plate 11	*Carcharhinus dussumieri*
Plate 12	*Carcharhinus falciformes*
Plate 13	*Carcharhinus fitzroyensis*
Plate 14	*Carcharhinus galapagensis*
Plate 15	*Carcharhinus hemiodon*
Plate 16	*Carcharhinus isodon*
Plate 17	*Carcharhinus leiodon*
Plate 18	*Carcharhinus leucas*
Plate 19	*Carcharhinus limbatus*
Plate 20	*Carcharhinus longimanus*
Plate 21	*Carcharhinus macloti*
Plate 22	*Carcharhinus melanopterus*
Plate 23	*Carcharhinus obscurus*
Plate 24	*Carcharhinus perezii*
Plate 25	*Carcharhinus plumbeus*
Plate 26	*Carcharhinus porosus*
Plate 27	*Carcharhinus sealei*
Plate 28	*Carcharhinus signatus*
Plate 29	*Carcharhinus sorrah*
Plate 30	*Carcharhinus tilstoni*
Plate 31	*Carcharhinus wheeleri*

Plate 1. ***Carcharhinus acronotus*****.** Teeth of a male of 119 cm TL caught on Felly Beach, South Carolina, USA.

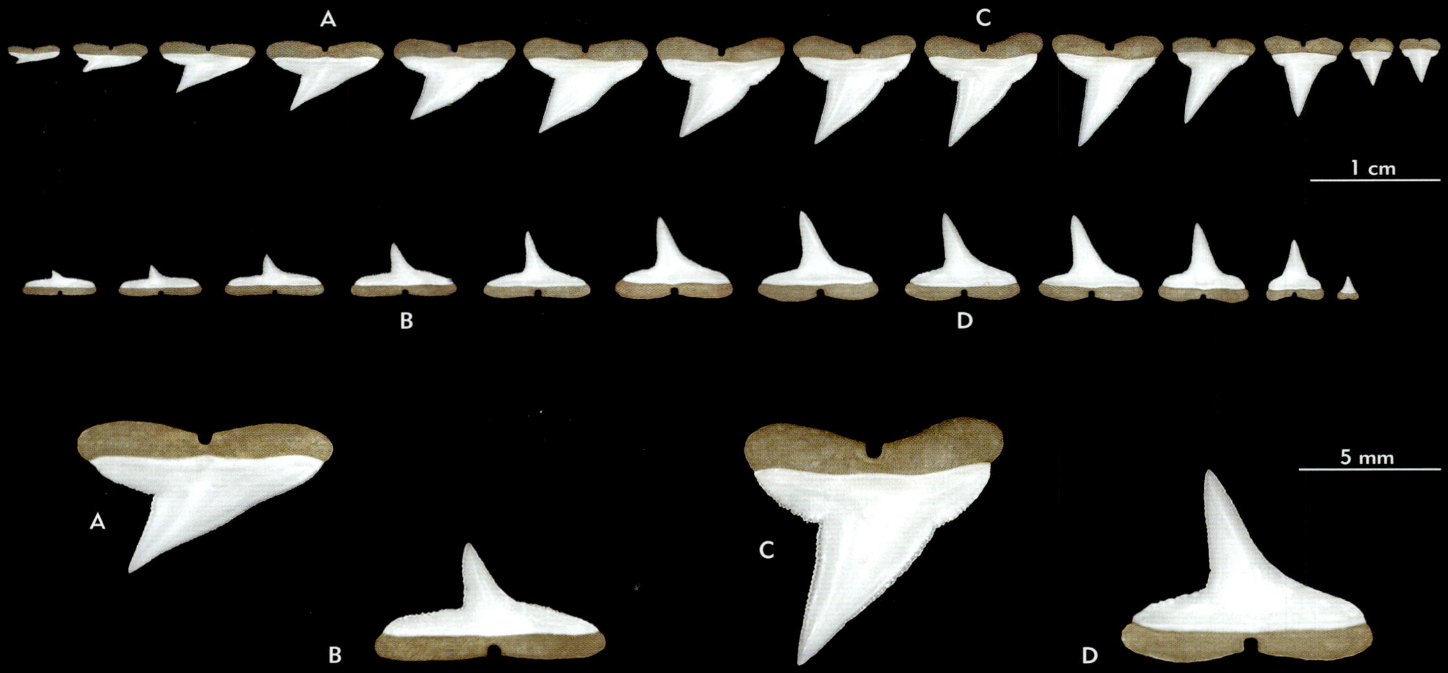

Plate 2. ***Carcharhinus albimarginatus*****.** Teeth of a female of 53 cm TL, from Madagascar.

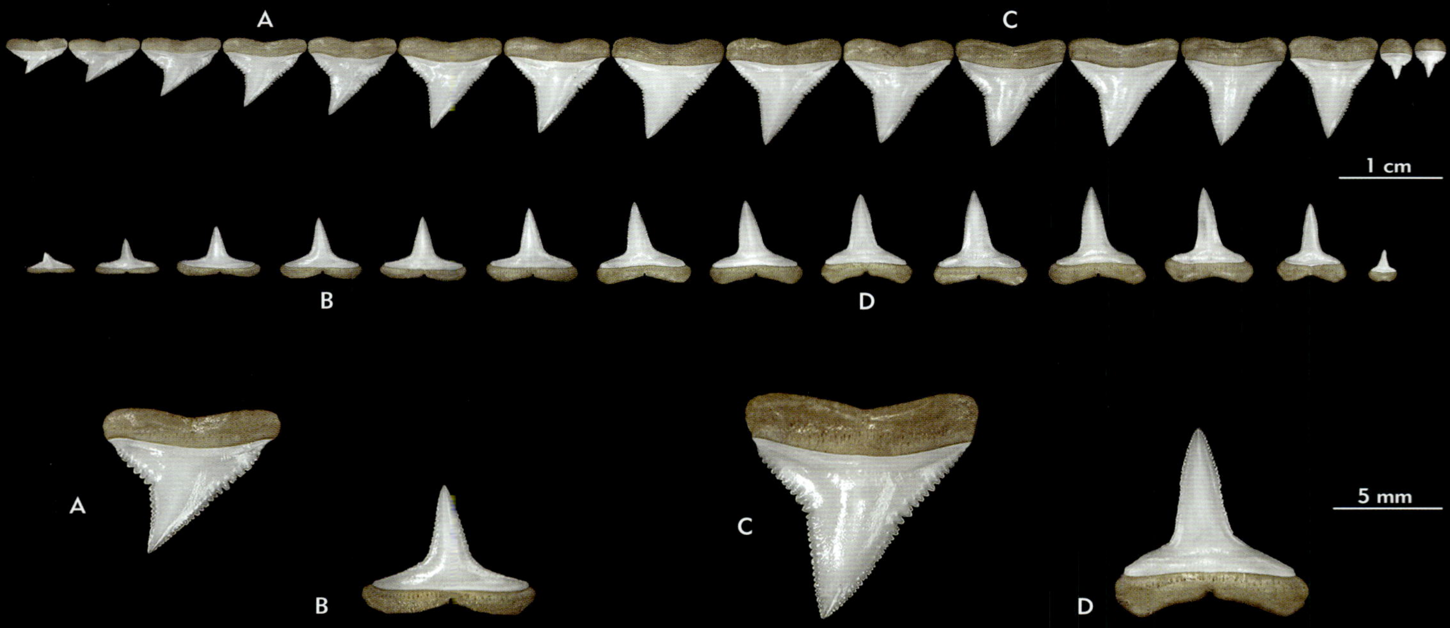

Plate 3. ***Carcharhinus altimus*****.** Teeth of a male of 234 cm TL, from Southeastern Florida, USA.

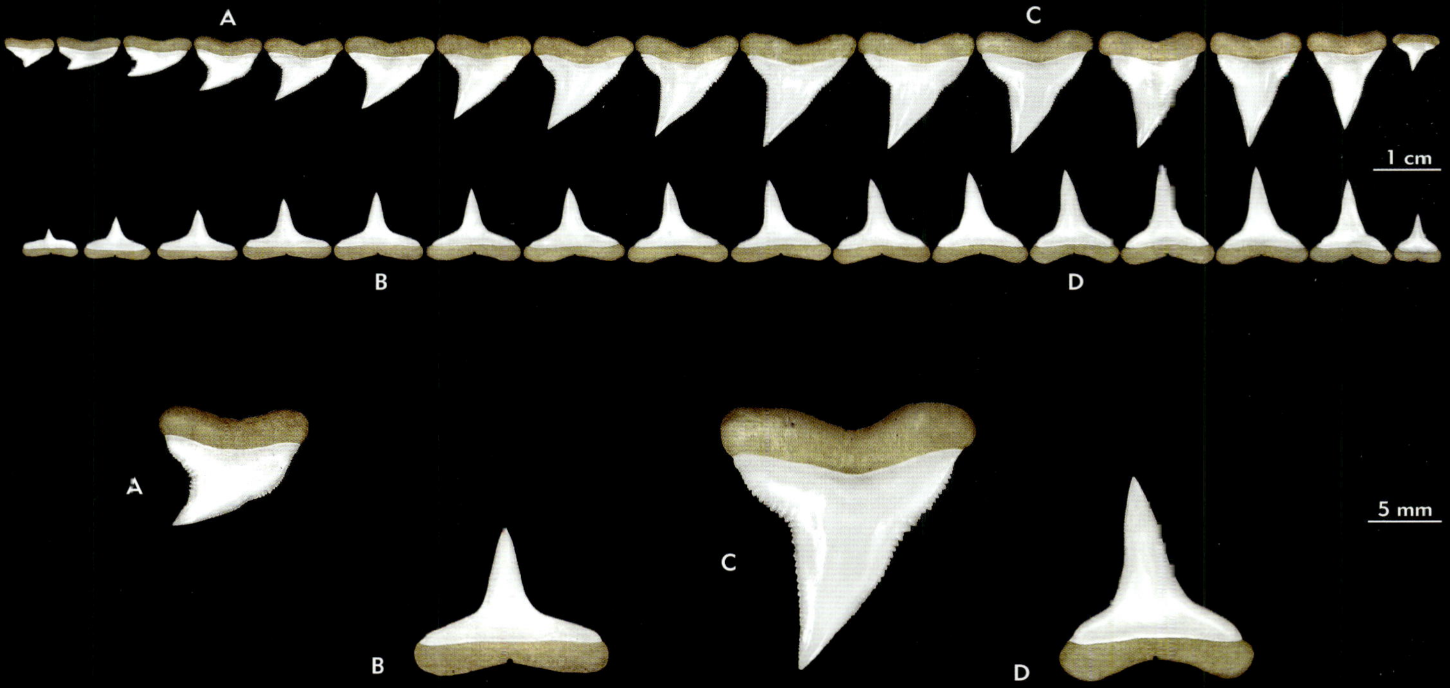

Plate 4. ***Carcharhinus amblyrhynchoides*****.** Teeth of a female of 230 cm TL, from Vietnam, South China Sea.

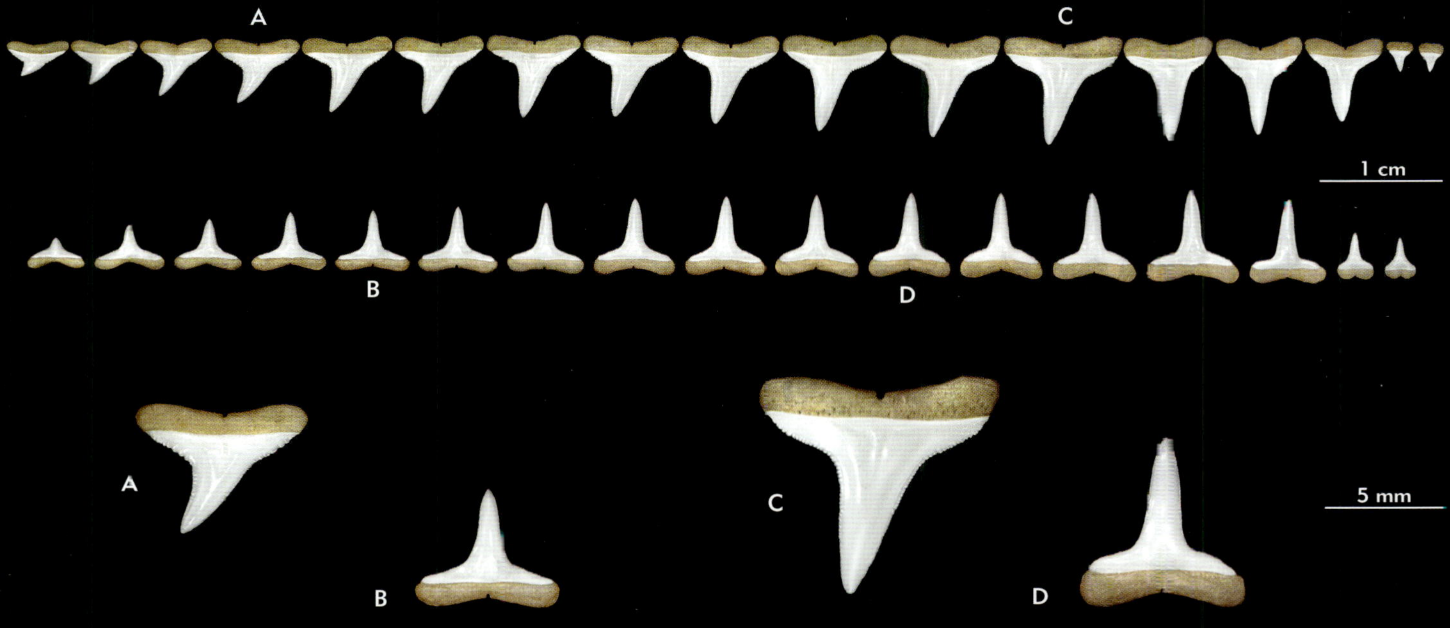

Plate 5. ***Carcharhinus amblyrhynchos*****.** Teeth of a male of 78 cm TL, from Mauritius.

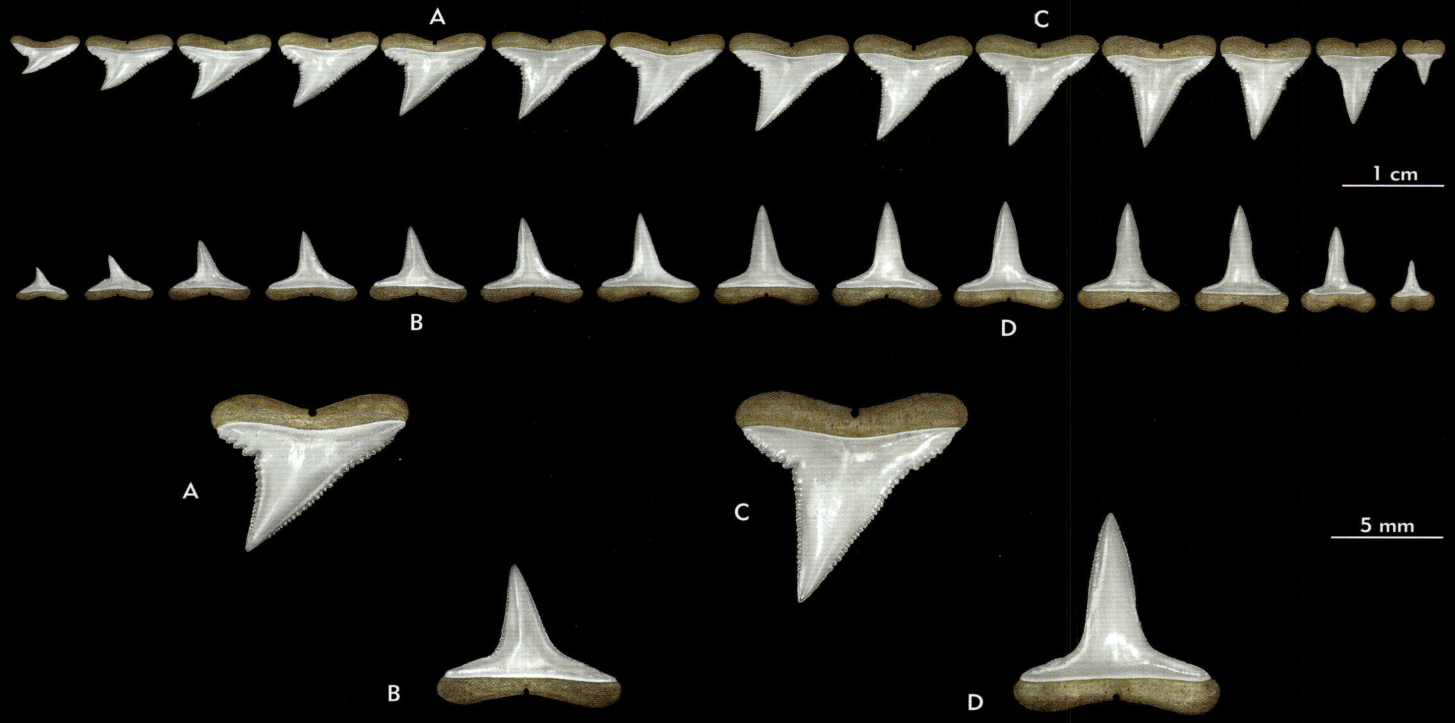

Plate 6. ***Carcharhinus amboinensis*****.** Teeth of a male of 70 cm TL, from Ambon, Moluccas, Indonesia (Holotype, RMNH D 2582).

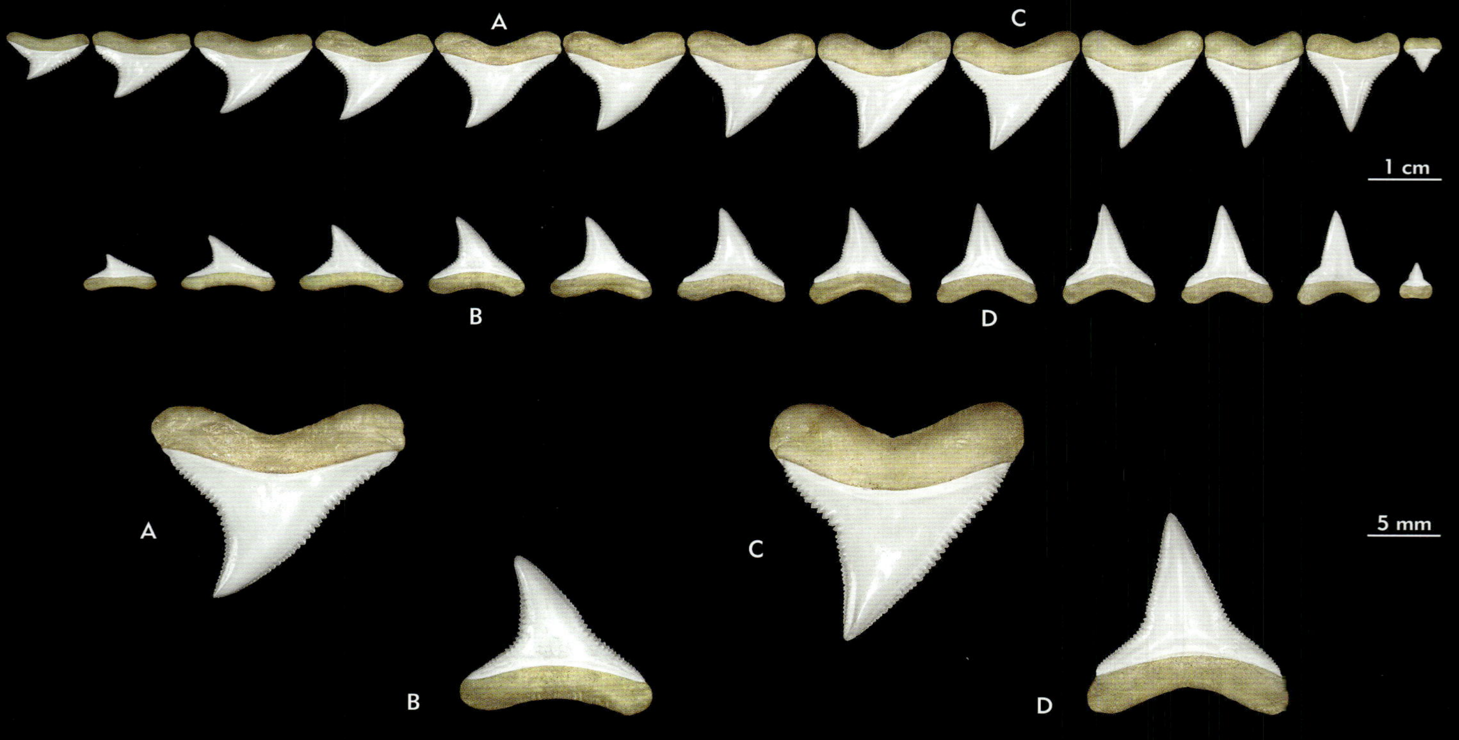

Plate 7. ***Carcharhinus borneensis*****.** Teeth of a female of 24 cm TL, from Sinkawang, Borneo, Indonesia (Holotype, RMNH 7386).

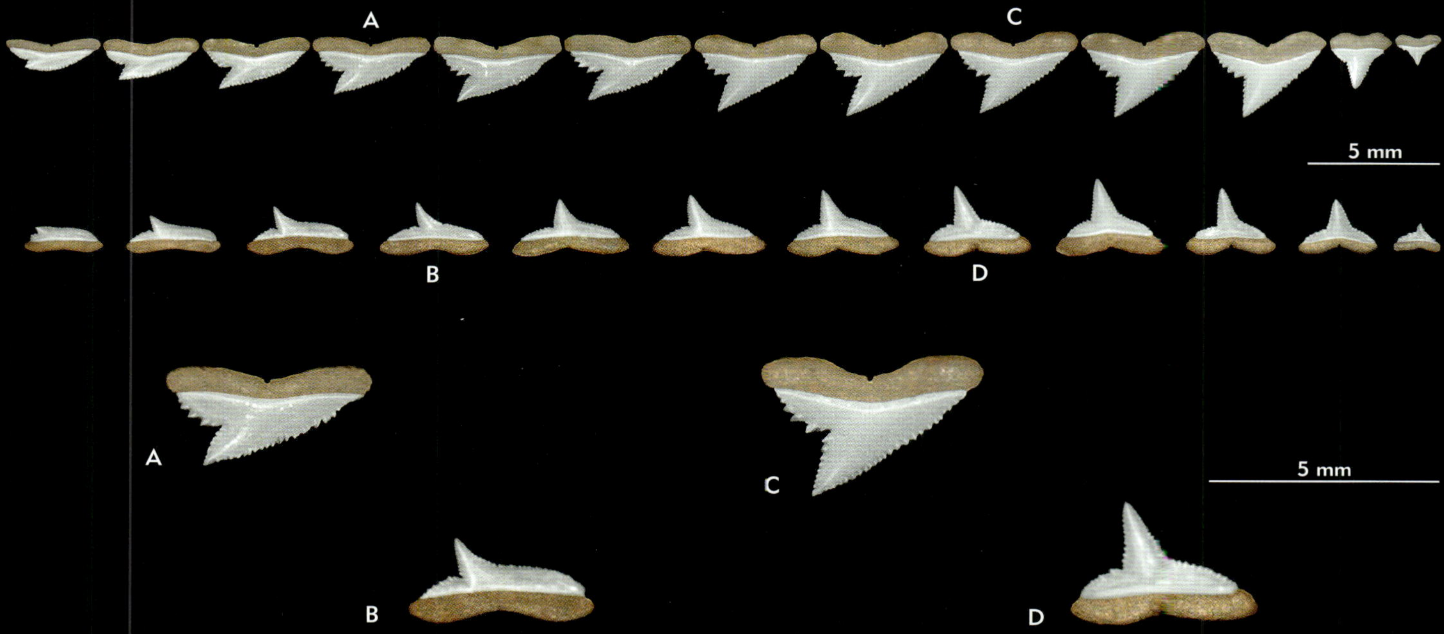

Plate 8. ***Carcharhinus brachyurus*****.** Teeth of a male of 40 cm TL, from the French Mediterranean coast (MNHN 1898-1228).

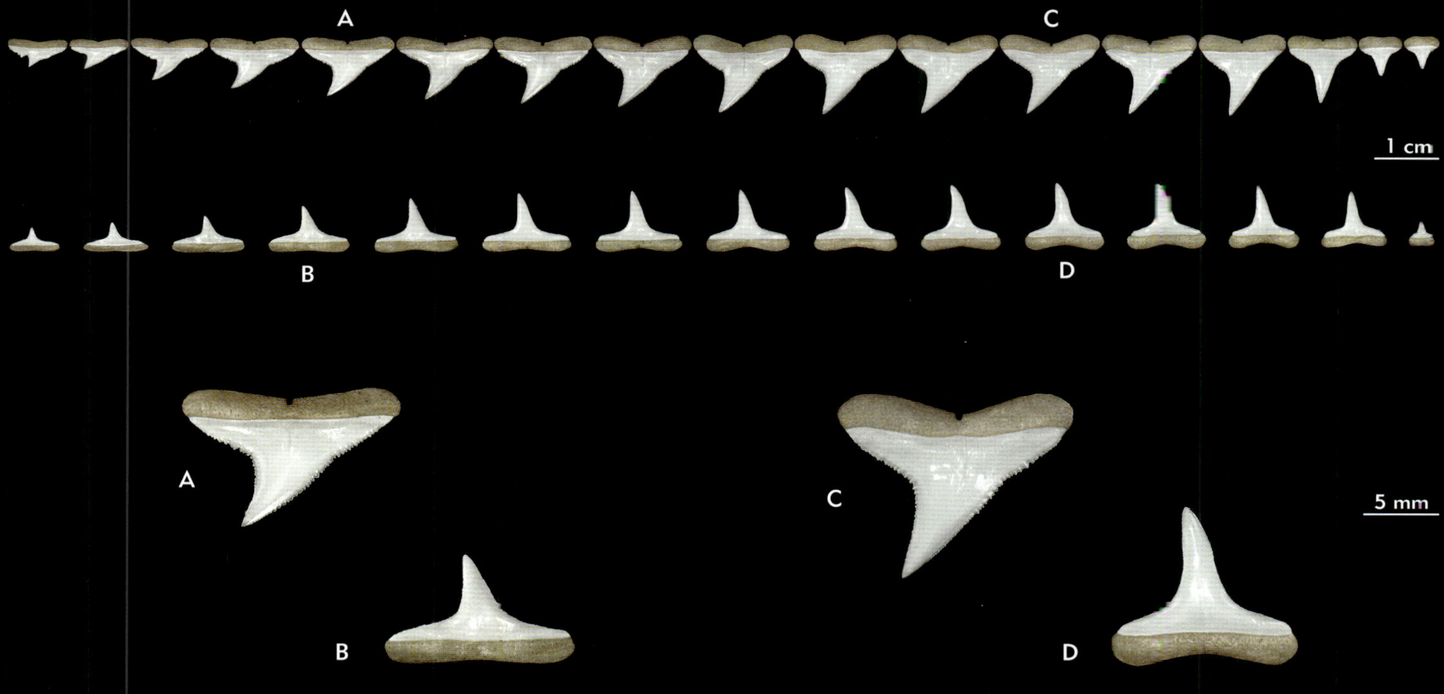

Plate 9. ***Carcharhinus brevipinna***. Teeth of a female of 75.5 cm TL, from Java, Indonesia (Holotype, RMNH-2525.1).

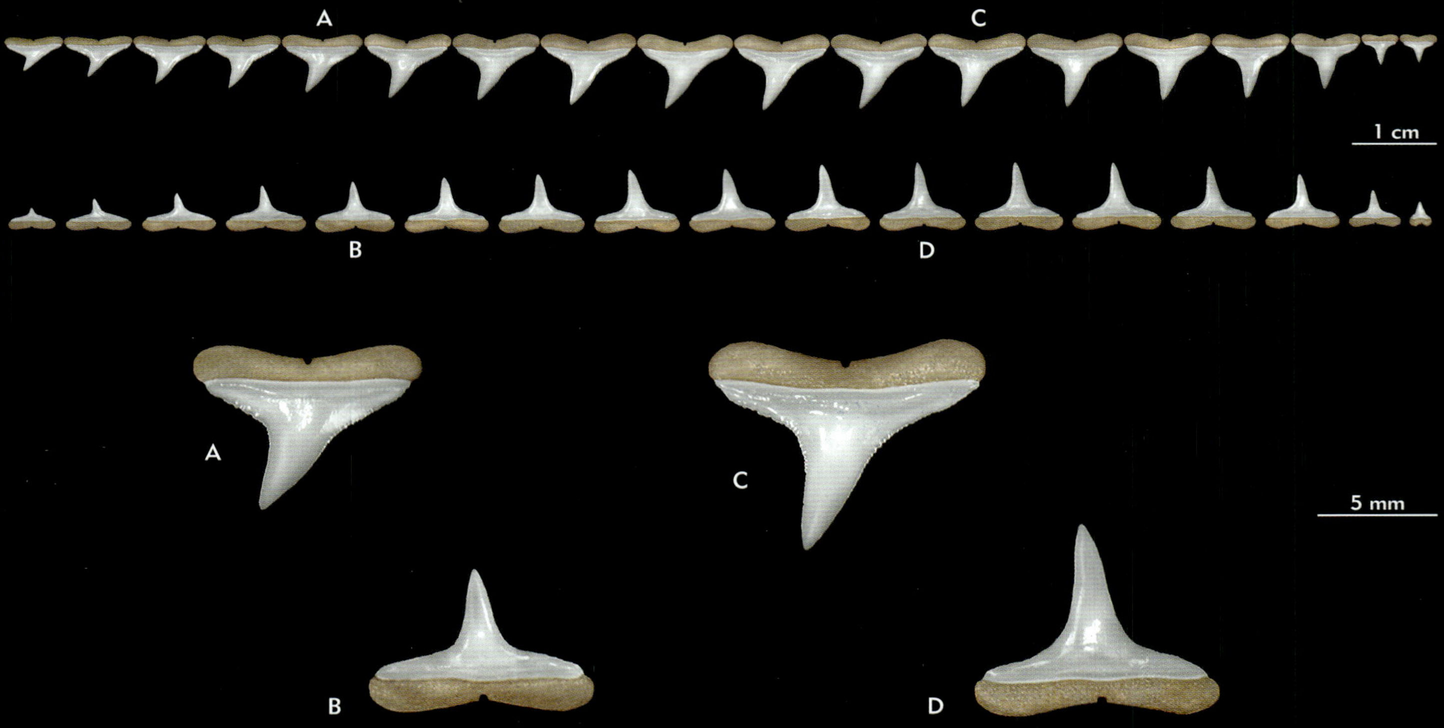

Plate 10. ***Carcharhinus cautus***. Teeth of a female of 92 cm TL, caught at Shark Bay, Western Australia (Holotype, AMS IB 1622).

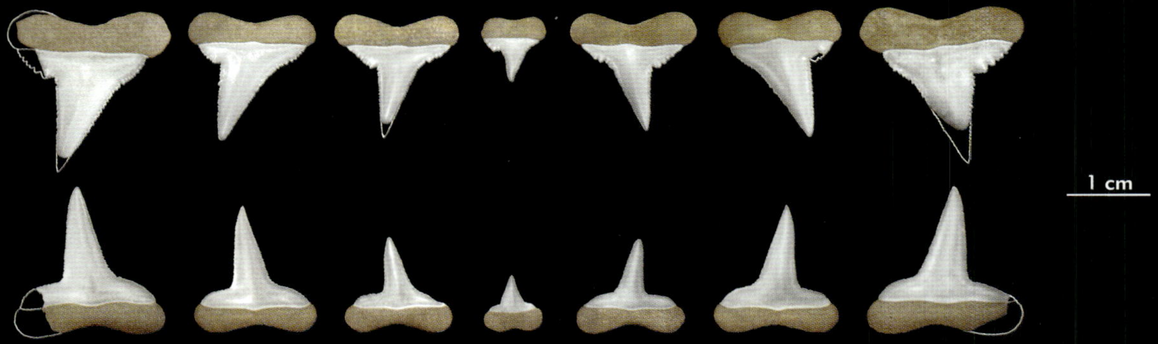

Plate 11. ***Carcharhinus dussumieri***. Teeth of a female of 61.5 cm TL, from Jakarta, Indonesia (Syntype, RMNH 7382).

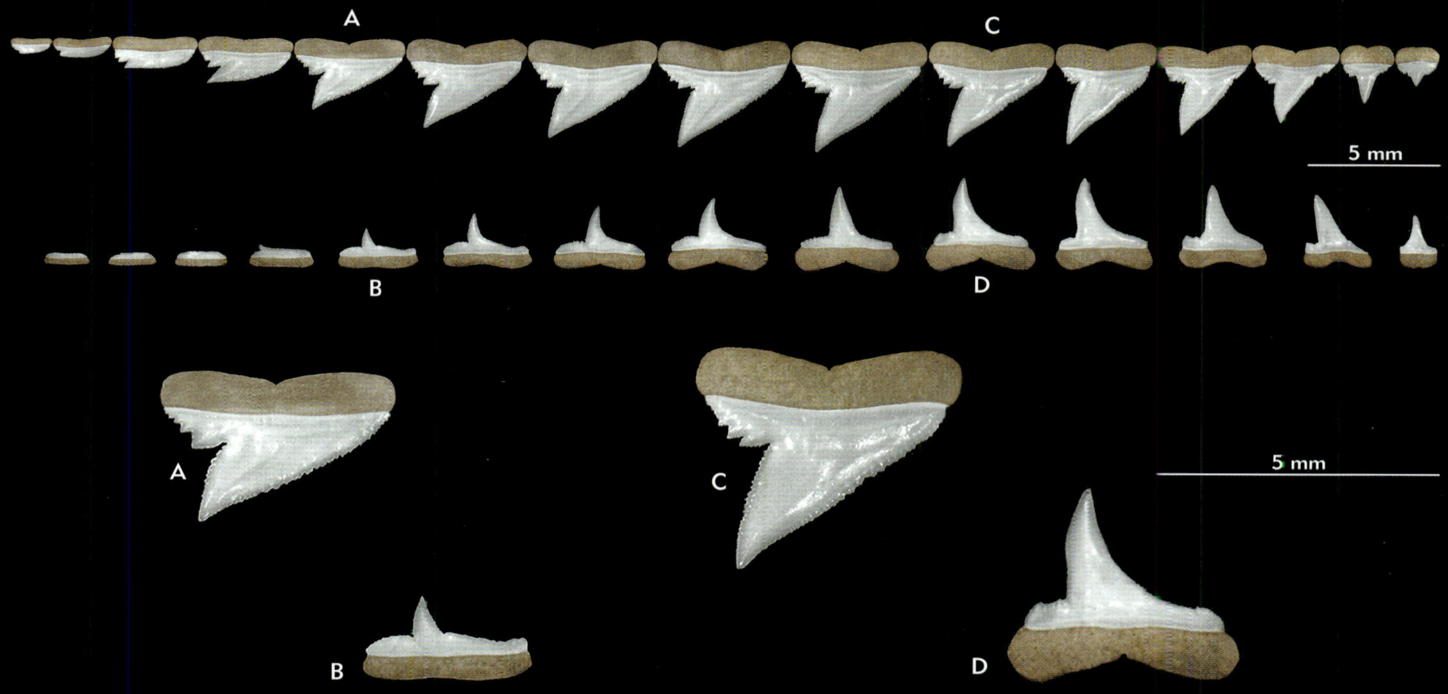

Plate 12. ***Carcharhinus falciformes***. Teeth of a male of 93.5 cm TL, from Madagascar (Syntype, Fish collection, MNHN 1132).

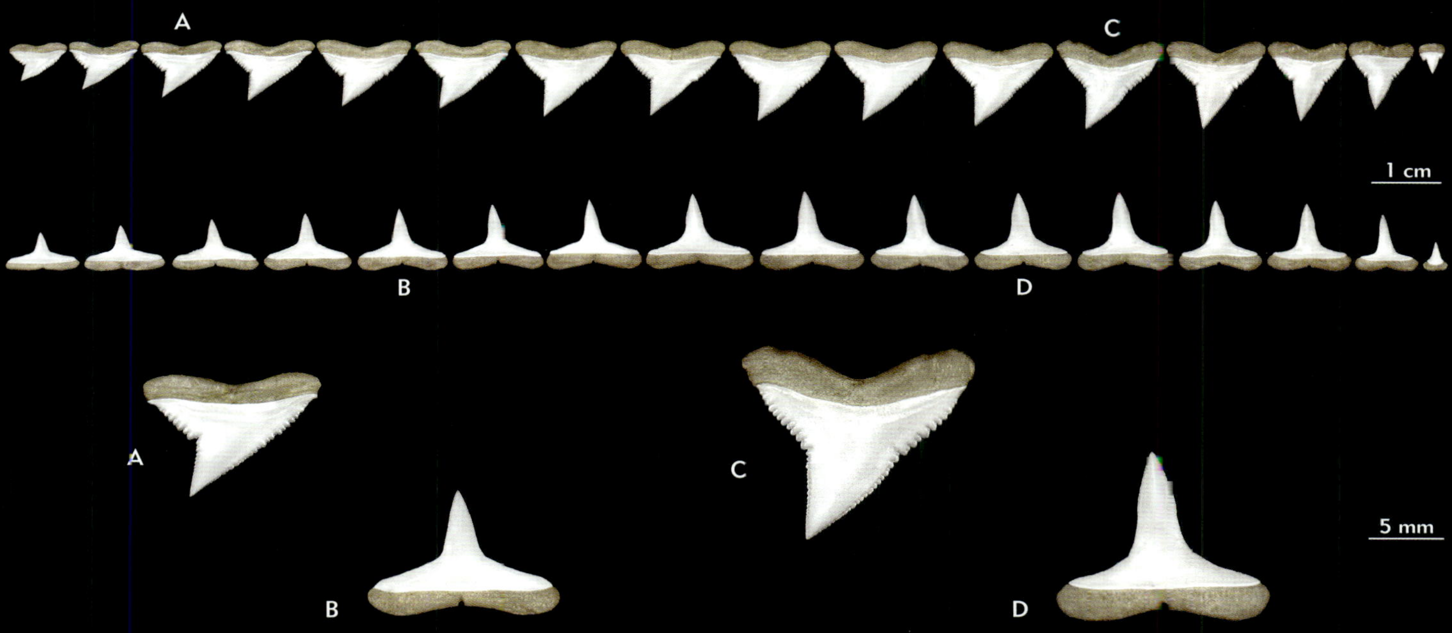

Plate 13. ***Carcharhinus fitzroyensis***. Teeth of a female of 118 cm TL, from Fitzroy River estuary, Queensland, Australia (Holotype, AMS IB1229).

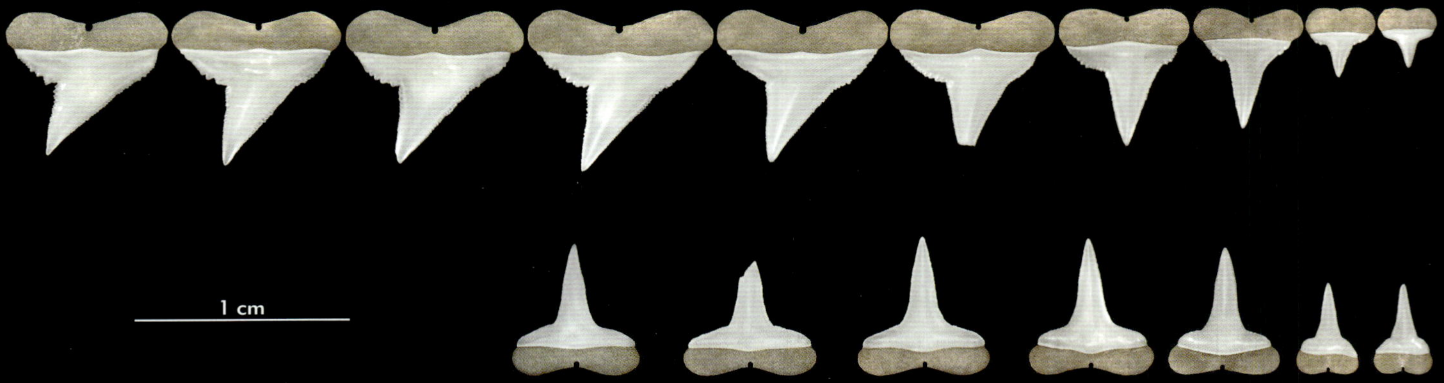

Plate 14. ***Carcharhinus galapagensis***. Teeth of a female of 224 cm TL, from Panama

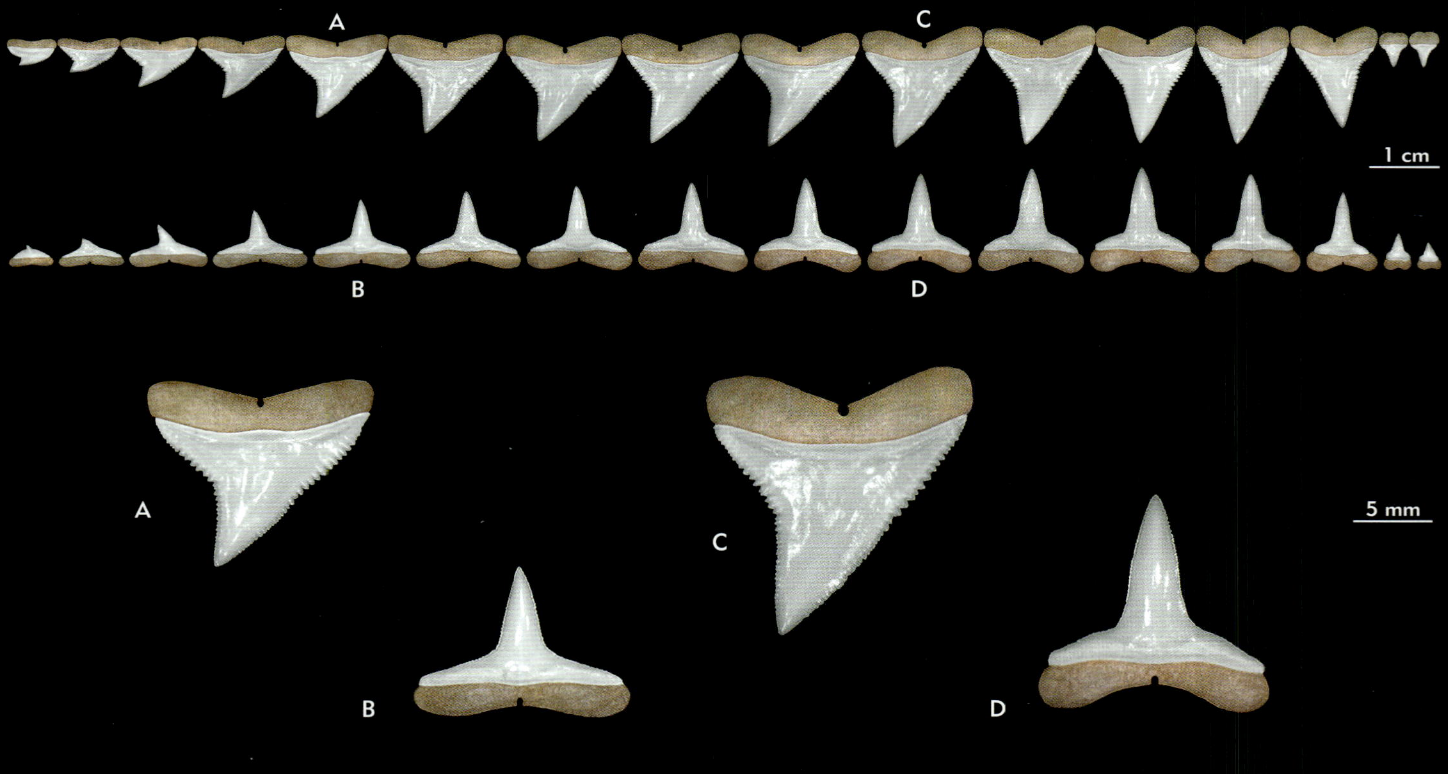

Plate 15. ***Carcharhinus hemiodon*****.** Teeth of a female of 54 cm TL, caught at Borneo, Indonesia (Paratype, NMV-50084).

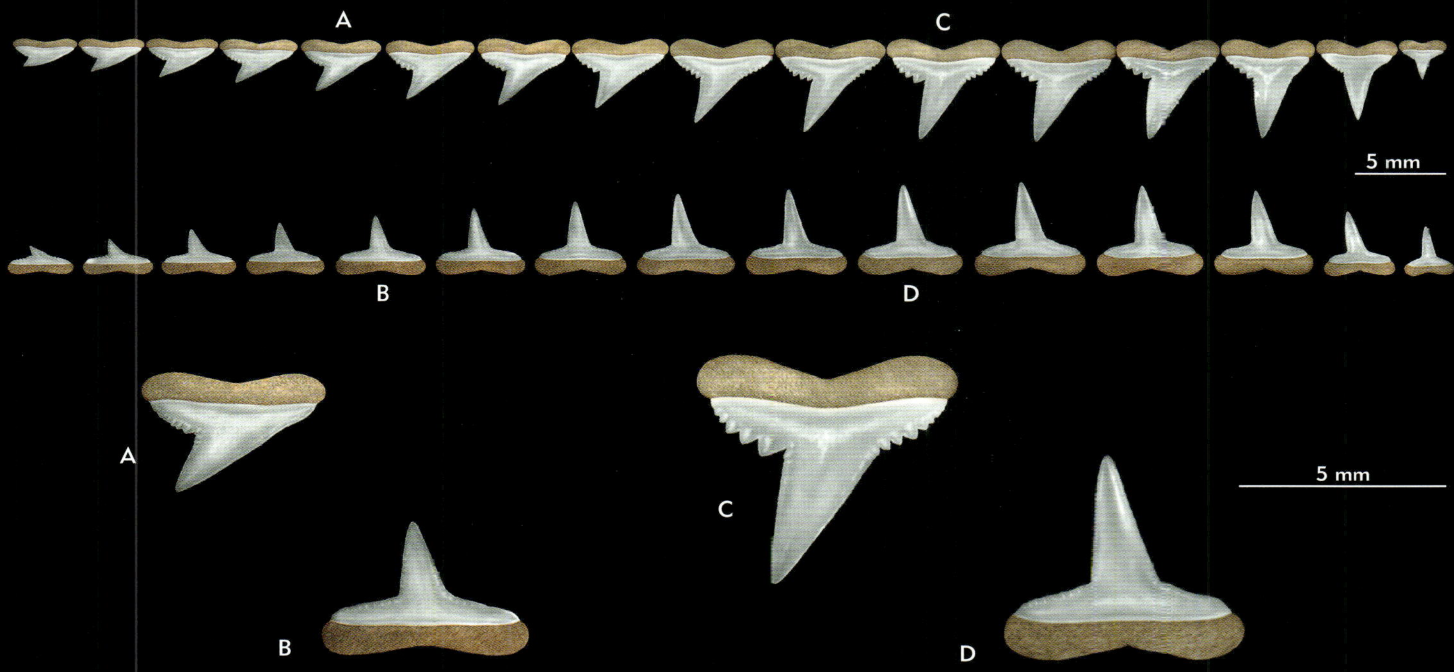

Plate 16. ***Carcharhinus isodon*****.** Teeth of a male of 63 cm TL, caught at New York, USA (Holotype, Fish collection, MNHN 0000-1037).

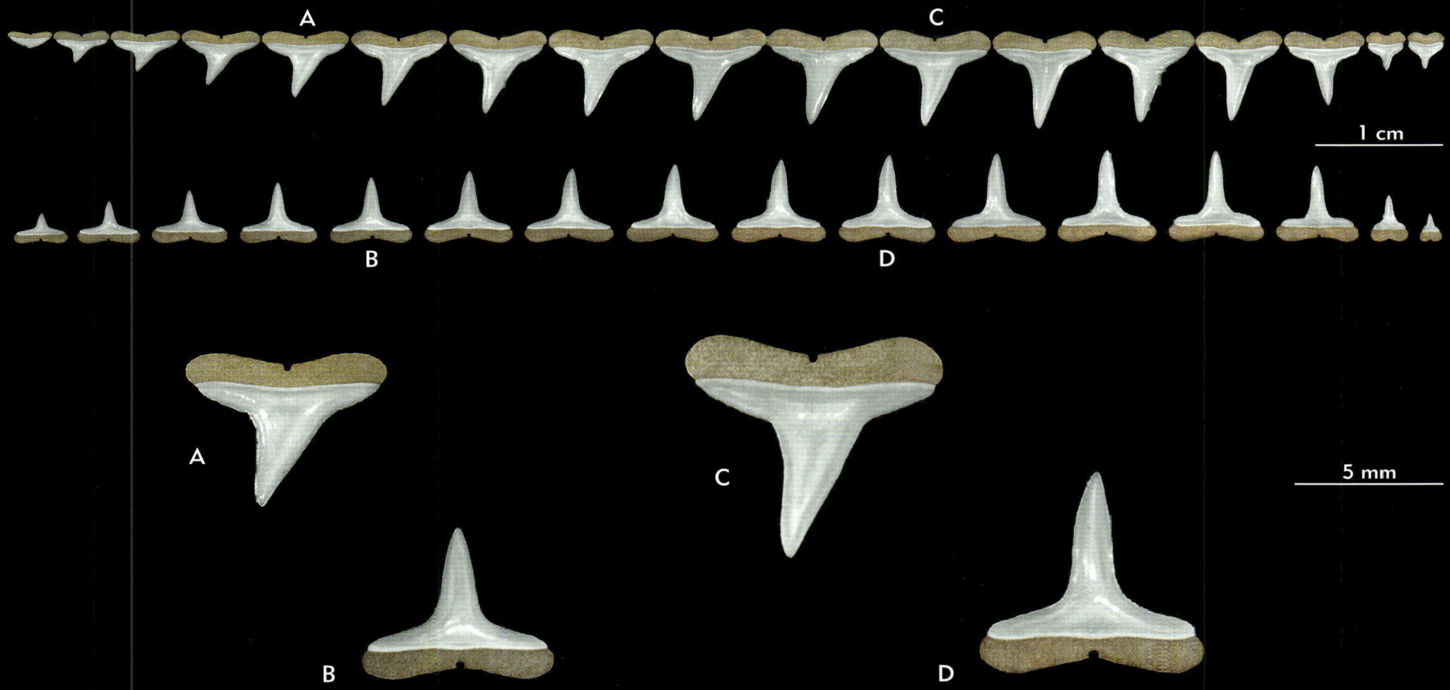

Plate 17. ***Carcharhinus leiodon***. Teeth of an immature male of 75 cm TL, caught at Quishn, Gulf of Aden (Holotype, NMV 61465).

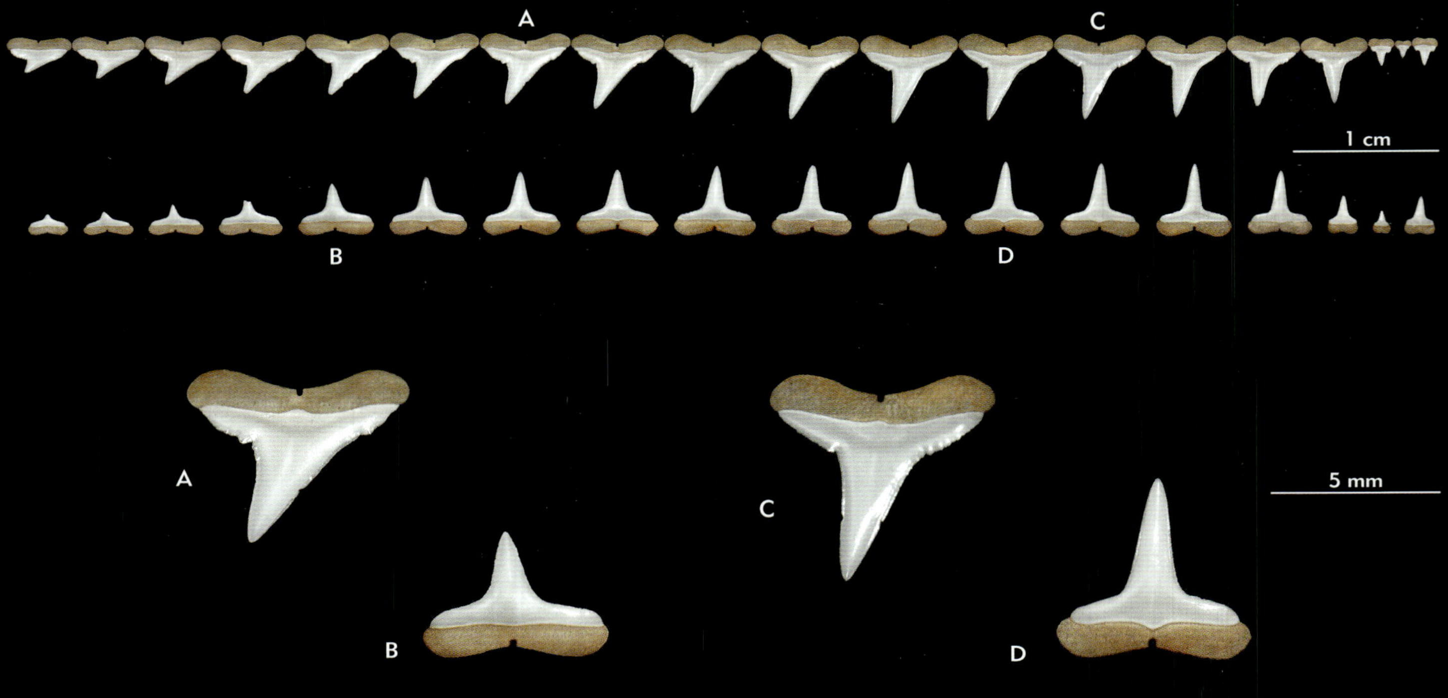

Plate 18. ***Carcharhinus leucas***. Teeth of a male of 161.5 cm TL, from the Antilles.

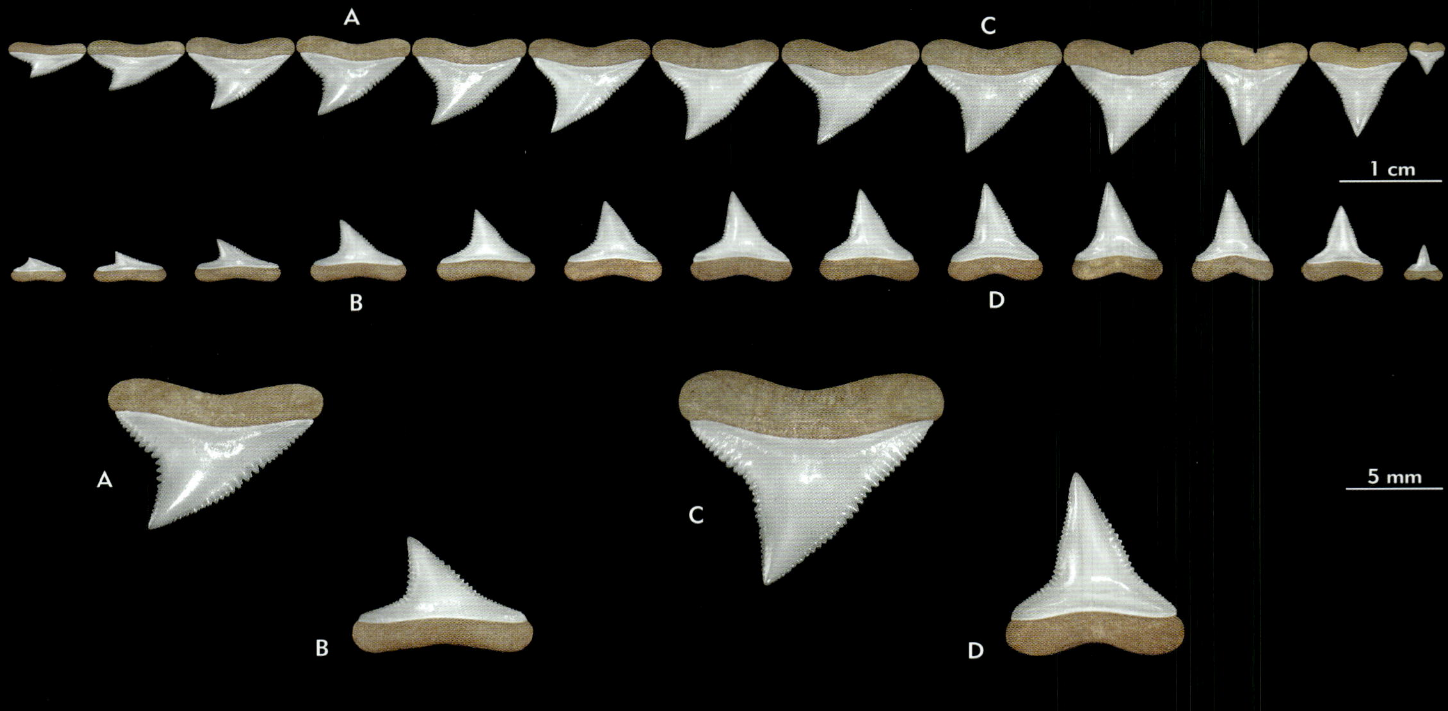

Plate 19. ***Carcharhinus limbatus*.** Teeth of a male of 72 cm TL, from Java, Indonesia.

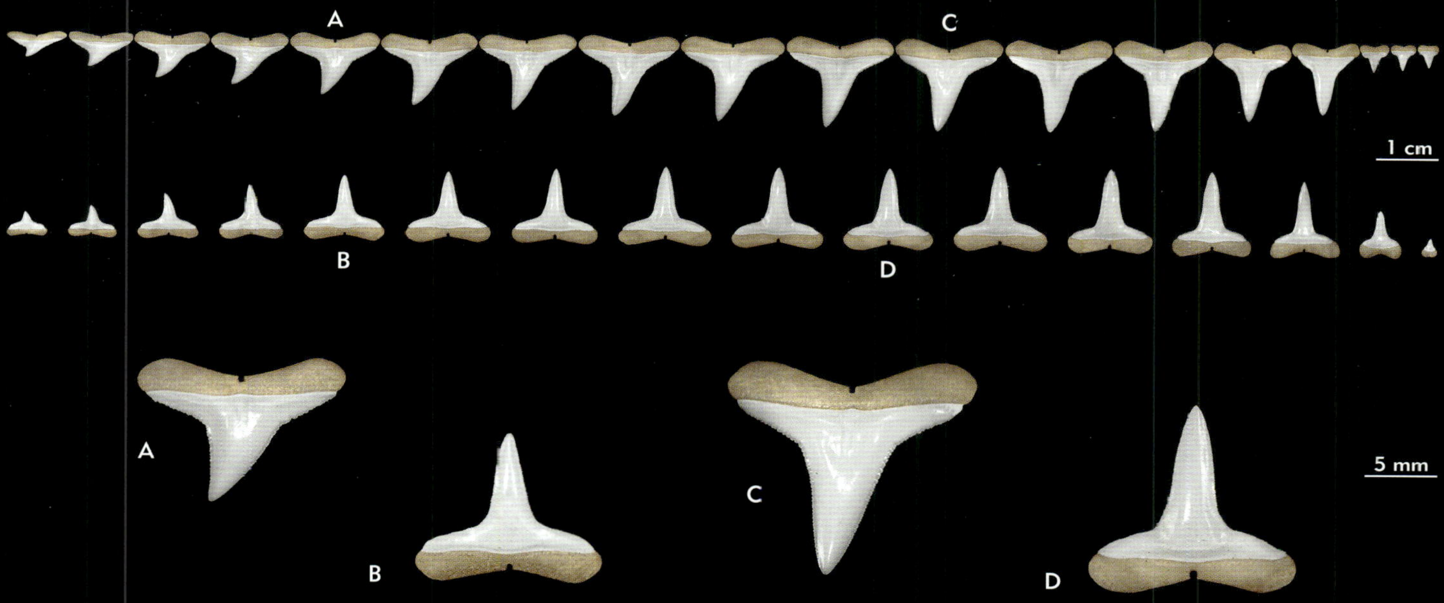

Plate 20. ***Carcharhinus longimanus*.** Teeth of a male of 52.5 cm TL, from the French Mediterranean coast.

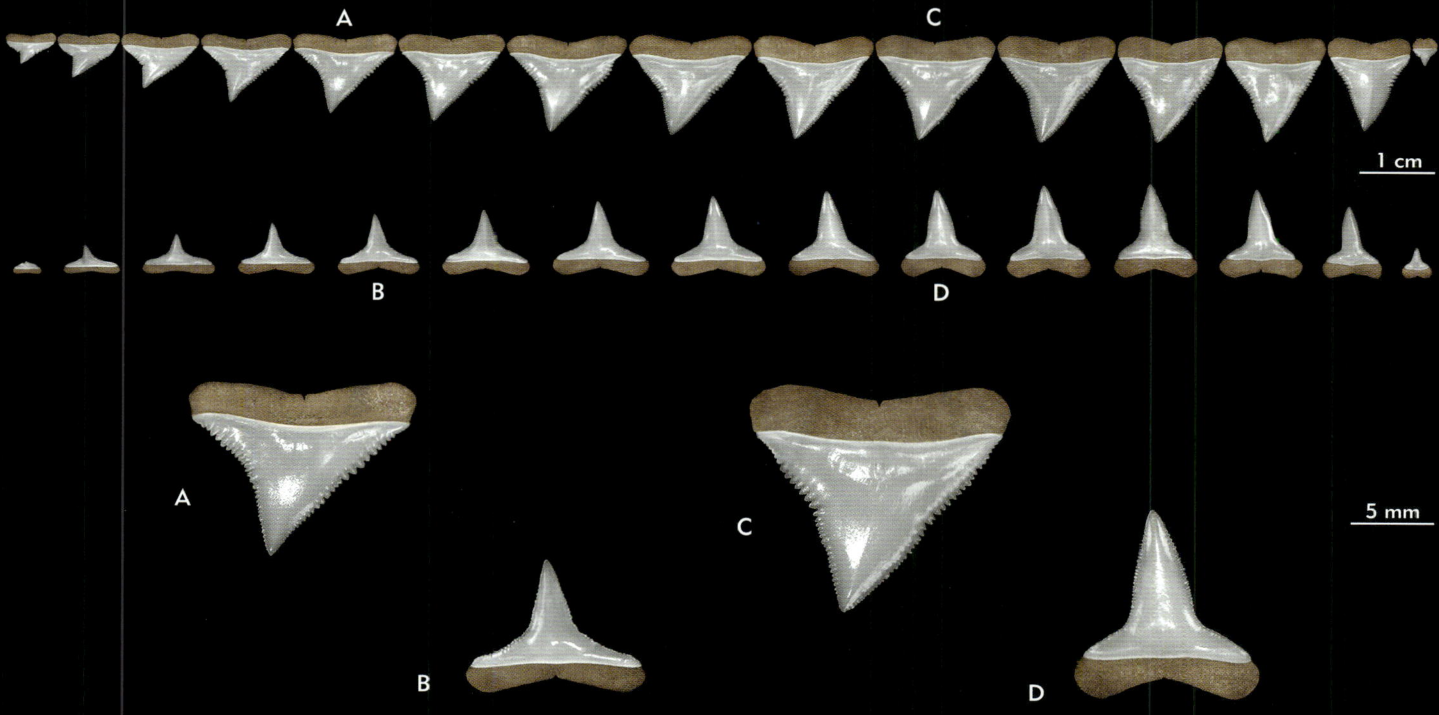

Plate 21. ***Carcharhinus macloti***. Teeth of a male of 71 cm TL, from Papua New Guinea (Holotype, RMNH-2591).

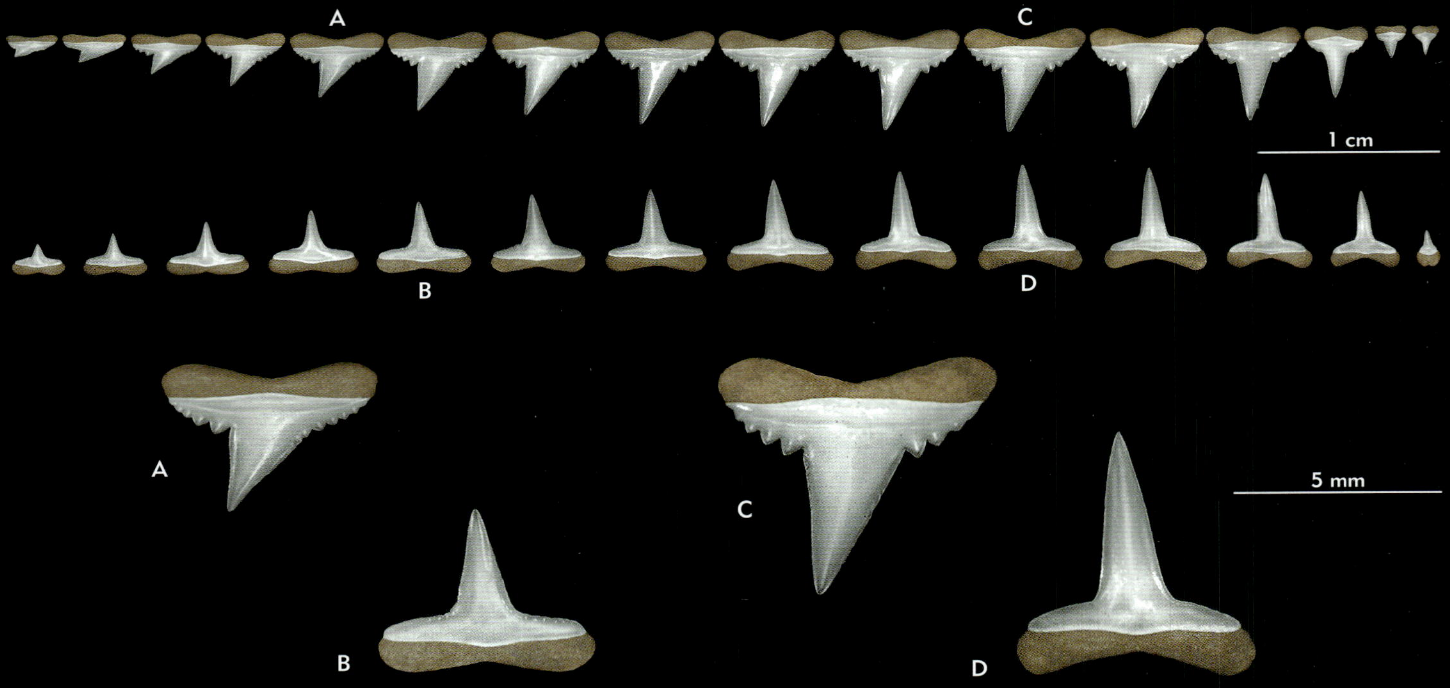

Plate 22. ***Carcharhinus melanopterus***. Teeth of a male of 53 cm TL, from Papua New Guinea.

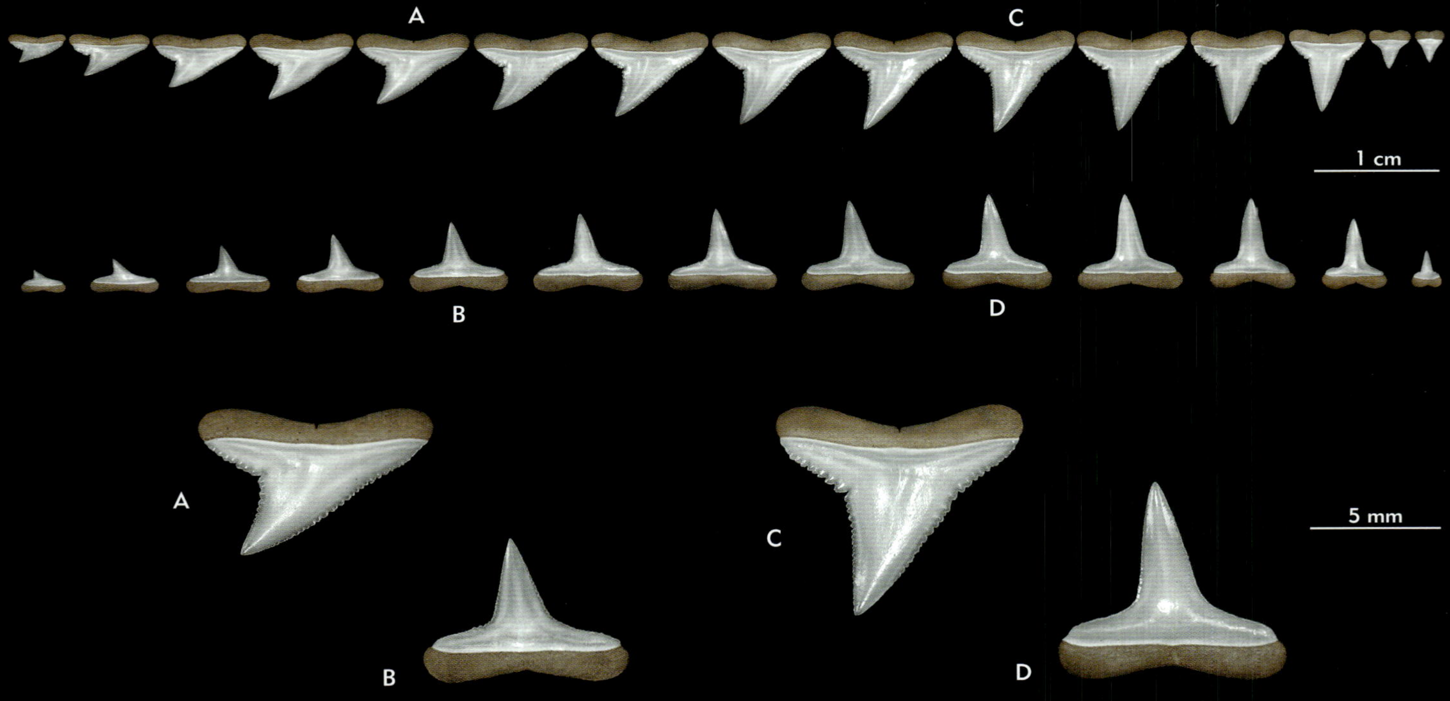

Plate 23. ***Carcharhinus obscurus*****.** Teeth of a female of 92.5 cm TL, from French Guinea.

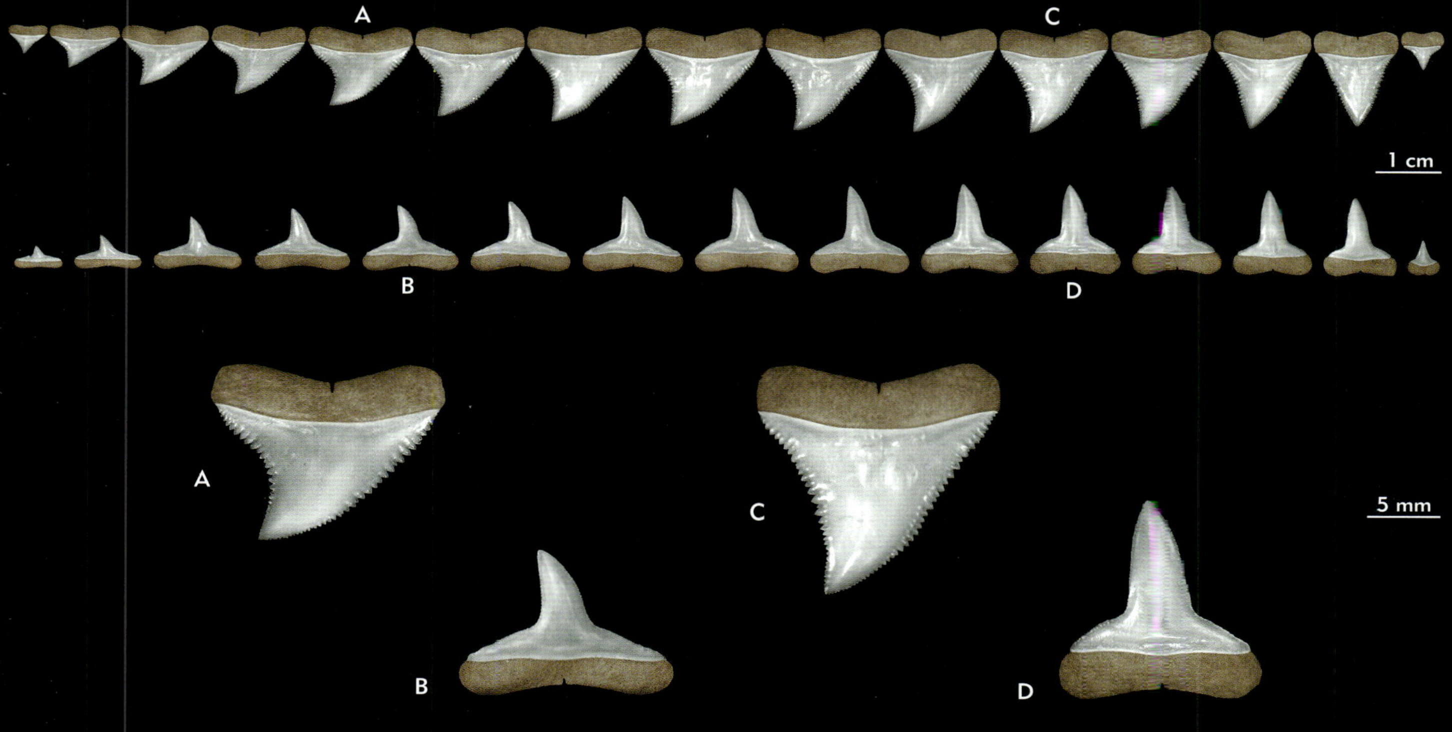

Plate 24. ***Carcharhinus perezii*****.** Teeth of a female of 73.5 cm TL, from Jamaica.

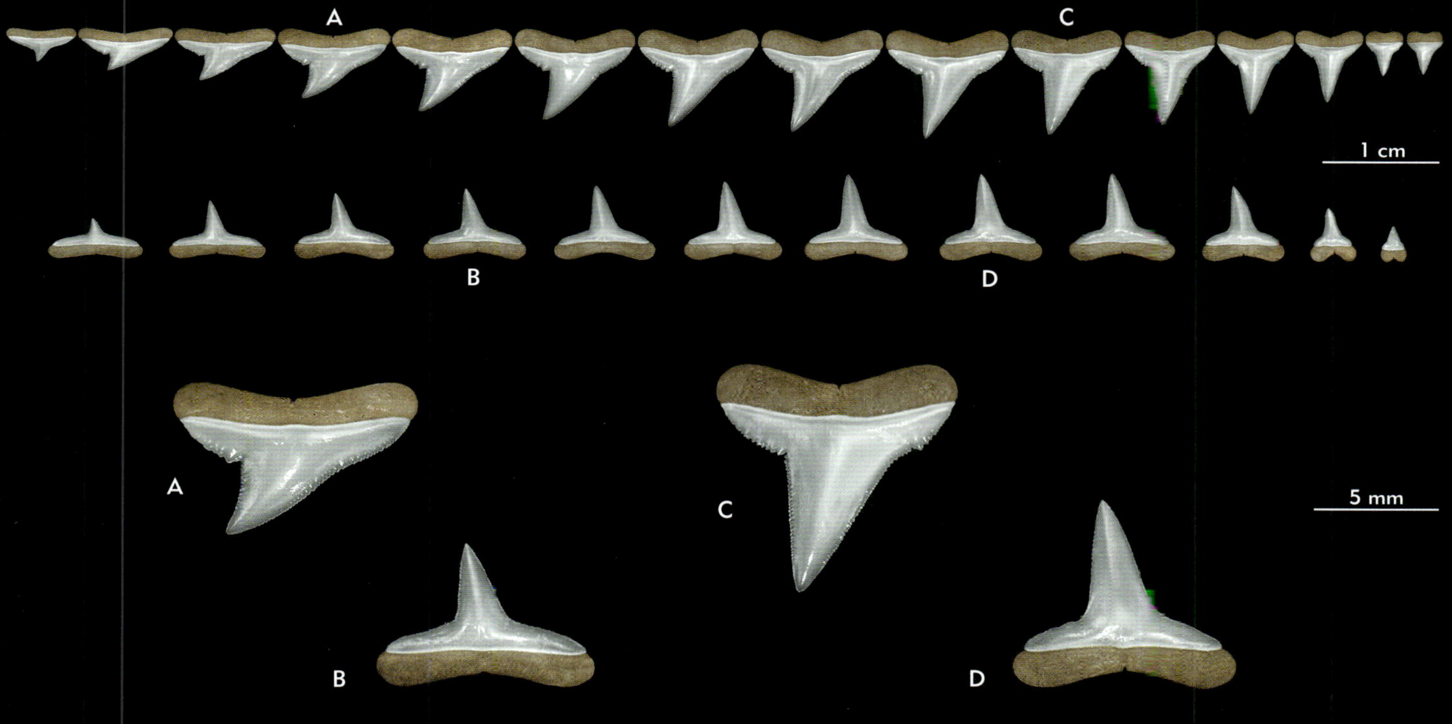

Plate 25. ***Carcharhinus plumbeus***. Teeth of a male of 178 cm TL, from Italy (Syntype, RMNH D-2555).

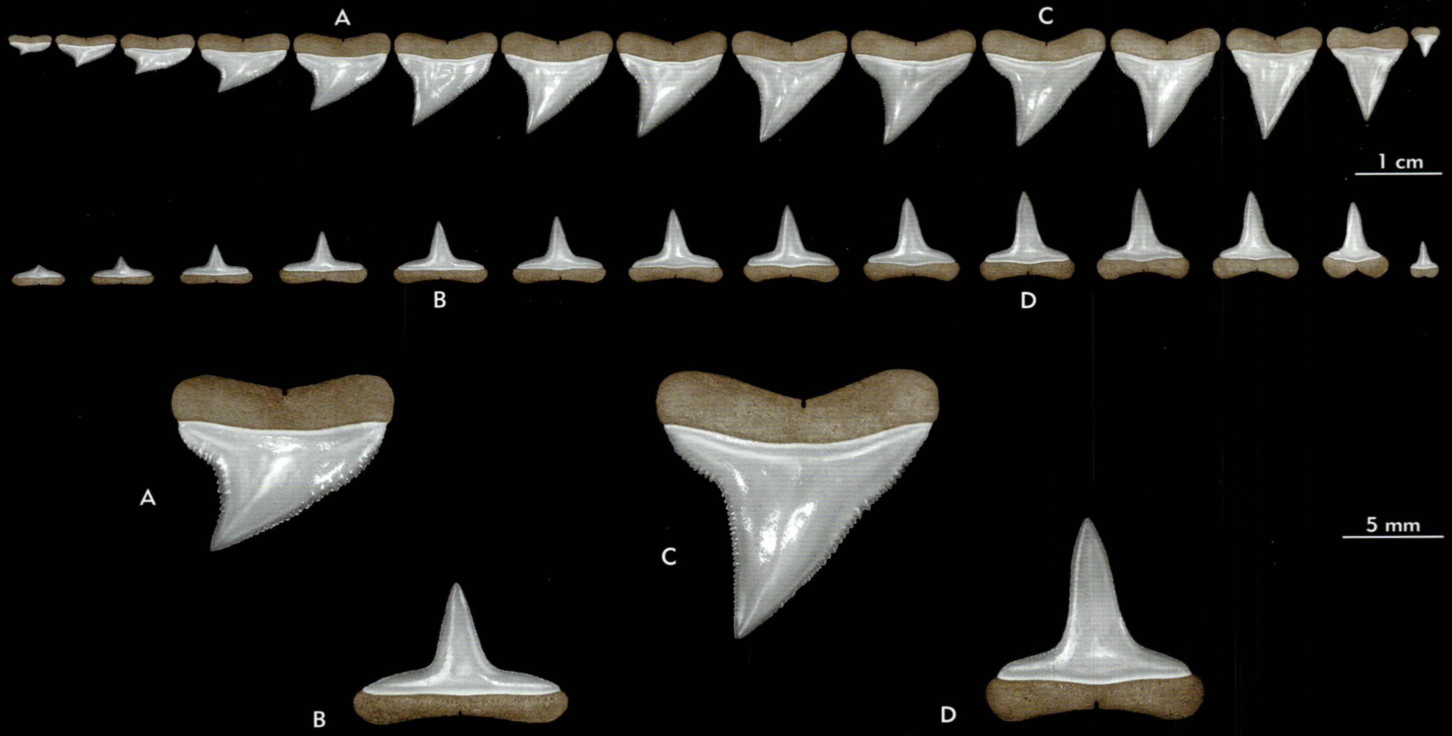

Plate 26. ***Carcharhinus porosus***. Teeth of a specimen (sex unknown) of 93.5 cm TL, from Manzinilla Bay, Trinidad (Original Photos: G. Hubbell).

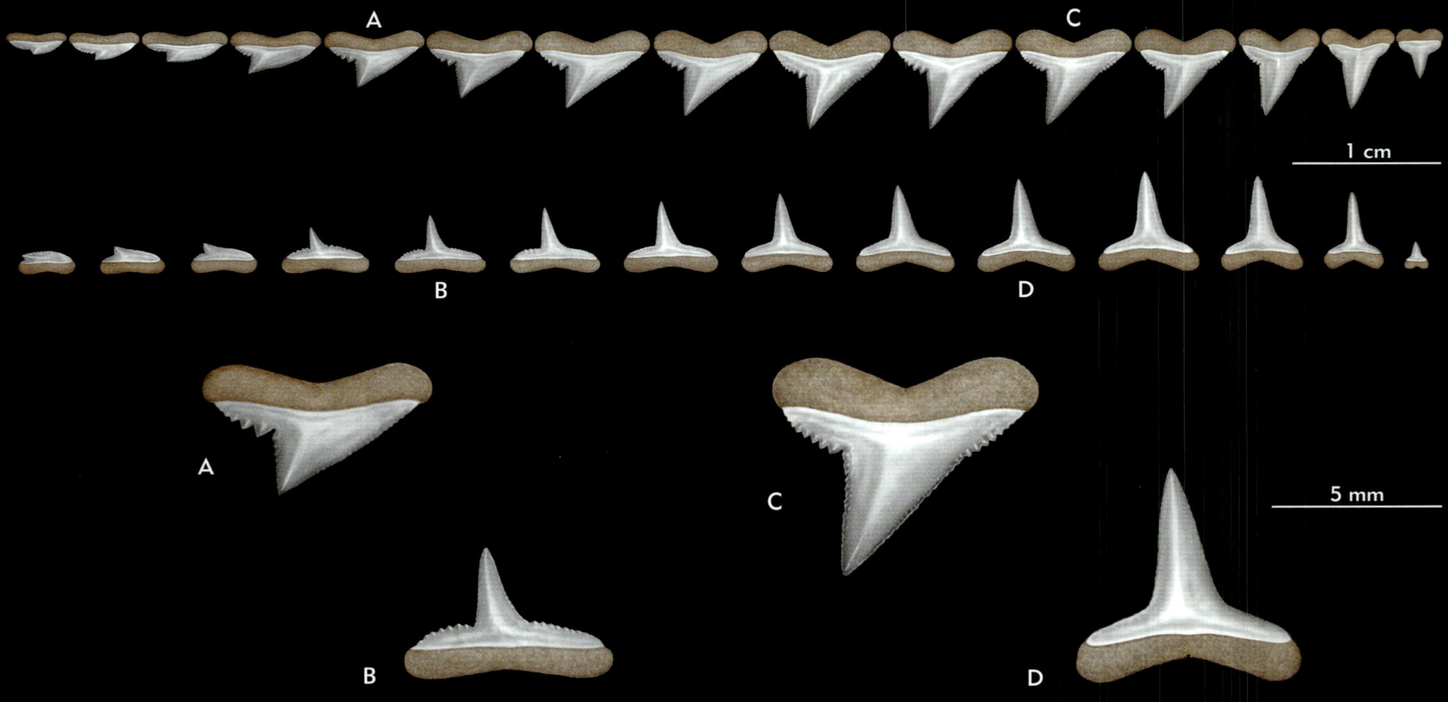

Plate 27. ***Carcharhinus sealei*****.** Teeth of a female of 60 cm TL, from Vietnam, South China Sea (Fish collection, MNHN a-7803).

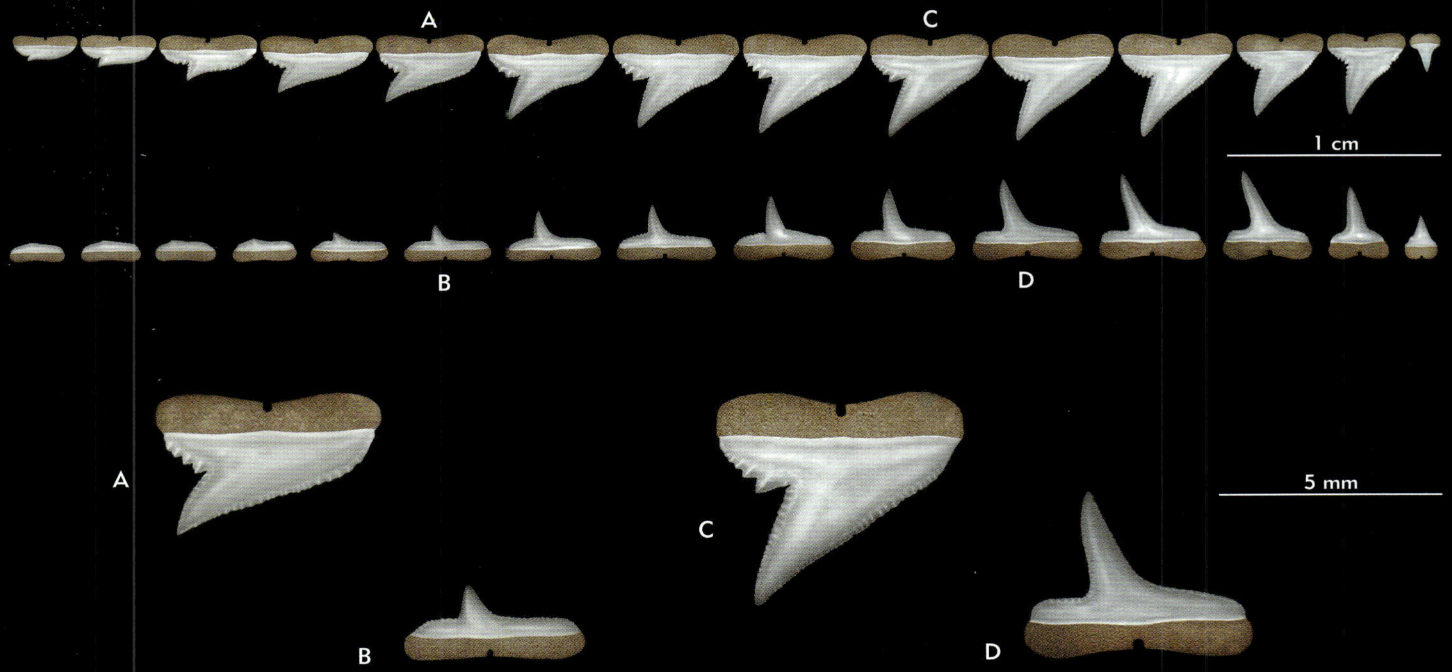

Plate 28. ***Carcharhinus signatus*****.** Teeth of a male of 203 cm TL caught on Daytona Beach, Florida, USA.

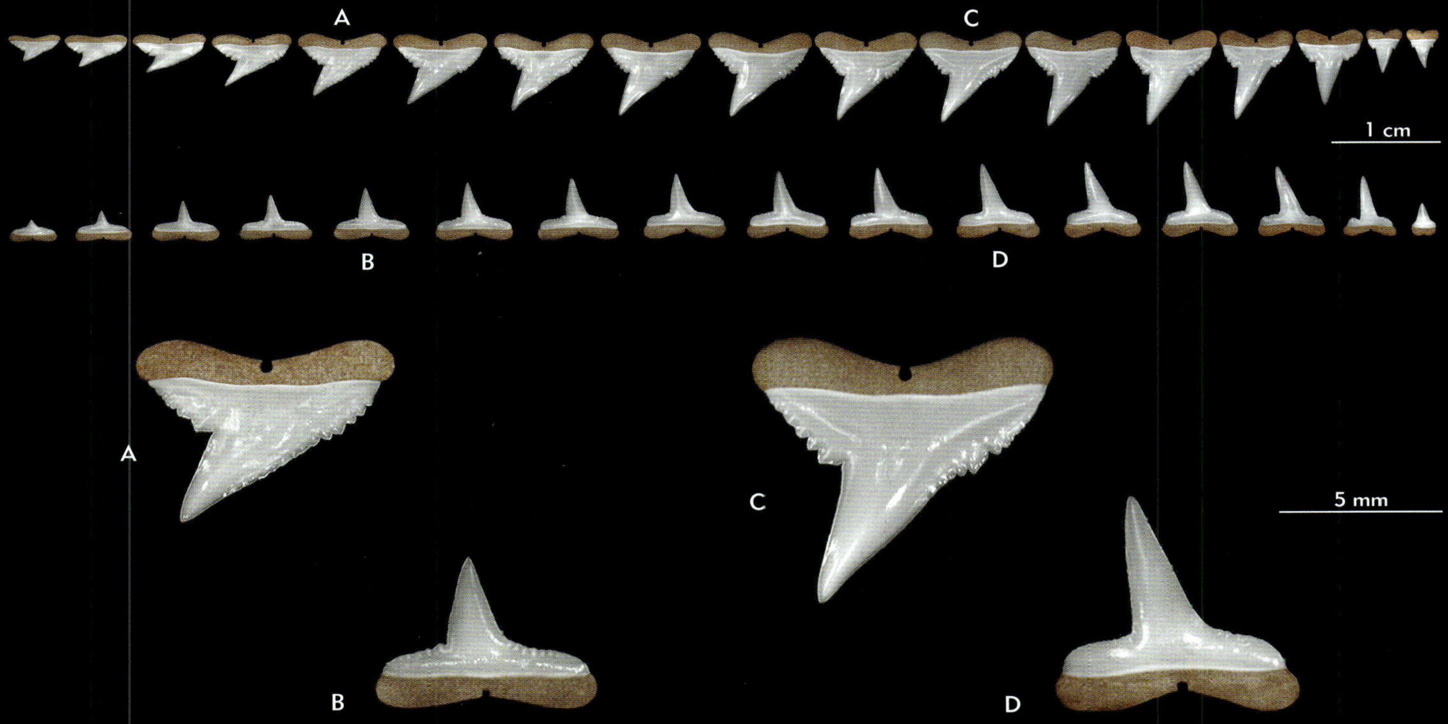

Plate 29. ***Carcharhinus sorrah*****.** Teeth of a female of 58.5 cm TL (Syntype, RMNH 4294.1).

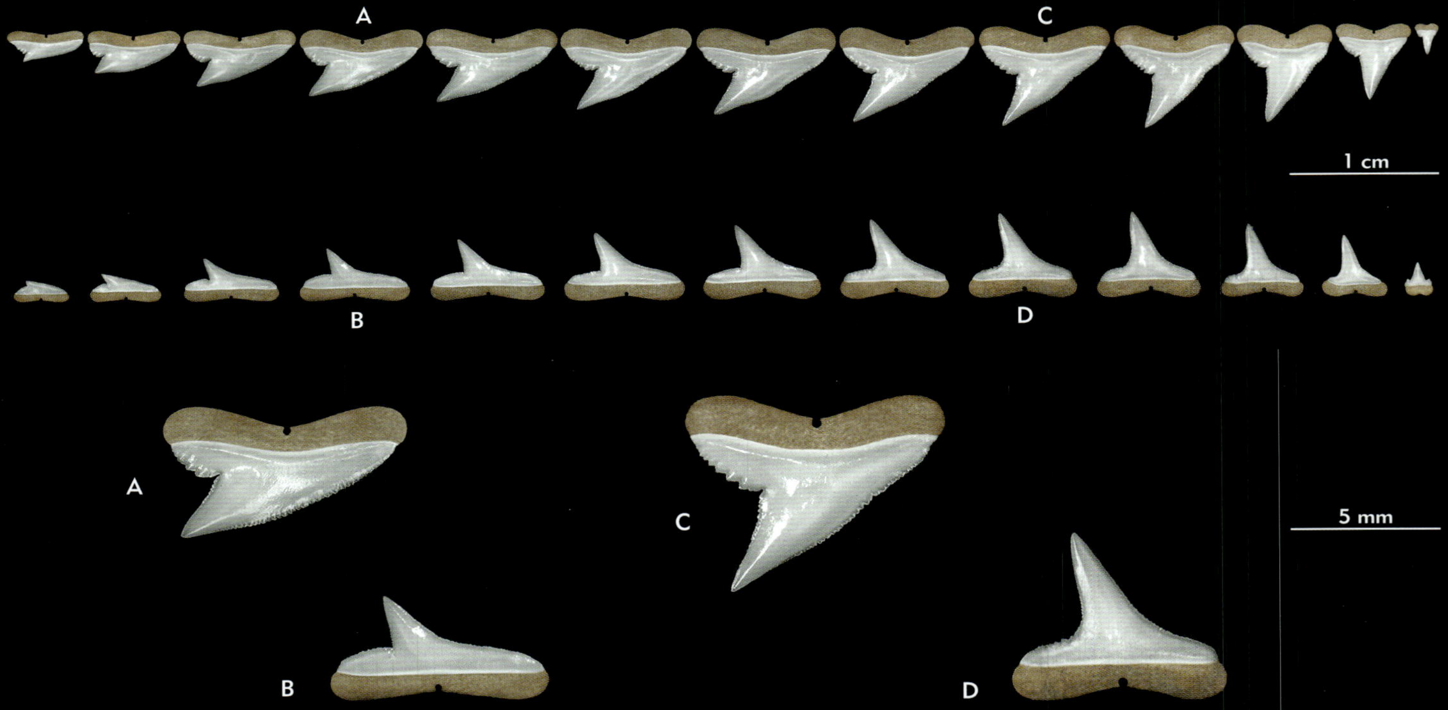

Plate 30. ***Carcharhinus tilstoni*****.** Teeth of a specimen (TL and sex unknown) (Paratype of *Galeolamna pleurotaenia tilstoni*, AMS IB-3803).

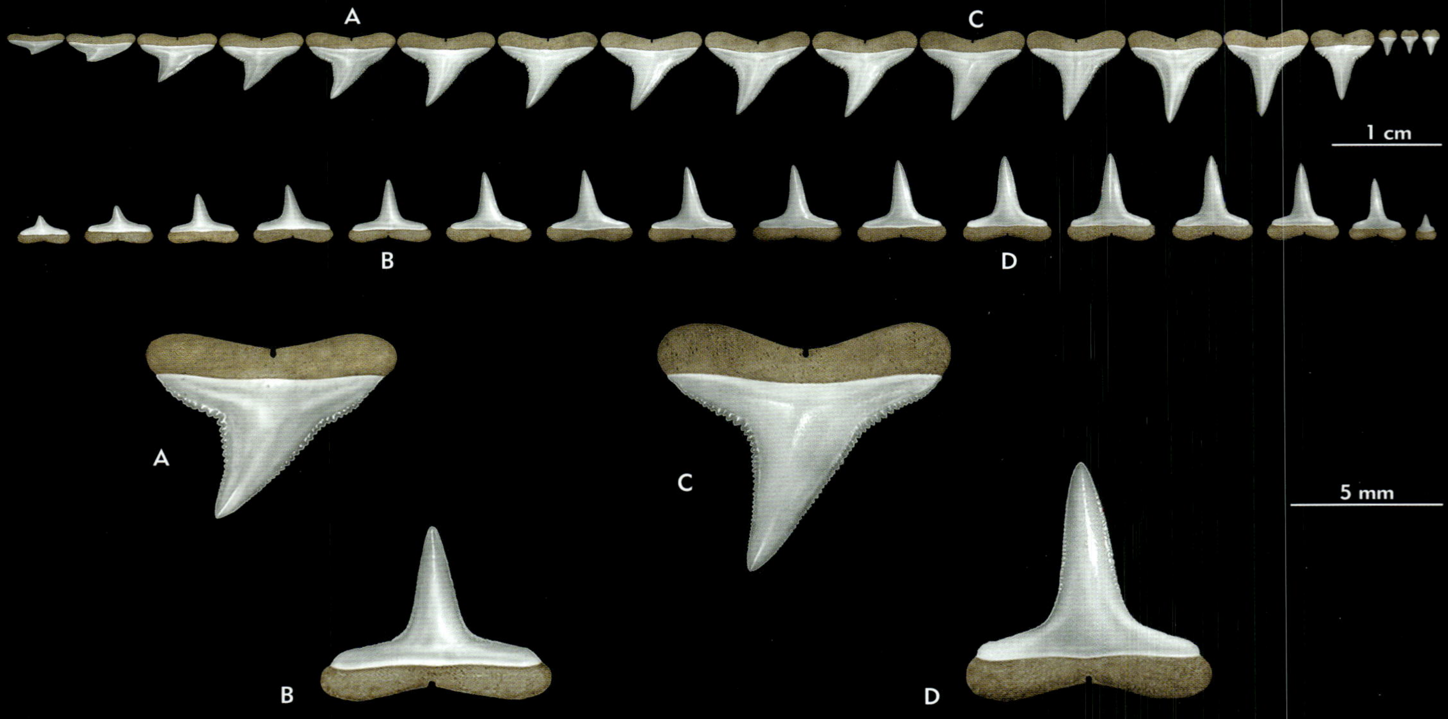

Plate 31. ***Carcharhinus wheeleri*.** Teeth of a male of 73 cm TL, from the Red Sea.

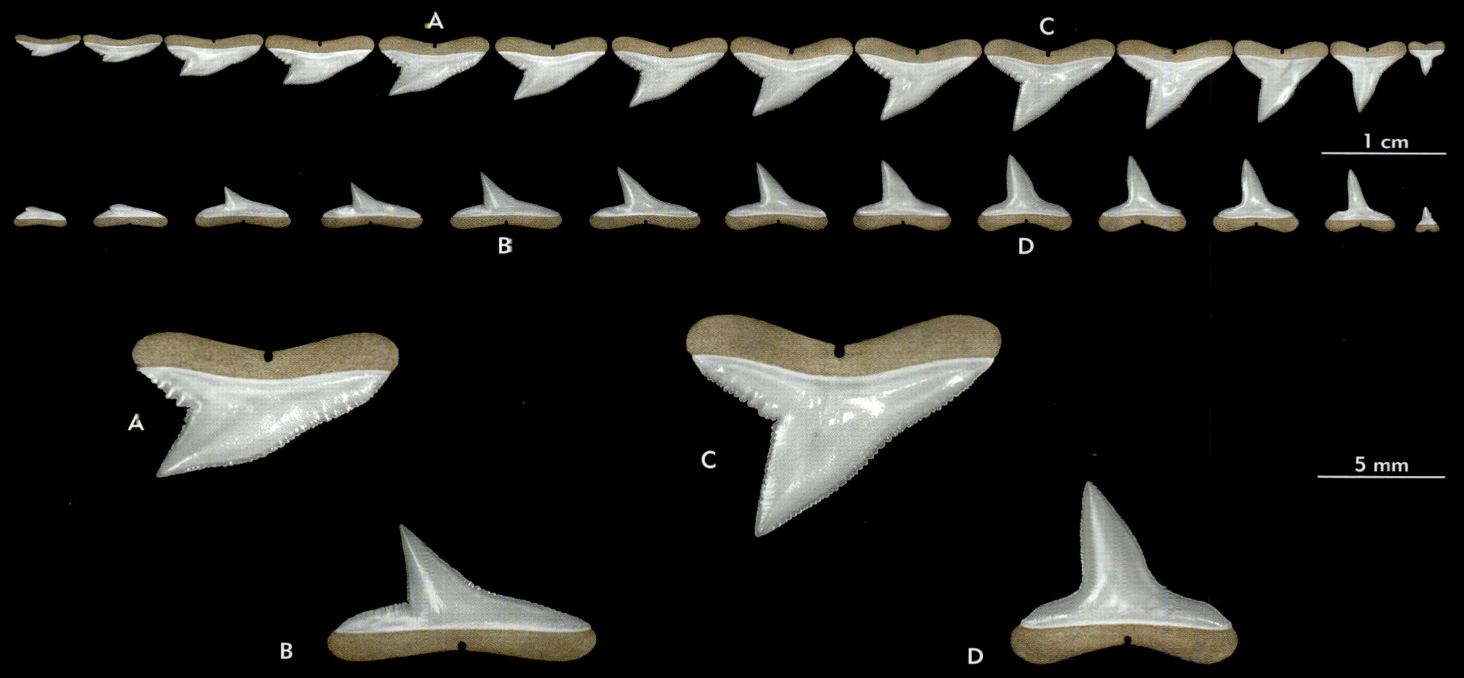

Abbreviations

The following abbreviations have been used for the names of museums in the text:

AMS	Australian Museum, Sydney, Australia
BMNH	British Museum of Natural History, London, UK
FMNH	Florida Museum of Natural History, Gainesville, FL, USA
MNHN	Muséum National d'Histoire Naturelle, Paris, France
NMNZ	National Museum of New Zealand, Wellington, New Zealand
NMV	Naturhistorisches Museum Vienna, Austria
QM	Queensland Museum, Brisbane, Queensland, Australia
RMNH	Nationaal Natuurhistorisch Museum, Leiden, the Netherlands
SMF	Naturmuseum Senckenberg, Frankfurt/M., Germany
SMNS	Staatliches Museum für Naturkunde, Stuttgart, Germany
SU	Stanford University, Natural History Museum, Stanford, CA, USA
USNM	U.S. National Museum of Natural History, Washington, USA
VBCM	Cátedra de Vertebrados, Facultad de Biológicas, Universitad Complutense de Madrid, Spain

Used abbreviations for the descriptions of fins

A	Anal fin
C	Caudal fin
D1	First dorsal fin
D2	Second dorsal fin
P	Pelvic fins
PP	Pectoral fins

Used abbreviations for the descriptions of tooth position in the jaw

A	anterior
AL	antero-lateral
C	commissural
CP	commissural-posterior
L	lateral
LP	latero-posterior
P	posterior
PS	pseudo- or parasymphysial position
S	symphysial
X	further rows

Other used abbreviations

IGFA	World Record Fish Game. International Game Fish Association, Florida, USA.
PSU	Part Salinity Unit
TL	Total length

Abbreviations of the IUCN Red List for species status

EX	Extinct
EW	Extinct in the Wild
CR	Critically Endangered
EN	Endangered
VU	Vulnerable
NT	Near Threatened
LC	Least Concern
LR/nt	Low Risk/near threatened
NE	Not Evaluated
DD	Data Deficient

Bibliography

ABELLA, A. (1972). Hallazgo de nueva especie de *Carcharhinus*, en la costas de Rocha, Uruguay. – Bol. Soc. Zool. Uruguay, 2: 102–106.

ACERO, P. A. & A. SANTOS MARTINEZ (1992). Confirmed presence of *Carcharhinus perezii* (POEY 1876) (Pisces: Carcharhinidae) in the Colombian Caribbean. – An. Inst. Cienc. Mar. Limnol. Univ. Nac. Auton. Mex., 19(2): 211–213.

AJAZUDDIN, S. & M. AHMED (2002). Some observations on occurrence and abundance of Finfish in the Miani Hor Lagoon, Balochistan (Northern Arabian Sea, Pakistan). – Pak. J. Zool., 34(2): 101–111.

ALLEN, B. R. & G. CLIFF (2000). Sharks caught in the protective gill nets off Kwazulu-Natal, South Africa. 9. The spinner shark *Carcharhinus brevipinna* (MUELLER & HENLE). – S. Afr. J. Mar. Sci./S. Afr. Tydskr. Seewet, 22: 199–215.

ANDERSON, R. C. & J. D. STEVENS (1996). Review of information on diurnal vertical migration in the bignose shark, *Carcharhinus altimus*. – Mar. Freshwat. Res., 47(4): 605–608.

APPLEGATE, S. P. (1965). Tooth terminology and variation in sharks with special reference to the Sand shark, *Carcharias taurus* (RAFINESQUE 1810). – Contr. Sci. (86): 3–18.

BACKUS, R. H., S. SPRINGER & E. L. ARNOLD, jr. (1956) Contributions to the natural history of the whitetip shark, *Pterolamiops longimanus* (POEY). Deep-Sea Res., 3: 178–188.

BANE, G. W., jr. (1966). Observations on the silky shark, *Carcharhinus falciformis*, in the Gulf of Guinea. – Copeia: 354–356.

BARANES, A. & A. BEN-TUVIA (1978a). Notes on *Carcharhinus altimus* (SPRINGER 1950) from the northern Red Sea. Cybium (3e Ser.) 4: 61–64.

– (1978b). Occurrence of the sandbar shark *Carcharhinus plumbeus* in the northern Red Sea. – Isr. J. Zool., 27: 45–51.

– (1978c). On sharks, skates and rays in the Gulf of Eilat. Israel land and nature. – Q. J. Soc. Protect. Nat. Isr., 4(1): 9 pp.

BARGAR, T. W. & T. B. THORSON (1992). scanning electron microscopies study of the dermal denticles of the bull shark, *Carcharhinus leucas*. – J. Aquaricult. Aquat. Sci., 7: 120–137.

BARRULL, J. & I. MATE (2002). Tiburones del Mediteráno. – 292 pp.; Liberia El Set-ciències, Arenys de Mar.

BASS, A. J., J. D. D'AUBREY & N. KISTNASAMY (1973). Sharks of the east coast of southern Africa. 1. The genus *Carcharhinus* (Carcharhinidae). – Invest. Rep. Oceanogr. Res. Inst., Durban, (33): 168 pp.

BASS, A. J., P. C. HEEMSTRA & L. J. V. COMPAGNO (1986). Families Hexanchidae, Carcharhinidae. – Pp. 49–62 in SMITH, M. M. & P. C. HEEMSTRA (eds.): Smith's Sea Fishes; Southern Book Publishers, Johannesburg.

BATISTA, V. S. & T. C SILVA (1995). Age and growth of juveniles of Junteiro shark, *Carcharhinus porosus*, in the coast of Maranhao, Brazil. – Rev. Bras. Biol., 55(1): 25–32.

BIBRON, B. (1839). – Pp. 39–102 in MÜLLER, J. & F. G. J. HENLE: Systematische Beschreibung der Plagiostomen; Berlin, Veit.

BIGELOW, H. B. & W. C. SCHROEDER (1944). New sharks from the Western North Atlantic. – Proc. New Engl. Zool. Club, 23: 21–36.

– (1961). *Carcharhinus nicaraquensis*, a synonym of the bull shark, *C. leucas*. – Copeia, 3: 359 pp.

BIGELOW, H.B., W. C. SCHROEDER & S. SPRINGER (1943). A new species of *Carcharhinus* from the western Atlantic. – Proc. New England Zool. Club, 22: 69–74.

BLAINVILLE, H. M. D. (1816). Prodrome d'une distribution systématique du regne animal. – Bull. Soc. Philo.-mat. Paris, 8: 105–124.

– (1825). Vertébrés. Classe V. Poissons. Poissons cartilagineux. – Faune Fr., (13–14): 96 pp.

BLANDON, I. R., R. WARD, W. J. KAREL & B. MOBLEY (2004). Use of biochemical and molecular genetic techniques to distinguish two closely related and morphologically similar carcharhinic shark species – *Carcharhinus brevipinna* and *Carcharhinus limbatus*. – Proc. Gulf Caribb. Fish. Inst., 55: 1015 pp.

BLEEKER, P. R. (1852). Bijdrage tot de kennis der Plagiostomen van den Indischen Archipel. – Verh. Batav. Genoot. Kunst. Wet., 24: 1–92.

– (1853). Diagnostische beschrijvingen van nieuwe of weinig bekende vischsoorten van Batavia. – Natuur. Tijdschr. Ned. Indie 4: 451–516.

– (1854). Vijde bijdrage tot de kennis der ichthyologische fauna van Amboina. – Natuur. Tijdschr. Ned. Indie 5: 455–508.

– (1856). *Carcharias (Prionodon) amblyrhynchos*, eene nieuwe haaisort, gevangen nabij het Eiland Solombo. – Natuur. Tijdschr. Ned. Indie, 6: 467 pp.

– (1858–59). Twaalfde bijdrage tot de kennis der vischfauna von Borneo. Vischen van Sinkawang. – Acta Soc. Sci. Indo.-Neerl., 5(7): 1–10.

BOESEMAN, M. (1956). A tragedy of errors; the status of *Carcharhinus* (BLAINVILLE 1816), *Galeolamna* (OWEN 1853), *Eulamia* (GILL 1861) and the identity of *Carcharhinus commersonii* (BLAINVILLE 1825). – Zool. Meded., 37(6): 81–100.

BONFIL, R. S. (1989). An abnormal embryo of the reef shark, *Carcharhinus perezii* (POEY), from Yucatan, Mexico. – Northeast Gulf Sci., 10(2): 153–155.

– (1994). Overview of world elasmobranch fisheries. – FAO, Fisheries Technical Paper No. 341: 119 pp.; Rome.

BONFIL, R. S., R. MENA & D. DE ANDA (1993). Biological parameters of commercially exploited silky sharks, *Carcharhinus falciformis*, from the Campeche Bank, Mexico. – NOAA, Tech. Rep., NMFS: 115 pp.

BORNATOWSKI, H. (2008). A parturition and nursery area for *Carcharhinus limbatus* (Elasmobranchii, Carcharhinidae) off the coast of Paraná, Brazil. – Brazilian J. Oceanography, 56(4): 317–319.

BRANSTETTER, S. (1981). Biological Notes on the sharks of the North Central Gulf of Mexico. – Contrib. Mar. Sci. Univ. Texas, 24: 13–34.

– (1982). Problems associated with the identification and separation of the spinner shark, *Carcharhinus brevipinna*, and the blacktip shark, *Carcharhinus limbatus*. – Copeia, 2: 461–465.

– (1987). Age and growth estimates for blacktip, *Carcharhinus limbatus*, and spinner, *C. brevipinna*, sharks from the north-western Gulf of Mexico. – Copeia, 4: 964–974.

BRANSTETTER, S. & R. L. SHIPP (1980). Occurrence of the finetooth shark, *Carcharhinus isodon*, off Dauphin Island, Alabama. – Fish. Bull., 78(1): 177–179.

BRANSTETTER, S. & R. STILES (1987). Age and growth estimates of the bull shark, *Carcharhinus leucas*, from the northern Gulf of Mexico. – Environmental Biology of Fishes. The Hague 20(3): 169–181.

BRITO, A. (1991). Catalogo de los pesces de las Islas Canarias. – 230 pp.; Francisco Lemus, La Laguna.

BRUM, J. M. M. & J. M. N. AZEVEDO (1993). First record of the Galapagos shark, *Carcharhinus galapagensis* (SNODGRASS & HELLER 1905) (Carcharhinidae) on the Azores. – Bol. Mus. Mun. Funchal, no. suppl. 4, pt. A.: 139–143.

BRUNNSCHWEILER, J. M. & L. J. V. COMPAGNO (2008). First record of *Carcharhinus leucas* from Tonga, South Pacific. – Mar. Biodiversity Rec., 1: 51 pp.

BYKOV, V. P. (1983). Marine fishes: chemical composition and processing properties. – 333 pp.; Amerind Publishing Co. Pvt. Ltd., New Delhi.

CABRERA CHAVEZ-COSTA, A. A. (2003). Feeding habits of the silky shark, *Carcharhinus falciformis* (BIBRON 1839) in the western coast of Baja California Sur. – Book Monographie, Dissertation.

CADENAT, J. (1956). Notes d' Ichtyologie ouest africaine. 11. Description dune espèce nouvelle de Requin appartenant au genre Hypoprion (MÜLLER

& HENLE). *Hypoprion bigelowi* sp. nov. – Bull. Inst. Fondam. Afr. Noire (Afr. Sci. Nat.), 18(2): 539–545.
CAPAPE, C., J. ZAOULI & M. DESOUTTER (1979). Occurrence of *Carcharhinus obscurus* (LESUEUR 1818) in Tunisia and identification key of Carcharhinidae from Tunisian coasts. – Bull. Off. Natl. Peches (Tunisie), 3(2): 171–182.
CAPAPE, C., A. A. SECK, Y. DIATTA, C. REYNAUD, F. HEMIDA & J. ZAOUALI (2004). Reproductive biology of the blacktip shark, *Carcharhinus limbatus* (Chondrichthyes: Carcharhinidae) off West and North African coast. – Cybium, 28(4): 275–284.
CARLSON, J. K. (1999). Occurrence of neonate and juvenile sandbar sharks, *Carcharhinus plumbeus*, in the north-eastern Gulf of Mexico. – Fish Bull., 97(2): 387–391.
CARLSON, J. K. & E. CORTES (2003). Gillnet selectivity of small coastal sharks off the south-eastern USA. – Fish Res., 60(2–3): 405–414.
CARLSON, J. K., E. CORTES & A. G. JOHNSON (1999). Age and Growth of the Blacknose shark, *Carcharhinus acronotus*, in the eastern Gulf of Mexico. – Copeia, 3: 684–691.
CARLSON, J. K., E. CORTES & D. M. BETHEA (2003). Life history and population dynamics of the finetooth shark, *Carcharhinus isodon* in the north-eastern Gulf of Mexico. – Fish. Bull., 101(2): 281–292.
CARLSON, J. K., E. CORTES, J. A. NEER, C. T. MCCANDLESS & L.R. BEERKIRCHER (2008). The status of the United States population of night shark, *Carcharhinus signatus*. – Mar. Fish. Rev., 70(1): 1–13 .
CASTRO, J. I. (1993). The biology of the finetooth shark, *Carcharhinus isodon*. The Hague, 36(3): 219–232.
– (1996). Biology of the blacktip shark, *Carcharhinus limbatus*, off the south-eastern US. – Bull. Mar. Sci., 59(3): 508–522.
CHAN, R. W. K., P. I. DIXON, J. G. PEPPERELL & D. D. REID (2003). Application of DNA-based techniques for the identification of whaler sharks (*Carcharhinus* spp.) caught in protective beach meshing and by recreational fisheries off the coast of New South Wales. – Fish. Bull., 101: 910–914.
CHAPMANN, D. D., B. FIRCHAU & M. S. SHIVJI (2008). Parthenogenesis in a large-bodied requiem shark, the blacktip shark, *Carcharhinus limbatus*. – J. Fish Biol., 73(6): 1473–1477.
CHAPMAN, D. D., D. PINHAL & M. S. SHIVJI (2009). Tracking the fin trade: genetic stock identification in western Atlantic scalloped hammerhead sharks *Sphyrna lewini*. – Endang. Species Res. (Reprint). DOI: 10.3354/esr00241: 1–8.
CHIARAMONTE, G. E. (1998). Shark fisheries in Argentina. – Mar. Freshwat. Res., 49(7): 601–609.
CHU, Y. T. (1960). Fishes of the East China Sea. – 642 pp.; Republic of China.
CLIFF, G. & S. F. J. DUDLEY (1991). Sharks caught in the protective gill nets off Natal, South Africa. 6. The copper shark, *Carcharhinus brachyurus* (GUENTHER). – S. Afr. J. Mar. Sci./S. Afr. Tydskr Seewet: 12 pp.
CLOQUET, H. (1817). Dictionnaire des sciences naturelles. Vol. 7. Paris and Strasbourg.
COMPAGNO, L. J. V. (1973) Carcharhinidae. – In: HUREAU, J.-C. & T. MONOD (eds.): CLOFNAM. Checklist of the fishes of the North Atlantic and Mediterranean. UNESCO, Paris, 1: 23–31.
– (1984). FAO Species Catalogue. Vol. 4: Sharks of the World. An annotated and illustrated catalogue of shark species known to date. Part 2: Carcharhiniformes. – FAO Fish. Synop., 125(4)2: 251–655.
– (1988). Sharks of the order Carcharhiniformes. – 486 pp.; Blackburn Press, Caldwell, New Jersey, USA.
– (1999). Checklist of living elasmobranchs. – In: HAMLETT, W. C. (ed.): Sharks, skates and rays. The biology of elasmobranch fishes. John Hopkins Press, Maryland: 471–478; Chapter 1, Systematics and body-form: 1–41, and Chapter 3, Endoskeleton: 69–92.
– (2000). Distributional note on the dusky shark, *C. obscurus*, from the Mediterranean sea, with a first record from the Maltese Islands. – Proc. 3rd Meet. Africain Elasmobranch Association, EEA, Boulogne-sur-Mer: 55–67.
– (2005). Checklist of living chondrichthyes. – In: HAMLETT, W. C. (ed.): Reproductive biology and philology of chondrichthyans: Sharks, Batoids and chimaeras. Sciences Publishers Inc., Enfield, USA.
COMPAGNO, L. J. V. & J. R. F. GARRICK (1983) *Nasolamia*, new genus, for the shark *Carcharhinus velox* (GILBERT 1898). – Zool. Publ. Vict. Univ. Wellington, 76: 16 pp.
COMPAGNO, L. J. V. & V. H. NIEM (1998). Carcharhinidae. Requiem sharks. – Pp. 1312–1360 in CARPENTER & NIEM (eds.): FAO Identification Guide for Fishery Purposes. The Living Marine Resources of the Western Central Pacific. – FAO, Rome.
COMPAGNO, L. J. V., D. A. EBERT & M. J. SMALE (1989). Guide to the sharks and rays of southern Africa. – 160 pp.; Struik Publishers, Cape Town, South Africa.
COMPAGNO, L. J. V., M. DANDO & S. FOWLER (2005). Sharks of the world. – Princeton Field Guides; 357 pp.
COSTANTINI, M. & M. AFFRONTE (2003). Neonatal and juvenile sandbar sharks in the northern Adriatic Sea. – J. Fish. Biol., 62(3): 740–743.
COPPO, M. (1992) Bronze whaler sharks in South Australia. – Sawfish., 17(1): 10–13.
CROW, G. L., C. G. LOWE & B. M. WETHERBEE (1996). Shark records from longline fishing programs in Hawaii with comments on Pacific Ocean distributions. Pac. Sci., 50(4): 382–392.
CUVIER, G. (1817). Le règne animal. – Paris, Deterville, Tome, 2: 532 pp.
DARRACOTT, A. (1977). Availability, morphometrics, feeding and breeding activity in a multi-species, demersal fish stock of the western Indian Ocean. – J. Fish Biol., 10(1): 1–16.
DAVENPORT, S. & J. D. STEVENS (1988). Age and growth of two commercially important sharks *(Carcharhinus tilstoni* and *Carcharhinus sorrah)* from northern Australia. – Aust. J. Mar. Freshwat. Res., 39 (4): 417–433.
DAY, F. (1873–78). The fishes of India, being a natural history of the fishes known to inhabit the seas and fresh waters of India, Burma and Ceylon. – London, B. Quaritch, Vol. 1: 778 pp., and Vol. 2: 779–816.
DENG SI-MING, XIONG GUO-QIANG & ZHAN HONG-XI (1981). On three new species of sharks of the genus *Carcharhinus* from China. – Acta Taxonom. Sin., 6(2): 216–220.
DE CASTELNAU, F. L., (1860). – In: BLEEKER, P.: Over eenige vischsoorten van de Kaap de Goede Hoop. Natuurk. – Tijdschr. Ned.-Indie, 21: 49–80.
DE CROSTA, M. A. (1981). Age determination and growth of three species of shallow-water carcharhinid sharks in Hawaii. – Pac. Sci., 35(3): 266–267.
DE KAY, J. E. (1842). Zoology of New York, or the New York fauna. Part IV. Fishes. – 415 pp.; D. Appleton & Co. and Wiley & Putnam, N.Y.
DE MADDALENA, A. & H. BÄNSCH (2005). Haie im Mittelmeer. – Kosmos Naturführer, Kosmos Verlag, Germany.
DIAZ, M. J. M. (1979). Teeth and dental replacement in some Carcharhinid sharks from the Colombian Pacific. – An. Inst. Invest. Mar. Punta de Betin., 11: 135–145.
DRIGGERS, W. B. (2002). Age, growth, reproduction and population genetics of the Blacknose shark, *Carcharhinus acronotus*, in the western North Atlantic. – Diss. Abst. Int. Pt. B – Sci. & Eng., 62 (12): 5585 pp.
DRIGGERS, W. B. & E. R. HOFFMAYER (2009). Variability in the reproductive cycle of finetooth sharks, *Carcharhinus isodon*, in the northern Gulf of Mexico. – Copeia, 2: 390–393 .
DRIGGERS, W. B., D. A. OAKLEY, G. ULRICH, J. K. CARLSON, B. J. CULLUM & J. M. DEAN (2004). Reproductive biology of *Carcharhinus acronotus* in the coastal waters of South Carolina. – J. Fish Biol., 64 (6): 1540–1551.
DUDLEY, S. F. J. & G. CLIFF (1993). Sharks caught in the protective gill nets off Natal, South Africa. 7. The blacktip shark *Carcharhinus limbatus* (VALENCIENNES). – S. Afr. J. Mar. Sci./S. Afr. Tydskr Seewet., 13: 237–254.
DULVY, N. K. & J. D. REYNOLDS (1997). Evolutionary transitions among egg-laying, live-bearing and maternal inputs in sharks and rays. – Proc. R. Soc. Lond. B 264: 1309–1315.

DUMERIL, A. (1865). Histoire naturelle des poissons ou ichthyologique générale. Tome premier. Elasmobranches, plagiostomes et holocéphales, ou chimères. Première partie. – 720 pp.; Paris, Librairie encyclopédique de Roret.
ECCLES, D. H. (1992). FAO species identification sheets for fishery purposes. Field guide to the freshwater fishes of Tanzania. Prepared and published with the support of the United Nations Development Programme (project URT/87/016). – 145 pp.; FAO, Rome
ECONOMAKIS, A. E. & P. S. LOBEL (1998). Aggregation behaviour of the gray reef shark, *Carcharhinus amblyrhynchos*, at Johnston Atoll, central Pacific Ocean. Environ. – Biol. Fish., 51 (2): 129–139.
EDWARDS, A. J. & H. R. LUBBOCK (1982). The shark population of Saint Paul's Rocks. – Copeia, 2: 223–225.
ENGELHARDT, R. (1912). Monographie der Selachier der Münchener Zoologischen Staatssammlung (mit besonderer Berücksichtigung der Haifauna Japans). 1. Teil. Tiergeographie der Selachier. – Abh. Bayer. Akad. Wiss. Math.-Phys. Kl. (Beitr. Naturgesch. Ostasiens), 4 (3): 110 pp.
ESCHMEYER, W. N. (1990). Catalogue of the genera of recent fishes. – 697 pp.; California Acad. Sci. San Francisco, CA (USA)
– (1998/2007). Catalogue of Fishes. – Spec. Pub. of the Centre for Biodiv. Res. & Information. Calif., Acad. of Sci., San Francisco, California, (vol. 1-3): 2905 pp.; online version available.
FANG, P. W. & K. F. WANG (1932). The elasmobranchiate fishes of Shantung coast. – Contrib. Biol. Lab. Sci. Soc China (Zool. Ser.), (8): 213–283.
FERGUSSON, I. K. & L. J. V. COMPAGNO (1999). Distributional note on the dusky shark, *Carcharhinus obscurus*, from the Mediterranean Sea, with a first record from the Maltese Islands. – Proc. of 3rd meeting of the European Elasmobranch Association, Boulogne sur-Mer.: 57–65.
FISCHER, W., I. SOUSA, C. SILVA, A. DE FREITAS, J. M. POUTIERS, W. SCHNEIDER, T. C. BORGES, J. P. FERAL & A. MASSINGA (1990). Fichas. FAO de identificaçao de espécies para actividades de pesca. Guia de campo das espécies comerciais marinhas e de águas salobras de Moçambique. Publicaçao preparada em collaboraçao com o Instituto de Investigaçao Pesquiera de Moçambique, com financiamento do Projecto PNUD/FAO MOZ/86/030 e de NORAD. – 424 pp.; Roma, FAO.
FORSTER, G., J. BADCOCK, M. LONGBOTTOM, N. MERRETT & K. THOMSON (1970). Results of the Royal Society Indian Ocean Deep Slope Fishing Expedition. – Proc. R. Soc. Lond. (B Taxon.), 175 (1041): 367–404.
FOURMANOIR, M. P. (1961). Requins de la côte ouest de Madagascar. – Mem. Inst. Sci. Madagascar (Ser. F), 4: 1–81.
– (1976a). New Caledonia sharks. (Requins de Nouvelle Calédonie). – Nature Calédonienne, 11–12: 23–29.
FOURMANOIR, M. P. & P. LABOUTE (1976b). Poissons de nouvelle Calédonie et des nouvelles Hébrides. – 376 pp.; Papeete, Tahiti, Editions du Pacifique.
FOWLER, H. W. (1940). *Eulamia (Platyodon) pleurotaenia*. – Bull. US. Nat. Mus., 100: 150 pp.
– (1944). New taxonomic names of fish-like vertebrates. – Notul. Nat. Acad. Nat. Sci. Philad., 187: 16 pp.
FROESE, R. & D. PAULY (eds.) (2010). Fish Base. – World Wide Web electronic publication (online available).
GARLA, R. C., D. D. CHAPMANN, B. M. WETHERBEE & M. SHIVJI (2006a). Movement patterns of young Caribbean reef sharks, *Carcharhinus perezii*, at Fernando de Noronha Archipelago Brazil: the potential of marine protected areas for conservation of a nursery ground. – Mar. Biol., 149: 189–199.
GARLA, R. C., D. D. CHAPMANN, M. S. SHIVJI, B. M. WETHERBEE & A. F. AMORIM (2006b). Habitat of juvenile Caribbean reef sharks, *Carcharhinus perezii*, at two oceanic insular marine protected areas in the south-western Atlantic Ocean: Fernando de Noronha Archipelago and Atol das Rocas, Brazil. – Fisheries Res., 81 (2–3): 236–241.
GARLA, R. C., D. D. CHAPMAN, M. S. SHIVJI, B. M. WETHERBEE & A. F. AMORIM (2006c). Habitat of juvenile Caribbean reef sharks, *Carcharhinus perezi*, at two oceanic insular marine protected areas in the south-western Atlantic Ocean: Fernando de Noronha Archipelago and Atol das Rocas, Brazil. – Fish. Res., 81 (2–3): 236–241.
GARRICK, J. A. F. (1962a). Reasons in favour of retaining the generic name *Carcharhinus* (BLAINVILLE), and a proposal for identifying its type species as the Indo-Pacific Black-tipper shark, *C. melanopterus*. – Proc. Biol. Soc. Wash., 75: 89–96.
– (1962b). *Carcharhinus* (BLAINVILLE 1816): proposed designation of *Carcharias melanopterus* (QUOY & GAIMARD) as type species under the Plenary Powers. – Bull. Zool. Nomencl., 19: 100–102.
– (1982). Sharks of the genus *Carcharhinus*. – NOAA Tech. Rep. NMFS Circ., 445: 194 pp.
– (1985). Additions to a revision of the shark genus *Carcharhinus*: synonymy of *Aprionodon* and *Hypoprion*, and description of a new species of *Carcharhinus* (Carcharhinidae). – NOAA, Tech. Rep., NMFS 34, National Marine Fisheries Service: 26 pp.
GARRICK, J. A. F., R. H. BACKUS & R. H. GIBBS, jr. (1964). *Carcharhinus floridanus*, the silky shark, a synonym of *C. falciformis*. – Copeia, 2: 369–375.
GILBERT, C. H. (1892). Descriptions of thirty-four new species of fishes collected in 1888 and 1889, principally among the Santa Barbara Islands and in the Gulf of California. – Proc. U.S. Nat. Mus., 14: 539–566.
– (1905) The deep-sea fishes of the Hawaiian Islands. Part II, Section 11. – Bull. U.S. Fish Comm., 23 (2): 575–713.
GILBERT, C. H. & E. C. STARKS (1904). The fishes of Panama Bay. – Mem. Calif. Acad. Sci., 4: 304 pp.
GILBERT, P. W. & D. A. SCHLERNITZAUER (1966). The placenta and gravid uterus of *Carcharhinus falciformis*. – Copeia, 3: 451–457.
GILL, T. (1862) Analytical synopsis of the order of squali, and revision of the nomenclature of the genera. – Ann. Lyceum Nat. Hist. N.Y., 7: 367–408 (413).
GILL, T. & J. F. BRANSFORD (1877). Synopsis of the fishes of Lake Nicaragua. – Proc. Acad. Nat. Sci. Phila., Ser. 3, 29: 175–191.
GOHAR, H. A. F. & F. M. MAZHAR (1964a). The Elasmobranches of the north-western Red Sea. – Publ. Mar. Biol. Stn Al-Ghardaqa, 13: 3–144.
– (1964b). Keeping elasmobranches in vivaria. – Publ. Mar. Biol. Stn. Al-Ghardaqa, 13: 241–250.
GÜNTHER, A. (1870). Catalogue of the fishes in the British Museum. – London, Brit. Mus. (Nat. History). 8: 369–549.
HAZIN, F. H. V., F. M. LUCENA, T. S. A. L. SOUZA, C. E. BOECKMAN, M. K. BROADHURST & R. C. MENNI (2000). Maturation of the night shark, *Carcharhinus signatus*, in the south-western equatorial Atlantic Ocean. – Bull. Mar. Sci., 66 (1): 173–185.
HAZIN, F. H. V., P. G. OLIVEIRA & M. K. BROADHURST (2002). Reproduction of the Blacknose shark, *Carcharhinus acronotus* in coastal waters off north-eastern Brazil. – Fish Bull., 100 (1): 143–148.
HEITHAUS, M. R., G. J. MARSHALL, B. M. BUHLEIER & L. M. DILL (2001). Employing Critter cam to study habitat use and behaviour of large sharks. – Mar. Ecol. Prog. Ser., 209: 307–310.
HEMIDA, F. & N. LABIDI (2002). Carcharinids of the Algerian basin. – 4th Meeting of the European Elasmobranch Association, Proceedings: 192 pp.
HEMPRICH, F. G. & C. G. EHRENBERG (1899) Symbolae physicae. – HILGENDORF, F. (ed.); Berlin, Paris, Zoologica.
HERMANN, J., M. HOVESTADT-EULER & D. C. HOVESTADT (1991). Contributions to the study of the comparative morphology of teeth and other relevant ichthyodorulites in living supraspecific taxa of Chondrichthyan fishes. STEHMANN, M. (ed.), Part A: Selachii. No. 2c: Order: Carcharhiniformes. Families: Proscylliidae, Hemigaleidae, Pseudotriakidae, Leptochariidae and Carcharhinidae.

Herre, A. W. C. T. (1934). Notes on fishes in the Zoological Museum of Stanford University. 2. Two new genera and species of Japanese sharks and a Japanese species of Narcetes. – Copeia, 3: 122–127.

Heupel, M. R. & R. E. Hueter (2002). Importance of prey density in relation to the movement patterns of juvenile blacktip sharks *(Carcharhinus limbatus)* within coastal nursery area. – Mar. Freshw. Res., 53: 543–550.

Heupel, M. R. & C. A. Simpfendorfer (2002). Estimation of mortality of juvenile blacktip sharks, *Carcharhinus limbatus*, within a nursery area using telemetry data. – Can. J. Fish Aquat. Sci., 59: 624–632.

Heupel, M. R., B. G. Yeiser, A. B. Collins, L. Ortega & C. A. Simpfendorfer (2010). Long-term presence and movement patterns of juvenile bull sharks, *Carcharhinus leucas*, in an estuarine river system. – Mar. & Freshwat. Res., 61 (1): 1–10.

Hilgendorf, F. M. (1899). Über eine Fischsammlung von Haiti, welche 2 neue Arten, *Poecilia* (subg. n. *Acropocilia*) *tridens* and *Eleotris maltzani*, enthält. – Sitzungsberichte der Gesellschaft naturforschender Freunde zu Berlin: 51–55.

Hobson, E. S. (1961). Feeding behaviour in three species of sharks. – MSc. thesis, University of Hawaii, Hawaii Marine Laboratory.

– (1964). Feeding behaviour in three species. – Pacific Sci., 17: 171–194.

Holtzhausen, J. A. (2002). Trans-boundary Bronze whaler, *Carcharhinus brachyurus* tagging program between Namibia and Angola. – Sci. Counc. Res. Doc. NAFO. 02/135: 4 pp.

Hubbs, C. L. (1951). Record of *Carcharhinus longimanus* with *Naucrates* and *Remora* from the east-central Pacific. – Pac. Sci., 5: 78–81.

Hussey, N. E., I. D. McCarthy, S. F. J. Dudley & B. Q. Mann (2009). Nursery grounds, movement patterns and growth rates of dusky sharks, *Carcharhinus obscurus*: a long-term tag and release study in South African waters. – Mar. & Freshwat. Res., 60 (6): 571–583.

IGFA (1991). World Record Game Fishes. – International Game Fish Association, Florida, USA.

– (2001). Database of IGFA angling records until 2001. – IGFA, Fort Lauderdale, USA.

Jensen, N. H. (1976). Reproduction of the bull shark, *Carcharhinus leucas*, in the Lake Nicaragua-Rio – San Juan system. – Pp. 539–559 in Thorson, T. B. (ed.): Investigation of the ichthyofauna of Nicaraguan lakes; Lincoln, Nebraska, Univ. of Nebraska Press.

Johnson, R. H. (1978). Sharks of tropical and temperate seas. – 170 pp.; Papeete, Tahiti, Les Editions du Pacifique.

Johnson, R. H. & D. R. Nelson (1973). Agonistic display in the gray reef shark, *Carcharhinus menisorrah*, and its relationship to attacks on man *(Carcharhinus amblyrhynchos)*. – Copeia: 76–84.

Jordan, D. S. & B. W. Evermann (1904). Descriptions of new genera and species of fishes from the Hawaiian Islands. – Bull. US Fish Comm., 22: 161–208.

– (1905). The aquatic resources of the Hawaiian Islands. Part 1. Section 1. The shore fishes of the Hawaiian Islands, with a general account of the fish fauna. – Bull. US Fish. Comm., 23 (1): 574 pp.

Jordan, D. S. & C. H. Gilbert (1882). Synopsis of the fishes of North America. – Bull. US Nat. Mus., 16: 1018 pp.

– (1883a). Description of four new species of sharks, from Mazatlan, Mexico. – Proc. US Nat. Mus., 5: 102–110.

– (1883b). Description of a new shark *(Carcharias lamielia)*, from San Diego, California. – Proc. US Nat. Mus., 5: 110–111.

Jordan, D. S. & A. Seale (1906). *Carcharias (Prionodon) pleurotaenia*. – Bull. US Bur. Fish, New Guinea [Error for Batavia].

Joung, S.-J. & C.-T. Chen (1995). Reproduction in the sandbar shark, *Carcharhinus plumbeus*, in the waters off north-eastern Taiwan. – Copeia, 3: 659–665.

Joung, S.-J., Y.-Y. Liao & C.-T. Chen (2004). Age and growth of sandbar shark, *Carcharhinus plumbeus*, in north-eastern Taiwan waters. – Fish. Res., 70 (1): 83–96.

Joung, S.-J., Liao Y.Y., Liu K.-M., Chen C.-T. & Leu L.-C. (2005). Age, growth and reproduction of the spinner shark, *Carcharhinus brevipinna* in the North eastern waters of Taiwan. – Zool. Stud., 44 (1): 102–110.

Joung, S.-J., C.-T. Chen, H.-H. Lee, K.-M. Liu (2008). Age, growth, and reproduction of silky sharks, *Carcharhinus falciformis*, in north-eastern Taiwan waters. – Fish. Res., 90 (1–3): 78–85.

Kailola, P. J., M. J. Williams, P. C. Stewart, R. E. Reichelt, A. McNee & C. Grieve (1993). Australian fisheries resources. – 422 pp.; Bureau of Resource Sciences, Canberra, Australia.

Kasim, H. M., K. M. S. A. Hamsa & S. Rajapackiam (1996). Age, growth, mortality, yield per recruit and stock assessment of *Carcharhinus sorrah* (Valenciennes, Muller and Henle). – The 4th Indian Fisheries Forum, Proc. Kochi, Kerala: 381–384.

Kato, S. (1964). Sharks of the genus *Carcharhinus* associated with the tuna fishery in the eastern tropical Pacific Ocean. – Circ. U.S. Fish Wildl. Serv., 172: 22 pp.

Klausewitz, W. (1960). Die Typen und Typoide des Naturmuseums Senckenberg, 23: Pisces, Chondrichthyes, and Elasmobranchii. – Senckenb. Biol., 41 (5/6): 289–296.

Klunzinger, C. B. (1871). Synopsis der Fische des Rothen Meeres. 2. Theil. – Verh. K. K. Zool.-Bot. Ges. Wien, 21: 441–688.

Kulkarni, G. N., S. L. Shanbhogue & K. S. Udupa (1988). Length-weight relationship of *Scoliodon laticaudus* (Muller & Henle) and *Carcharhinus limbatus* (Muller & Henle), from Dakshina Kanada coast. – Indian journal of Fisheries, 35 (4): 300–301.

Lafrance, S. (1994). Archipel des Bijjagos: ichtyofauna et élements d'ecologie marine. – Doc. Sci. CIPA No. 1: 66 pp.

Last, P. R. & J. D. Stevens (1994). Sharks and rays of Australia. – 513 pp.; CSIRO, Australia.

Lavery, S. & J. B. Shaklee (1989). Population genetics of two tropical sharks, *Carcharhinus tilstoni* and *C. sorrah*, in Northern Australia. – Aust. J. Mar. Freshwater Res., 40: 541–557.

– (1991). Genetic evidence for separation of two sharks, *Carcharhinus limbatus* and *C. tilstoni*, from northern Australia. – Mar. Biol., 108 (1): 1–4.

Lessa, R. & Z. Almeida (1997). Analysis of stomach contents of the smalltail shark, *Carcharhinus porosus*, from northern Brazil. – Cybium, 21 (2): 123–133.

Lessa, R. & F. M. Santana (1998). Age determination and growth of the smalltail shark, *Carcharhinus porosus*, from northern Brazil. – Mar. Freshwat. Res., 49 (7): 705–711.

Lessa, R. & F. M. Santana & R. Paglerani (1999a). Age, growth and stock structure of the oceanic whitetip shark, *Carcharhinus longimanus*, from the south-western equatorial Atlantic. – Fish Res., 42 (1–2).

Lessa, R., R. Paglerani & F. M. Santana (1999b). Biology and morphometry of the oceanic whitetip shark, *Carcharhinus longimanus* (Carcharhinidae), off north-eastern Brazil. – Cybium, 23 (4): 353–368.

Lessa, R., F. Santana, R. Menni & Z. Almeida (1999c). Population structure and reproductive biology of the smalltail shark, *Carcharhinus porosus* off Maranhao (Brazil). – Mar. Freshwat. Res., 50 (5): 383–388.

Lesson, M. (1831). Zoologie. Voyage autour du Monde, exécuté par Ordre du Roi, sur la Corvette de la Majesté, « La Coquille », pendant les années 1822, 1823, 1824 et 1825. – 471 pp.; Paris, Arthus Bertrand.

LeSueur, C. A. (1818). Descriptions of several new species of North American fishes. – J. Acad. Nat. Sci. Philad., 1 (2): 222–235.

– (1822). Description of a squalus, of a very large size, which was taken on the coast of New Jersey. – J. Acad. Nat. Sci. Philad., 2: 343–352.

Lieske, E. & R. Myers (1994). Collins Pocket Guide. Coral reef fishes. Indo-Pacific & Caribbean including the Red Sea. – 400 pp.; Haper Collins Publishers.

Lipej, L., B. Mavric, Z. Dobrajc & C. Capape (2008). On the occurrence of the sandbar shark, *Carcharhinus plumbeus* (Chondrichthyes: Carcharhinidae) off the Slovenian coast (northern Adriatic). – Acta Adriatica, 49 (2): 137–145.

Lowe, R. T. (1839). A supplement to a synopsis of the fishes of Madeira. – Proc. Zool. Soc. London, 7: 76–92.

– (1840). On new species of fishes from Madeira. – Proc. Zool. Soc. Lond., 8: 36–39.

– (1843). Notices of fishes newly observed or discovered in Madeira during the years 1840–42. – Proc. Zool. Soc. Lond., 11: 81–95.

– (1860). A history of the fishes of Madeira. – 196 pp.; London, Bernard Quaritch.

Lucifora, L. O. (2003). Ecology and conservation of large coastal sharks from Anegada Bay, Buenos Aires province, Argentina. – Physical medium: Printed matter; Author's Degree, Thesis. Available at INIDEP Library, Mar del Plata (Argentina).

Lucifora, L. O., R. C. Menni & A. H. Escalante (2003). Reproduction and seasonal occurrence of the copper shark, *Carcharhinus brachyurus* from North Patagonia, Argentina. – ICES J. Mar. Sc., 62(1): 107–115.

Lyle, J. M. (1984). Mercury concentrations in four carcharhinid and three hammerhead sharks from coastal waters of the Northern Territory. – Aust. J. Mar. Freshwat. Res., 35(4): 441–451.

– (1987). Observations on the biology of *Carcharhinus cautus* (Whitley), *C. melanopterus* (Quoy & Gaimard) and *C. fitzroyensis* (Whitley) from northern Australia. – Aust. J. Mar. Freshwat. Res., 38(6): 701–710.

Lyle, J. M. & G. J. Timms (1987). Predation on aquatic snakes by sharks from northern Australia. – Copeia, 3: 802–803.

McAuley, R. B., C. A. Simpfendorfer, G. A. Hyndes & R. C. J. Lenanton (2007). Distribution and reproductive biology of the sandbar shark, *Carcharhinus plumbeus* (Nardo), in western Australian water. – Mar. and Freshwater Res., 58(1): 116–126.

McKibben, J. N. & D. R. Nelson (1986). Patterns of movement and grouping of gray reef sharks, *Carcharhinus amblyrhynchos*, at Eniwetok, Marshall Islands. – Bull. Mar. Sci., 38(1): 89–110.

Meek, S. E. & S. F. Hildebrand (1923). The marine fishes of Panama. – Field Mus. Nat. Publ., 215, Zool. Ser., 15: 1–330.

Merson, R. R. & H. L. Pratt, jr. (2001). Distribution, movements and growth of young sandbar sharks, *Carcharhinus plumbeus*, in the nursery grounds of Delaware Bay. Environ. – Biol. Fishes, 61: 13–24.

Moore, A. B. M., W. T. White, G. J. P. Naylor, R. D. Ward, L. V. J. Compagno & R. Peirce (2010). Rediscovery and conservation status of *Carcharhinus leiodon* (Carcharhinidae) from Kuwait, a geographically restricted whaler shark. Abstract presented at the Sharks Int. Conf., Cairns, Queensland, Australia.

Moreau, H. (1881). Histoire naturelle des poissons de la France. – Paris, G. Masson, 1: 478 pp.

Moreno, J. A. (1995). Guia de los tiburonesde Atlantico Nororiental y Mediterrraneo. – 310 pp.; Pirámide, Madrid.

Moreno, J. A. & A. Hoyos (1983a). *Carcharhinus acarenatus*, n. sp, nouveau requin carcharhinidé de l'Atlantique nord Oriental et de la Méditerranée Occidentale. – Cybium, 7(1): 57–64.

– (1983b). Première capture en aux Espagnoles et en Méditerranée de *Carcharhinus altimus* (Springer 1950). – Cybium, 7(1): 65–70.

Morey, G. & E. Massuti (2003). Record of the copper shark, *Carcharhinus brachyurus*, from the Balearic Islands (Western Mediterranean). – Cybium, 27(1): 53–56.

Müller, J. & F. G. J. Henle (1837a). Gattungen der Haifische und Rochen, nach ihrer Arbeit: Ueber die Naturgeschichte der Knorpelfische. – Ber. K. Preuss. Akad. Wiss. Berl., 2: 111–118.

– (1837b). Ueber die Gattungen der Plagiostomen. – Arch. Naturgesch., 1837: 394–401.

– (1838a). On the generic characters of cartilaginous fishes, with descriptions of new genera. – Ann. Maq. Nat. Hist., 2: 33–91.

– (1838b). Poissons cartilaginoux. – J. Gen. Soc. Trav. Sci. Fr. Etranger, 6(224): 63–65.

– (1838c). Ueber die Gattungen der Plagiostomen. – Arch. Naturgesch., 4: 83–85.

– (1838d). Systematische Beschreibung der Plagiostomen. – Pp. 1–38; Berlin, Veit.

– (1838e). Systematische Beschreibung der Plagiostomen. – Pp. 27–28; Berlin, Veit.

– (1839). Systematische Beschreibung der Plagiostomen. – Pp. 39–102; Berlin, Veit.

– (1841). Systematische Beschreibung der Plagiostomen. – Pp. 103–200; Berlin, Veit.

Munoz-Chapuli, R. (1984). Reproduction behaviour of some sharks from the north-eastern Atlantic. Cybium, (3rd Ser.), 8(3): 1–14.

Musick, J. A., M. M. Harbin, S. A. Berkeley, G. H. Burgess, A. M. Eklund, L. Findley, R. G. Gilmore, J. T. Golden, D. S. Ha, G. R. Huntsman, J. C. McGovern, S. J. Parker, S. G. Poss, E. Sala, T. W. Schmidt, G. R. Sedberry, H. Weeks & S. G. Wright (2000). Marine, estuarine, and diadromous fish stocks at risk of extinction in North America (exclusive of Pacific salmonids). – Fisheries, 25(11): 6–30.

Myers, R. F. (1991). Micronesian reef fishes. – Second ed., 298 pp.; Coral Graphics, Barrigada, Guam.

– (1999). Micronesian reef fishes: a comprehensive guide to the coral reef fishes of Micronesia. – 3rd revised and expanded edition, 330 pp.; Coral Graphics, Barrigada, Guam.

Nakaya, K. (1984). Carcharhinidae. – Pp. 5–7 in Masuda, Amaoka, Araga, Uyeno & Yoshino (eds.): Fishes of the Japanese Archipelago. – Tokai Univ. Press, Tokyo, Japan.

Nardo, J. D. (1827). Prodromus Observationum et Disquisitionum Ichthyologiae Adriaticae. – Isis (von Oken), 20(6): 474–625.

– (1847a). Prospetto delle faune marine volgare del Vento Estuario con cenni sulle principali specie commestibili del'Àdriatico, ecc. Venezia: 45 pp.

– (1847b). Sinonimia moderna delle specie registrate nell'opera inititolata: Descizione de'Crostacei, de'Testacei e dei Pesci che abitano le Lagune e Golfo Veneto rappresentati in figure, a chiaroscuro ed a colori dall'Abate Stefano Chiereghini ven clodiense applicata per commissione governativa dal Dr. Gio Domenico Nardo. Parte terza. Indice dei Pesci della Laguna e Golfo Veneto, Venezia: 108–127.

Natanson, L. J. & N. E. Kohler (1996). A preliminary estimate of age and growth of the dusky shark *Carcharhinus obscurus* from the South-West Indian Ocean, with comparisons to the western North Atlantic population. – S. Afr. J. Mar. Sci./S. Afr. Tydskr. Seewet., 17: 217–224.

Natanson, L. J., J. G. Casey & N. E. Kohler (1995). Age and growth estimates for the dusky shark, *Carcharhinus obscurus*, in the western North Atlantic Ocean. – Fish. Bull., 93(1): 116–126.

Naylor, G. J. P. & L. F. Marcus (1994). Identifying isolated shark teeth of the genus *Carcharhinus* to species: Relevance for tracking phyletic change through the fossil record. – Novitates, Amer. Mus. Nat. Hist, N.Y., 3109: 53 pp.

Ogilby, J. D. (1910a). On new or insufficiently described fishes. – Proc. R. Soc. Queensl., 23: 1–55.

– (1910b). On new genera and species of fishes. – Proc. R. Soc. Queensl., 2: 3–26.

– (1911). On new or insufficiently described fishes of Queensland waters. – Ann. Queensl. Mus., 10: 36–58.

Ortega, L. A., M. R. Heupel, P. van Beynen & P. J. Motta (2009). Movement patterns and water quality preferences of juvenile bull sharks *(Carcharhinus leucas)* in a Florida estuary. Environ. – Biol. Fishes, 84(4): 361–373.

Ovenden, J. R., T. Kashiwagi, D. Broderick, J. Giles & J. Salini (2009). The extent of population genetic subdivision differs among four co-distributed shark species in the Indo-Australian archipelago. – BMC, Evolutionary Biol., 9: 40–54.

OVENDEN, J. R., J. A. T. MORGAN, T. KASHIWAGI, D. BRODERICK & J. SALINI (2010). Toward better management of Australia's shark fishery: genetic analyses reveal unexpected ratios of cryptic blacktip species *Carcharhinus tilstoni* and *C. limbatus*. – Mar. & Freshwat. Res., 61: 253–262.

OWEN, SIR RICHARD (1853). Descriptive catalogue of the osteological series contained in the Museum of the Royal College of Surgeons of England. – Vol. 1. Taylor & Francis, London.

PAPASTAMATIOU, Y. P., B. M. WETHERBEE, C. G. LOWE & G. L. CROW (2006). Distribution and diet of four species of carcharhinid shark in the Hawaiian Islands: evidence for resource partitioning and competitive exclusion. – Mar. Ecol. Prog. Ser., 320: 239–251.

PAPASTAMATIOU, Y. P., J. E. CASELLE, A. M. FRIEDLANDER & C. G. LOWE (2009). Distribution, size frequency, and sex ratios of blacktip reef sharks *Carcharhinus melanopterus* at Palmyra Atoll: a predator-dominated ecosystem. – J. Fish Biol., 75(3): 647–654.

PETERS, W. C. H. (1852). Hr. Peters legte einige neue Säugethiere und Flussfische aus Mossambique vor. – Ber. Bekanntmachung geeigneten Verh. Königl. Preuss. Akad. Wiss. Berlin: 273–276.

PHILIPPI, R. A. (1887). Sobre los tiburones y algunos otros peces de Chile. – An. Univ. Chile, 71: 3–31.

– (1893). Abbildung und Beschreibung einiger chilenischer Fische. – An. Mus. Nac. Chile, 1893: 1–19.

PIETSCHMANN, V. (1916). Fische des Wiesbadener Museums. – Jahrb. Nassau. Ver. Nat., 66: 171–201.

PIKITCH, E. K., D. D. CHAPMAN, E. A. BABCOCK & M. S. SHIVJI (2005). Habitat use and demographic population structure of elasmobranches at a Caribbean atoll (Glover's Reef, Belize). – Mar. Ecol. Prog. Ser., 302: 187–197.

PILLIANS, R. D., J. P. GOOD, W. G. ANDERSON, N. HAZON & C. E. FRANKLIN (2005). Freshwater to seawater acclimation of juvenile bull sharks *(Carcharhinus leucas)*: plasma osmolytes and Na^+/K^+-ATPase activity in gill, rectal gland, kidney and intestine. – J. Comp. Phys. B-Biochem. Sys. And Environm. Phys., 175(1): 37–44.

POEY, F. (1858–61). Memorias sobre la historia natural de la Isla de Cuba, acompanadas de sumarios Latinos y extractos en Frances. Tomo 2. La Habana. Sections have subtitles. – Mem. Hist. Nat. Cuba, v. 2, 1960: 1–96; 1861: 337–442.

– (1865). Repertorio fisico-natural de la isla de Cuba. Havana, 1: 383 pp.

– (1868). Repertorio fisico-natural de la isla de Cuba. Havana: 468 pp.

– (1876). Enumeratio piscium cubensium. – An. Soc. Esp. Hist. Nat., 5: 177–218.

POLL, M. (1951). Poissons. 1. Generalités. 2. Sélaciens et Chimères. – Result. Sci. Exped. Oceanogr. Belgie, 4⁺(1): 154 pp.

PORCHER, I. F. (2005). On the gestation period of the Blackfin reef shark, *Carcharhinus melanopterus*, in the waters off Moorea, French Polynesia. – Mar. Biol. ISSN: 0025-3162: 1207–1211.

PURDY, R. W. (1990). A key to the common genera of neogene shark teeth. – Nat. Mus. Nat. Hist., WA, USA.

QUOY, J. R. C. & P. GAIRMARD (1824). Zoology. – In DE FREYCINET, L.: Voyage autour de monde exécuté sur les corvettes de S.M. l'Uranie et la Physicienne, pendant les années 1817, 1818, 1819, et 1820. – 712 pp.; Pillet Ainé, Paris.

RAFINESQUE, C. S. (1810). Caratteri d'alcuni nuovi generi e nuove specie d'animali e piante delta Sicilia. – 105 pp.; Palermo.

RAMSAY, E. P. & J. D. OGILBY (1887). On an undescribed shark from Port Jackson. – Proc. Linn. Soc. N.S.W., Ser. 2, 2: 163–164.

RANDALL, J. E. (1963a). Dangerous sharks of the western Atlantic. – Pp. 339–361 in GILBERT, P. W., J. A. F. GARRICK & L. P. SCHULTZ (eds.): Sharks and survival. Boston, D.C. Heath and Co.

– (1963b). A fatal attack by the shark *Carcharhinus galapagensis* at Saint Thomas, Virgin Islands. – Caribb. J. Sci., 3: 201–205.

RANDALL, J. E. & G. S. HELFMAN (1973). Attacks on humans by the blacktip reef shark *Carcharhinus melanopterus*. – Pac. Sci., 27(3): 226–238.

RANDALL, J. E., G. R. ALLEN & R. C. STEENE (1990). Fishes of the Great Barrier Reef and Coral Sea. – 506 pp.; University of Hawaii Press, Honolulu, Hawaii.

RASCHI, W., J. A. MUSICK & L. J. V. COMPAGNO (1982). *Hypoprion bigelowi*, a synonym of *Carcharhinus signatus* (Pisces: Carcharhinidae), with a description of ontogenetic heterodonty in this species and notes on its natural history. – Copeia, 1: 102–109.

RANZANI, C. (1839). De novis speciebus piscium. – Acad. Sci. Inst. Bononiensis, Novi comment, 4: 65–83.

RECHISKY, E. L. & B. M. WETHERBEE (2003). Short-term movements of juvenile and neonate sandbar sharks, *Carcharhinus plumbeus*, on their nursery grounds in Delaware Bay. – Environ. Biol. Fish, 68(2): 113–128.

RIEDE, K. (2004). Global register of migratory species – from global to regional scales. Final report of the R & D-Project 808 05 081. – 329 pp.; Federal Agency for Nature Conservation, Bonn, Germany.

ROSENBLATT, R. H. & W. J. BALDWIN (1958). A review of the eastern Pacific sharks of the genus *Carcharhinus*, with a redescription of *C. malpeloensis* (FOWLER) and California records of *C. remotus* (DUMERIL). – Calif. Fish Game, 44(2): 137–159.

RÜPPEL, W. P. E. S. (1837). Neue Wirbelthiere zu der Fauna von Abyssinien gehörig. – Frankfurt/M., 2(11): 64 pp.

SADOWSKY, V. (1967). The adult stage of the shark *Carcharhinus remotus* (DUMERIL 1865). – Senckenberg. Biol., 48: 327–334.

– (1971). Notes on the bull shark *Carcharhinus leucas* in the lagoon region of Cananeia, Brazil. – Bol. Inst. Oceanogr. S. Paulo, 20(2): 71–78.

SAIDÏ, B., M. N. BRADAÏ, A. BOUAÏN & C. CAPAPE (2007). Feeding habits of the sandbar shark *Carcharhinus plumbeus* (Chondrichthyes: Carcharhinidae) from the Gulf of Gabès, Tunisia. – Cahiers de Biologie Marine, 48: 139–144.

SALINI, J. P., S. J. M. BLABER & D. T. BREWER (1991). Diets of sharks from estuaries and adjacent waters of the north-eastern Gulf of Carpentaria, Australia. – Aust. J. Mar. Freshwat. Res., 43(1): 87–96.

SALINI, J. P., M. TONKS, S. J. M. BLABER & J. ROSS (1999). Feeding of captive, tropical carcharhinid sharks from the Embley River estuary, northern Australia. – Mar. Ecol. Prog. Ser., 184: 309–314.

SANCHES, J. G. (1991). Catálogo dos principais peixes marinhos da República de Guiné-Bissau. – Publicações avulsas do I.N.I.P. No. 16: 429 pp.

SANTANA, F. M. & R. LESSA (2004). Age determination and growth of the night shark *(Carcharhinus signatus)* off the north-eastern Brazilian coast. – Fish. Bull., 102(1): 156–167.

SAZIMA, I. & R. L. MOURA (2000). Shark *(Carcharhinus perezii)*, cleaned by the Goby *(Elacatinus randalli)*, at Fernando de Noronha Archipelago, western south Atlantic. – Copeia, 1: 297–299.

SCHLUESSEL, V., M. B. BENNETT, H. BLECKMANN, S. BLOMBERG & S. P. COLLIN (2008). Morphometric and ultra structural comparison of the olfactory system in elasmobranches: The significance of structure-function relationships based on phylogeny and ecology. – J. Morphol., 269: 1365–1386.

SCHWARTZ, F. J. (1984). Occurrence, abundance, and bio ogy of the Blacknose shark, *Carcharhinus acronotus* in North Carolina. – Inst. Mar. Sci., Univ. N.C. Northeast Gulf Sci., 7(1): 29–48.

SEALE, A. (1910). Fishes of Borneo, with descriptions of four new species. – Philipp. J. Sci. Ser. D, 5: 263–289.

SEKI, T., T. TANIUCHI, H. NAKANO & M. SHIMIZU (1998). Age, growth and reproduction of the oceanic whitetip shark from the Pacific Ocean. – Fish. Sci., 64(1): 14–20.

SETNA, S. B. & P. N. SARANGDHAR (1946). Selachian fauna of the Bombay waters. – Proc. Nat. Inst. Sci. India, 12(5): 243–259.

SIMPFENDORFER, C. A. (2000). Growth rates of juvenile dusky sharks, *Carcharhinus obscurus* (LESUEUR 1818), from south-western Australia estimated from tag-recapture data. – Fish Bull., 98(4): 811–822.

SIMPFENDORFER, C. A. & N. E. MILWARD (1993). Utilisation of a tropical bay as a nursery area by sharks of the families Carcharhinidae and Sphyrnidae. The Hague, 37(4): 337–345.
SIMPFENDORFER, C. A., R. B. MCAULEY, J. CHIDLOW & P. UNSWORTH (2002). Validated age and growth of the dusky shark, *Carcharhinus obscurus*, from Western Australian waters. – Mar. Freshwat. Res., 53(2): 567–573.
SMALE, M J. (1991). Occurrence and feeding of three shark species, *Carcharhinus brachyurus*, *C. obscurus* and *Sphyrna zygaena*, on the eastern Cape coast of South Africa. – S. Afr. J. Mar. Sci., 11: 31–42.
SMITH, J. L. B. (1952). Two chondrichthyan fishes new to South Africa. – Ann. Mag. Nat. Hist. (Ser. 12), 5: 760–765.
– (1958). A new shark from South Africa. – S. Afr. J. Sci., 53(10): 261–264.
SNELSON, F. F. jr. & S. E. WILLIAMS (1981). Notes on the occurrence, distribution, and biology of Elasmobranch fishes in the Indian River lagoon system, Florida. – Estuaries, 4(1): 110–120.
SNELSON, F. F. jr., T. J. MULLIGAN & S. E. WILLIAMS (1984). Food habits, occurrence and population structure of the bull shark, *Carcharhinus leucas*, in Florida coastal lagoons. – Bull. Mar. Sci., 34(1): 71–80.
SNODGRASS, R. E. & E. HELLER (1905). Papers from the Hopkins-Stanford Galapagos Expedition, 1898–1899. XVII. Shore fishes of the Revillagigedo, Clipperton, Cocos and Galapagos Islands. – Proc. Wash. Acad. Sci., 6: 333–427.
SNYDER, J. O. (1904). A catalogue of the shore fishes collected by the steamer "ALBATROSS" about the Hawaiian Islands in 1902. – Bull. U.S. Fish. Comm., 22: 513–538.
SOMMER, C., W. SCHNEIDER & J.-M. POUTIERS (1996). FAO species identification field guide for fishery purposes. The living marine resources of Somalia. – 376 pp.; FAO, Rome.
SPRINGER, S. (1950). A revision of North American sharks allied to the genus *Carcharhinus*. – Am. Mus. Novit., 9(13): 1451 pp.
– (1951). Correction for a revision of North American sharks allied to the genus *Carcharhinus*. – Copeia, 3: 244 pp.
– (1990). Carcharhinidae. – In QUERO, J. C., J. C. HUREAU, C. KARRER, A. POST, & L. SALDANHA (eds.): Check-list of the fishes of the eastern tropical Atlantic (CLOFETA). JNICT, Lisbon; SEI, Paris; and UNESCO, Paris. Vol. 1: 101–108.
SPRINGER, V. G. (1964). A revision of the Carcharhinid shark genera *Scoliodon*, *Loxodon*, and *Rhizoprionodon*. – Proc. U.S. Nat. Mus., 115(3493): 559–632.
STEAD, D. G. (1963) Sharks and rays of Australian seas. – 211 pp.; Sydney, Angus and Robertson.
STEINDACHNER, F. (1867). Ichthyologische Notizen (VI.). – Sitzungsber. Math.-Naturwiss. Cl. Kais. Akad. Wiss. Wien, 56(1): 307–376.
STEVENS, J. D. (1984). Biological observations on sharks caught by sport fishermen off New South Wales. – Aust. J. Freshwat. Res., 35(5): 573–590.
– (2004). Taxonomy and field techniques for identification, with listening of available regional guides. – In MUSICK, J. A. & R. BONFIL (eds.): Elasmobranch fisheries management techniques. 84 pp.; CSIRO, Hobart, Tasmania.
STEVENS, J. D. & P. D. WILEY (1986). Biology of two commercially important carcharhinid sharks from northern Australia. – Aust. J. Freshw. Res., (37): 671–688.
STOBUTZKI, I. C., M. J. MILLER, D. S. HEALES & D. T. BREWER (2002). Sustainability of elasmobranches caught as bycatch in a tropical prawn (shrimp) trawl fishery. – Fish Bull., 100(4): 800–821.
SULIKOWSKI, J. A., W. B. DRIGGERS, T. S. FORD, R. K. BOONSTRA & J. K. CARLSON (2007). Reproductive cycle of the Blacknose shark *Carcharhinus acronotus* in the Gulf of Mexico. – J. Fish Biol., 70(2): 428–440.
TAN, H. H. & K. K. P. LIM (1998). Freshwater elasmobranches from the Batang Hari basin of central Sumatra, Indonesia. – Raffles Bull. Zool., 46(2): 425–429.
TANIUCHI, T. (1971). Reproduction of the sandbar shark, *Carcharhinus milberti*, in the East China Sea. – Jap. J. Ichthyol., 18(2): 94–98.
TANIUCHI, T., H. TACHIKAWA, Y. KURATA & Y. HOSE (1985). Galapagos shark, *Carcharhinus galapagensis*, from the Ogasawara Islands, Japan. – Jap. J. Ichthyol., 31(4): 449–452.
TANIUCHI, T., H. ISHIHARA, S. TANAKA, S. HYODO, M. MURAKAMI & B. SERET (2003). Occurrence of two species of elasmobranches, *Carcharhinus leucas* and *Pristis microdon*, in Betsiboka River, West Madagascar. – Cybium, 27(3): 237–241.
TAVARES, R. & R. F. PROVENZANO (2000). Feeding habits of the blacktip shark juveniles, *Carcharhinus limbatus* (VALENCIENNES 1839), at the Archipelago Los Roques National Park, Venezuela. – Acta Biol. Venez., 20(1): 59–67.
TEMMINCK, C. & H. SCHLEGEL (1850). Pisces. – In VON SIEBOLD, P. V. (ed.): Fauna Japonica, 323 pp.
TESHIMA, K. (1973). Studies on sharks. III. The stage of placentation and the umbilical stalk in *Carcharhinus dussumieri*. – J. Shimonoseki Univ. Fish, 21(3): 25–30.
TESHIMA, K. & K. MIZUE (1972). Studies on sharks, I. Reproduction in the female Sumitsuki shark, *Carcharhinus dussumieri*. – Mar. Biol., 14: 222–231.
THOMERSON, J. E., T. B. THORSON & R. L. HEMPEL (1977). The bull shark, *Carcharhinus leucas*, from the upper Mississippi River near Alton, Illinois. – Copeia: 166–168.
THORSON, T. B., D. E. WATSON & C. M. COWAN (1966a). The status of the freshwater shark of Lake Nicaragua. – Copeia, 3: 385–402.
THORSON, T. B., M. COWAN & D. E. WATSON (1966b). Sharks and sawfish in the Lake Isabel-Rio Dulce system, Guatemala. – Copeia: 620–622.
THORSON, T. B., D. E. WATSON & C. M. COWAN (1971). Movement of bull sharks, *Carcharhinus leucas*, between Caribbean Sea and Lake Nicaragua demonstrated by tagging. – Copeia: 336–338.
Thorson, T. B., D. E. Watson & C. M. Cowan (1976). The status of the Lake Nicaragua shark: an updated appraisal. – Pp. 561–574 in THORSON, T. B. (ed.): Investigation of the ichthyofauna of Nicaraguan lakes; Univ. of Nebraska Press.
THORSON, T. B. & E. J. LACY, jr. (1982). Age, growth and longevity of *Carcharhinus leucas* estimated from tagging and vertebral rings. – Copeia, 1: 110–116.
TORRES, F. S. B., jr. (1991). Tabular data on marine fishes from Southern Africa, Part I. Length-weight relationships. – Fishbyte 9(1): 50–53.
TUMA, R. E. (1976). An investigation of the feeding habits of the bull shark, *Carcharhinus leucas*, in the Lake Nicaragua Rio San Juan system. – Pp. 533–538 in THORSON, T. B. (ed.): Investigation of the ichthyofauna of Nicaraguan lakes; Univ. of Nebraska Press.
VACCHI, M., F. SERENA & V. BIAGI (1996). Capture of *Carcharhinus brachyurus* (GUENTHER 1870) (Pisces, Selachii, Carcharhinidae) in the northern Tyrrhenian Sea. – Biol. Mar. Mediterr., 3(1): 389–390.
VALENCIENNES, A. (1839). – Pp. 39–102 in: MÜLLER & HENLE (ed.): Systematische Beschreibung der Plagiostomen; Berlin, Veit.
VILLAVICENCIO-GARAYZAR, C. J. (1996). Reproduction of *Carcharhinus obscurus* (Pisces: Carcharhinidae), in the northeast Pacific. – Rev. Biol. Trop., 44(1): 287–289.
WALTER, J. P. & D. A. EBERT (1991). Preliminary estimates of age of the Bronze whaler *Carcharhinus brachyurus* (Chondrichthyes: Carcharhinidae) from southern Africa, with a review of some life history parameters. – S. Afr. J. Mar. Sci./S. Afr. Tydskr. Seewet., 10: 37–44.
WASS, R. C. (1973). Size, growth, and reproduction of the sandbar shark, *Carcharhinus milberti*, in Hawaii. – Pac. Sci., 27(4): 305–318.
WETHERBEE, B. M., G. L. CROW & C. G. LOWE (1996). Biology of the Galapagos shark, *Carcharhinus galapagensis*, in Hawaii. – Environ. Biol. Fish., 45(3): 299–310.

WETHERBEE, B. M., G. L. CROW & C. G. LOWE (1997). Distribution, reproduction and diet of the gray reef shark *Carcharhinus amblyrhynchos* in Hawaii. – Mar. Ecol. Prog. Ser., 151 (1–3): 181–189.

WHEELER, J. F. G. (1960). Sharks of the western Indian Ocean. 3. *Carcharhinus menisorrah* (MÜLLER & HENLE, 1839). – East Afr. Agric. J., 25 (4): 271–273.

– (1962). Notes on the three common species of sharks in the Mauritius-Seychelles area. – Proc. R. Soc. Arts Sci. Mauritius, 2 (2): 146–160.

WHITE, W. T., N. G. HALL & I. C. POTTER (2002). Size and age compositions and reproductive biology of the nervous shark, *Carcharhinus cautus*, in a large subtropical embayment, including an analysis of growth during pre- and postnatal life. – Mar. Biol., 141 (6): 1153–1164.

WHITE, T. W., P. R. LAST & A. P. K. LIM (2010). Rediscovery of the rare and endangered Borneo shark *Carcharhinus borneensis* (BLEEKER 1858) (Carcharhiniformes: Carcharhinidae). – In LAST, WHITE & POGONOSKI (eds.): Descriptions of new sharks and rays from Borneo; CSIRO Mar. Atmosph. Res. Pap. 32: 17–28.

WHITEHEAD, D. L. (2002). Ampullary organs and electroreception in freshwater *Carcharhinus leucas*. – J. Physiology, Paris, 96 (5–6): 391–395.

WHITLEY, G. P. (1934). Notes on some Australian sharks. – Mem. Queensl. Mus., 10 (4): 180–200.

– (1939). Taxonomic notes on sharks and rays. – Aust. Zool., 9 (3): 227–262.

– (1940). The fishes of Australia. Part 1. The sharks, rays, devilfish, and other primitive fishes of Australia and New Zealand. Australian zoology handbooks. – 280 pp.; Mozman, Royal Zool. Soc. N.S.W.

– (1943a). Ichthyological descriptions and notes. – Proc. Linn. Soc. N.S.W., 68: 114–144.

– (1943b). Ichthyological notes and illustrations. Part 2. – Aust. Zool, 10 (2): 167–169.

– (1943c). A new Australian shark. – Rec. S. Aust. Mus., 7 (4): 397–399.

– (1944a). New sharks and fishes from Western Australia. – Aust. Zool., 10 (3): 252–273.

– (1944b). Illustrations of some Western Australian fishes. – Proc. R. Zool. Soc. N.S.W.: 25–29.

– (1945). New sharks and fishes from Western Australia. Part 2. – Aust. Zool., 11 (1): 1–41.

– (1947). New sharks and fishes from Western Australia. – Aust. Zool., 11 (2): 129–150.

– (1949). New sharks and fishes from Western Australia. Part 4. – Aust. Zool., 11 (3): 259–276.

– (1950). A new shark from north-western Australia. – West. Aust. Nat., 2 (5): 100–105.

– (1951). New fish names and records. – Proc. R. Zool. Soc. N.S.W.: 61–68.

– (1964a). Presidential address. A survey of Australian ichthyology. – Proc. Linn. Sac. N.S.W., 89 (1): 11–127.

– (1964b). Fishes from the Coral Sea and the Swain Reefs. – Rec. Aust. Mus., 26: 145–195.

– (1967). Sharks of the Australasian region. – Aust. Zool., 14 (2): 173–188.

WINTNER, S. P. & G. CLIFF (1996). Age and growth determination of the blacktip shark, *Carcharhinus limbatus*, from the east coast of South Africa. – Fish Bull., 94 (1): 135–144.

YANO, K. & J. F. MORRISSEY (1999). Confirmation of blacktip shark, *Carcharhinus limbatus*, in the Ryukyu Islands and notes on possible absence of *C. melanopterus* in Japanese waters. – Ichthyol. Res., 46 (2): 193–198.

YEISER, B. G., M. R. HEUPEL & C. A. SIMPFENDORFER (2008). Occurrence, home range and movement patterns of juvenile bull *(Carcharhinus leucas)* and lemon *(Negaprion brevirostris)* sharks within a Florida estuary. – Mar. & Freshwat. Res., 59 (6): 489–501.

ZEITZ, A. (1888). Notes on *Carcharias hemiodon* as an Australian species. Trans. Proc. Rep., R. Soc. S. Aust., 10: 303 p.

Internet

AMS – Australian Museum, Sydney, Australia
http://www.amonline.net.au/about/

BMNH – Natural History Museum, London, UK (formerly the British Museum of Natural History)
www.curator.org/LegacyVMNH/VMNH_Publications/ExpeditionReports/Placereviews/the_bmnh.htm

ESCHMEYER W. N. (1998/2007) Catalogues of fishes, Vol. 1–3.
www.calacademy.org/research/ichthyology

FMNH – Florida Museum of Natural History
Carcharhinus site: www.flmnh.ufl.edu

Fish Base
www.fishbase.org

German Elasmobranch Society (D.E.G.)
www.elasmo.de

ICZN – International Commission on Zoological Nomenclature
www.iczn.org

IUCN – International Union for Conservation of Nature
http://cms.iucn.org/

MNHN – Muséum National d'Histoire Naturelle, Paris, France
www.mnhn.fr/museum/foffice/transverse/transverse/accueil.xsp?cl=en

NMNH – National History Natural Museum, Washington, USA
www.nmnh.si.edu/paleo/sharkteeth/index.html

NMNZ – National Museum of New Zealand, Wellington, New Zealand
www.tepapa.govt.nz/Tepapa/English/

NMV – Naturhistorisches Museum Vienna, Austria
www.nhm-wien.ac.at/

QM – Queensland Museum, Queensland, Australia
www.qm.qld.gov.au/

RNMH – Nationaal Natuurhistorisch Museum, Leiden, the Netherlands (formerly Rijksmuseum van Natuurlijke Historie)
www.naturalis.nl

SMF – Naturmuseum Senckenberg, Frankfurt/M., Germany
www.senckenberg.de

SMNS – Staatliches Museum für Naturkunde, Stuttgart, Germany
www.naturkundemuseum-bw.de/

SU – Stanford University, Natural History Museum, Stanford, CA, USA
www.stanford.edu/

Swizz Shark Foundation
www.sharkinfo.ch

USNM – U.S. National Museum of Natural History, Washington, USA
http://nis.gsmfc.org/nis_instinfo.php?instcode=USNM

VBCM – Cátedra de Vertebrados, Facultad de Biológicas, Universitad Complutense de Madrid, Spain
www.ucm.es/centros/webs/fbio/

Index of Scientific and Vernacular Names

Ordered by genera or higher-level taxa

Ordered by species or subspecies names